Late for Tea at the Deer Palace

TAMARA CHALABI

Late for Tea at the Deer Palace

The Lost Dreams of My Iraqi Family

HARPER ● PERENNIAL

NEW YORK ● LONDON ● TORONTO ● SYDNEY

HARPER ● PERENNIAL

First published in Great Britain in 2010 by
HarperPress, an imprint of HarperCollins Publishers.

A hardcover edition was published in 2011 by HarperCollins Publishers.

HarperCollins books may be purchased for educational, business, or sales promotional use. For information please write: Special Markets Department, HarperCollins Publishers, 10 East 53rd Street, New York, NY 10022.

FIRST HARPER PERENNIAL EDITION PUBLISHED 2012.

The Library of Congress has catalogued the hardcover edition as follows:

Chalabi, Tamara.
Late for tea at the deer palace : the lost dreams of my Iraqi family / Tamara Chalabi.—
1st Harper Perennial ed.
p. cm.
"First published in Great Britain in 2010 by HarperPress, an imprint of HarperCollins Publishers"—T.p. verso.
ISBN 978-0-06-124039-3
1. Chalabi, Tamara—Family. 2. Chalabi family. 3. Iraq—Biography.
4. Iraq—History—1921- 5. Iraq—Social conditions—20th century. I. Title.
DS79.6.A2C48 2012
956.704092'2--dc22
[B]
2011013083

ISBN 978-0-06-124040-9 (pbk.)

12 13 14 15 16 OFF/RRD 10 9 8 7 6 5 4 3 2 1

To my dearest ammooooo, Hassan Chalabi

Contents

Maps   x
Family tree   xiv
Chronology   xvii
Prologue   xxv

BOOK ONE: Fallen Pomegranates

١ DECEMBER 2007   3

1 Duty Calls: *A Busy Day for Abdul Hussein* (1913)   5
2 Stacking Rifles: *Hadi and the War* (1914–1916)   23
3 All That is Good Will Happen: *A Marriage   35
   Prospect* (1916)
4 Sugared Almonds and Jasmine: *Bibi and Hadi's   46
   Wedding* (1916)
5 A Giant Broken: *The End of the Ottomans* (1917–1918) 58

٢ NOVEMBER 1999, BEIRUT   69

BOOK TWO: Replanting Eden

٣ SEPTEMBER 2005   73

6 Café Chantant: *The British in Baghdad* (1918)   77

7 Rebellion: *Fighting for Freedom* (1919–1920)   86

8 A New King for a New Country: *From Mesopotamia*   95
   to Iraq (1920–1921)

9 *Fesanjoon, a Royal Luncheon: Faisal Visits*   102
   Kazimiya (1921)

10 Banished: *Out of Kazimiya* (1922–1924)   108

11 Accidents of Nature: *The Baghdad Boil* (1925–1926)   117

12 In Between: *A Home Between Two Cities* (1926–1929)   126

13 Stolen Hopes: *A Young Life Lost* (1928–1929)   136

14 Bursting Energy: *Hadi's Growing Empire* (1931–1933)   142

15 Prison: *Uninvited Guests at a Feast* (1935–1936)   149

16 Carefree: *Growing Up in the Golden Age* (1936–1938)   161

17 A Dark Cloud: *The End of a Generation* (1938–1939)   173

18 A New Home: *The Shadow of Death* (1937–1939)   179

❦ OCTOBER 2006   185

BOOK THREE: A Dangerous Garden

☙ MAY 1993   189

19 Mountains and Floods: *Domestic Changes* (1939–1941)   191

20 Blood and Salons: *Mounting Tensions* (1941)   202

21 An Education Overseas: *Mixed Fortunes* (1941–1945)   212

22 Love in Strange Quarters: *Of Marriage and*   223
   Other Unions (1946–1947)

23 The Girl on the Bridge: *Anger on the Streets* (1947–1949)   229

24 Precious Things: *Towards a New World* (1950–1951)   235

25 Storm Clouds Gathering: *Family Feuds and*   245
   Revolution (1952–1956)

26 Defiance: *A Crisis and a Key* (1956)   255

27 Revolution: *Slaughter of a Family* (1958)   260

❧ FEBRUARY 2005, SADR CITY   279

CONTENTS

BOOK FOUR: Fields of Wilderness

Ⅴ DECEMBER 2007   283

28 Lost Lands: *Seeking Shelter* (1958)   285
29 Migration: *Precious Cargo* (1958)   289
30 Hunger Pangs: *Yearning for Home* (1958)   296
31 Arrivals and Departures: *The Importance of Contacts* (1958–1959)   298
32 Escape to Nowhere: *The Threat of the Clown Court* (1959)   308
33 A Temporary Home: *Visits to the Park* (1959)   312
34 Return to the Shrine: *A Life by the Sea* (1959–1963)   316
35 Of Carpets and New Blood: *The Emergence of New Patterns* (1967)   325
36 The Ruins of Kufa: *A Coup and a Birth* (1968–1972)   333
37 Civil War: *A Shattered Sanctuary* (1975–1982)   338
38 Creased Maps: *A Move to a Different Land* (1980s)   343
39 Lessons in Humility: *The Loss of Everything Precious* (1980s)   350
40 The Mortality of Gods: *Burials of the Banished* (1988)   357
41 The Lost Talisman: *When Everything is Taken* (1989–1992)   363
42 A Question of Identity: *In Search of a Way to Be* (1990–2009)   370

Ⅷ 30 JANUARY 2005, ELECTION DAY IN BAGHDAD   383

Epilogue   385
Glossary of Iraqi Terms   389
List of Illustrations   397
Acknowledgements   399
Index   403

CITY OF BAGHDAD

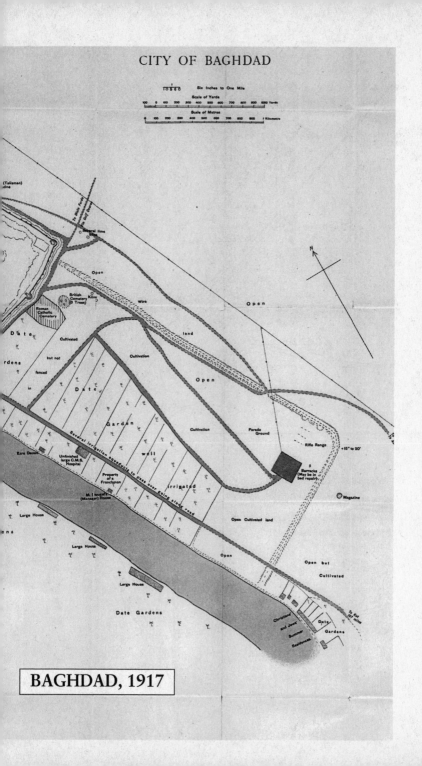

BAGHDAD, 1917

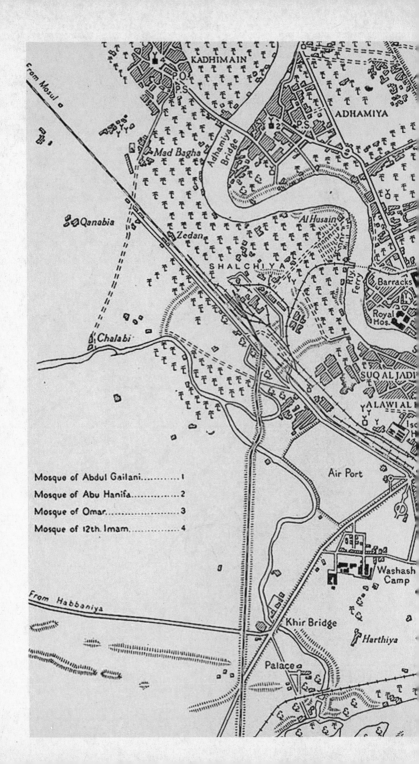

KADHIMAIN

From Mosul

ADHAMIYA

Mad Bagha

Adhamiya Bridge

Qanabia

Al Husain

Zedan

SHALCHIYA

Ferry

Barracks

Royal Hos.

Chalabi

SUQ AL JADI

ALAWI AL

Mosque of Abdul Gailani...........1

Mosque of Abu Hanifa...............2

Mosque of Omar.......................3

Mosque of 12th. Imam...............4

Air Port

Washash Camp

From Habbaniya

Khir Bridge

Harthiya

Palace

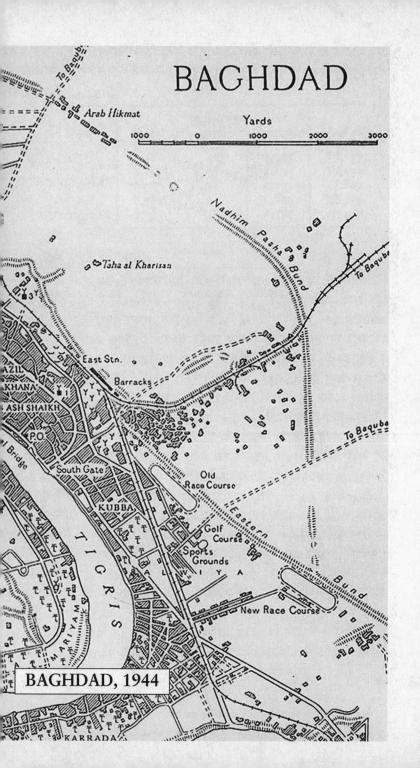

BAGHDAD

Yards

1000　　0　　1000　　2000　　3000

To Baquba

Arab Hikmat

Nadhim Pasha's Bund

To Baquba

Taha al Kharisan

East Stn.

Barracks

To Baquba

AZII

KHANA

ASH SHAIKH

P.O.

To Baquba

l Bridge

South Gate

Old Race Course

KUBBA

Eastern

Golf Course

Sports Grounds

A L W I Y A

Bund

TIGRIS

New Race Course

MARIYAM

BAGHDAD, 1944

KARRADA

CHALABI FAMILY TREE

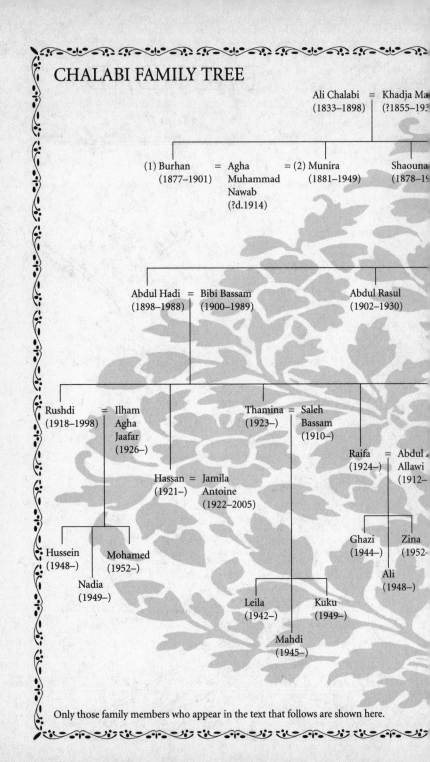

Ali Chalabi = Khadja Ma
(1833–1898)   (?1855–19?

(1) Burhan = Agha = (2) Munira   Shaouna
(1877–1901)   Muhammad   (1881–1949)   (1878–19
              Nawab
              (?d.1914)

Abdul Hadi = Bibi Bassam   Abdul Rasul
(1898–1988)   (1900–1989)   (1902–1930)

Rushdi = Ilham   Thamina = Saleh
(1918–1998)   Agha   (1923–)   Bassam
             Jaafar   (1910–)
             (1926–)
                                          Raifa = Abdul
                                          (1924–)   Allawi
             Hassan = Jamila   (1912–
             (1921–)   Antoine
                       (1922–2005)
                                          Ghazi   Zina
                                          (1944–)   (1952–
Hussein   Mohamed   Ali
(1948–)   (1952–)   (1948–)
      Nadia
      (1949–)
                    Leila   Kuku
                    (1942–)   (1949–)

                    Mahdi
                    (1945–)

Only those family members who appear in the text that follows are shown here.

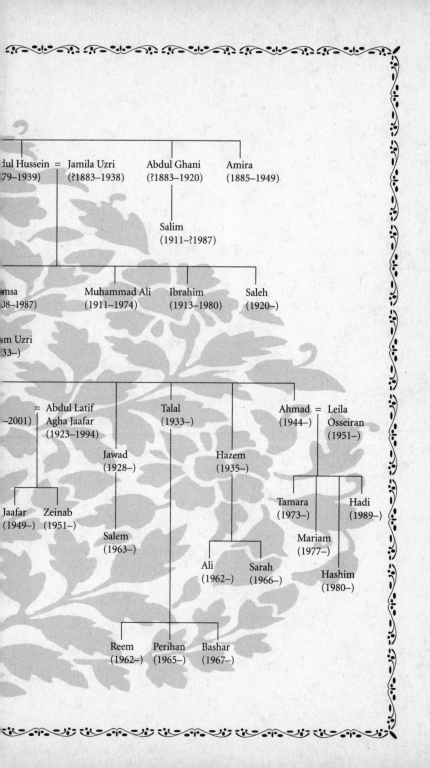

dul Hussein = Jamila Uzri   Abdul Ghani   Amira
79–1939)   (?1883–1938)   (?1883–1920)   (1885–1949)

Salim
(1911–?1987)

msa
08–1987)   Muhammad Ali   Ibrahim   Saleh
                   (1911–1974)   (1913–1980) (1920–)

m Uzri
33–)

              = Abdul Latif   Talal   Ahmad = Leila
–2001)   Agha Jaafar   (1933–)   (1944–) Osseiran
                (1923–1994)   (1951–)

                  Jawad   Hazem
                  (1928–)   (1935–)

Jaafar Zeinab   Tamara   Hadi
(1949–) (1951–)   (1973–)   (1989–)

                  Salem   Mariam
                  (1963–)   (1977–)

                                    Ali   Sarah   Hashim
                                    (1962–) (1966–)   (1980–)

              Reem   Perihan Bashar
              (1962–) (1965–) (1967–)

Chronology

1833   Ali Chalabi is born.

1869–72 Midhat Pasha, Ottoman governor in Baghdad, launches a series of much needed reforms aimed at modernization.

1879   Abdul Hussein is born.

1898   Hadi is born.

1900   Bibi is born.

1908   The Young Turks or CUP (Committee of Union and Progress) Revolt occurs.

1909   Persian Constitutional Revolution takes place in Iran.

1911   Bibi's father, Sayyid Hassan al-Bassam, dies.

1912   The Turkish Petroleum Company is formed for oil exploration. Its main partners are British and Dutch.

1914   World War I begins.

1916   T. E. Lawrence is sent by the British to Sharif Hussein of Mecca to foment an Arab revolt against the Ottomans, aimed at weakening the Ottoman position near the Persian Gulf to Britain's advantage.

       Faisal, the son of Sharif Hussein, leads the revolt alongside Lawrence.

       Signature of the Sykes–Picot Agreement, a contract between Britain and France carving up the Ottoman Empire, including the Arab provinces, into spheres of control.

       Gertrude Bell, traveller and archaeologist, arrives in Basra to serve as the only female British political officer with knowledge of the terrain and the Arab tribes of Iraq.

Bibi and Hadi are married.

1917 Baghdad is captured by British troops marching from the south.

1918 Word War I ends and Rushdi is born.

The British occupy the provinces that make up Iraq.

The French occupy Syria and Lebanon.

1919 The Paris Peace Settlement is signed. Faisal heads an Arab delegation to the conference, accompanied by T. E. Lawrence, to lobby unsuccessfully for Arab independence.

1920 The Treaty of San Remo is decided at a meeting to determine fate of former Arab Ottoman territories, based on the previous Sykes–Picot agreement and overlooking British promises to Sharif Hussein for an Arab Kingdom.

Gertrude Bell becomes Oriental Secretary to British Commissioner Sir Percy Cox and plays a key role in supporting Arab rule in Iraq.

Faisal is crowned King of Syria but is driven out by the French after the battle of Maysaloun in June.

Revolt breaks out in southern and central Iraq among Shi'a tribes and there is joint Sunni–Shi'i resistance to the British in Baghdad.

The idea of the India Office running Iraq as an administration is abandoned by the British in favour of national rule.

1921 In April, the Cairo Conference is called by Winston Churchill to decide the fate of Iraq, which is now too costly for the British to run. Lawrence and Bell attend. Faisal is chosen as King of Iraq.

King Faisal arrives in Basra in June, then through the main Iraqi cities to Baghdad. He receives a cool reception, but among those welcoming him is Abdul Hussein Chalabi.

Faisal is crowned at a ceremony in Baghdad in August. Gertrude Bell is charged with setting up his court.

Hassan is born.

1922 Abdul Hussein becomes Minister of Education, breaking with the Shi'a *fatwa* forbidding participation in government. He is banished from Kazimiya.

The Chalabi family moves to the Deer Palace late November.

1923 Thamina is born.

1924   Raifa is born.
        Hassan loses his sight.
        Tensions that will later explode under Saddam Hussein
            begin to brew in the Ministry of Education, under
            Sunni Arab nationalist Director of Education Sati'
            al-Husri.

1925–8 Abdul Rasul studies at Cambridge.

1926   Gertrude Bell founds the Iraqi Museum in June, then
            commits suicide in July.

1929   Abdul Rasul is diagnosed with cancer and travels to Europe
            with Hadi to seek treatment.
        The Turkish Petroleum Company becomes the Iraqi
            Petroleum Company. Majority shares are still held by
            European partners, with Iraq benefiting from only 20 per
            cent of revenue.

1930   Abdul Rasul dies of a brain tumour. Hadi becomes a Member
            of Parliament for the town of Diwaniya.
        Hadi becomes the agent for Andrew Weir & Co., supplying
            wheat and barley – Iraq's main agriculture exports along
            with dates.

1932   Faisal struggles with a tension between local demands and
            Britain's intervention in what has until now remained a
            British mandate. This year, Iraq officially becomes
            independent.
        Iraq joins the League of Nations.
        Rebellion breaks out in the north and is dealt with brutally
            by the British. 4000 die.

1933   Umm Kalthoum holds her first concert in Baghdad. Bibi
            attends with Shamsa.
        Talal is born. King Faisal dies unexpectedly. Ghazi becomes
            king.

1935   Hazem is born. Hadi is arrested on charges of treason,
            imprisoned and sentenced to death, but released three
            months later.

1936   Abdul Hussein serves his final parliamentary term and
            becomes a senator.
        A military *coup d'état* is attempted.
        Hadi becomes the first head of the Iraqi stock exchange.

1937   Rushdi travels to London to study, then transfers to the
            American University of Beirut.

1938   The Chalabi family move from the Deer Palace to new house built by Hadi, now called the Sif Palace.
        Jamila dies.

1939   Abdul Hussein dies.
        King Ghazi dies in car crash. Popular opinion holds that the British killed him because he was a nationalist and popular with the military.
        The Second World War breaks out while the Chalabi family is travelling back from holiday in Lebanon, then under French rule.

1940   Hassan starts his degree at the law college and is introduced to Jamila.
        Minister of Finance and former adviser to King Faisal Rustum Haidar is killed in his office. The murder is widely seen as a sectarian attack against the Shi'a.

1941   A successful pro-Nazi *coup d'état* is led by Rashid 'Ali Gailani. The royal family flees.
        British and Iraqi forces clash at Habbaniya.
        Baghdad is rocked by the *Farhud* – the Great Loot – during which Jewish shops and homes are attached by angry anti-British mobs.
        The royal family returns to Baghdad with British help. Some *coup d'état* leaders are arrested, others flee. The British temporarily reoccupy Baghdad.
        Thamina marries Bibi's cousin Saleh Bassam.
        Rushdi becomes involved in Freya Stark's anti-Nazi Brotherhood of Freedom.
        Four officers behind the coup are executed by order of the Prince regent, in a move that proves unpopular.

1942   Thamina gives birth to Leila, Bibi's first granddaughter.

1943   Iraq enters World War II, fighting against the Axis powers.
        Raifa marries Abdul Amir Allawi.
        Rumia dies.

1944   Hadi makes his first trip to America as part of a business delegation.
        Ahmad is born.
        Hassan goes to university in Egypt to pursue a doctorate in law. Jamila goes with him.
        Raifa and Thamina give birth one after the other to Ghazi and Mahdi, respectively.

1946   The Communist leader 'Fahd' is imprisoned for inciting workers to protest against the government.

1947   Rushdi marries Ilham and becomes MP for Kazimiya.

Hadi becomes a senator.

Hassan briefly returns to Baghdad before moving to Paris to continue his doctorate at the Sorbonne. Jamila follow him.

1948   Raifa's second son, Ali, is born. Rushdi first child, Hussein, is born.

The Arab Israeli war – what the Israelis call the War of Independence and the Palestinians call the Catastrophe (*Nakba*) – breaks out with the abrupt ending of the British mandate in Palestine. As part of the Arab Legion that included forces from Jordan, Syria, Egypt, Lebanon, Saudi Arabia and Yemen, Iraq initially sends 3,000 men to fight, later increasing that number to 21,000. Pressure on Iraq's Jewish population increases.

1949   Bibi travels to Europe for the first time to visit Hassan in Paris.

Fahd is executed.

Salim Chalabi temporarily becomes secretary general of the Iraqi Communist Party.

1950   Najla marries Abdul Latif Agha Jaafar, Ilham's brother.

1951   Hassan is hospitalized in Paris.

Hadi, Bibi and their three youngest boys take a family trip to Europe.

1952   Hassan returns to Cairo to defend his PhD.

Revolution in Egypt topples the monarch. Jamal Abdul Nasser, a military officer, takes over.

Hassan begins a career as a professor at Baghdad University.

1953   18-year-old King Faisal II is crowned in Baghdad.

1954   Hadi buys Latifiyyah Estate from Andrew Weir.

Rushdi becomes Minister of Agriculture.

1955   The Baghdad Pact is signed, including Iran, Turkey, Pakistan and Britain in a political and economic bloc aimed at countering Soviet influence in the region.

Hadi becomes deputy head of the Senate.

1956   During the Suez crisis, Britain, France and Israel attack Egypt in response to the nationalization of the Suez Canal.

1957   Frank Lloyd Wright visits Baghdad before designing the Baghdad Opera House. Le Corbusier and Walter Gropius are likewise commissioned to design public buildings.

1958   Rushdi participates in successful negotiations with British
        Petroleum. Iraq's share of its oil revenue increases to 80 per
        cent.

        In a *coup d'etat* on 14 July, the entire royal family is
        murdered, except for Princess Badiya and her children.

        Prime Minister Nuri Said hides in Thamina's house, then is
        taken to her in-laws in Kazimiya.

        On 15 July, Rushdi is arrested with Abdul Amir Allawi and
        Fadhil Jamali.

        Nuri's location is discovered. He seeks refuge with the
        Istrabadi family but is shot dead alongside Bibi's old friend,
        Bibi Istrabadi.

        Thamina's husband Saleh is arrested for helping Nuri Said.

        Britain officially recognizes new Iraq government on 1
        August.

        Hadi arrives in London in September and is joined a month
        later by Ahmad and Ghazi.

        Rushdi is released from jail and put under house arrest.

1959   In January Bibi joins Hadi in London.

        In March, Rushdi and his remaining siblings try to escape to
        Jordan by car. Their attempt is thwarted.

        Rushdi's house arrest ends in July. He leaves for London to
        join the rest of the family.

        Thamina's husband, Saleh, is released from jail.

        The Mahdawi court – known as 'the clown court' – is set up
        to try officials of the old regime. Four are executed.

        Saeeda dies.

1960   Bibi and Hadi move to Beirut. The rest of the family will
        follows in following years. Only Hassan and Jawad remain
        in Baghdad.

1961   Ahmad is admitted to MIT.

1962   Hadi is given permission to visit Baghdad briefly. It was his
        last visit.

1963   In a Ba'ath coup, President Abdul Karim Qassim is executed
        in Baghdad. Shortly thereafter, General Abdul Salam Arif
        becomes president.

1967   Arab countries suffer a major defeat in the Arab–Israeli War.

1968   A second Ba'ath *coup d'etat* takes place, leaving the
        Ba'athists in charge.

        Ni'mati dies.

1969   Hassan is the last family member to leave for Beirut.

Ahmad returns from the U.S. with a Phd in Mathematics, then travels with Hassan to Iran to meet with Mulla Mustafa Barzani, a Kurdish leader fighting the Ba'athists for Kurdish autonomy. Ahmad takes teaching post at the American University of Beirut.

1975   The Persian Gulf Treaty is signed in Algiers, resolving a dispute between Iran and Iraq over the Persian Gulf and reneging on a promise to allow Kurds autonomy in Iraq.

Civil war breaks out in Lebanon.

1978   Ahmad establishes the Petra Bank in Jordan.

1979   The Iranian revolution overthrows the Shah. An Islamic government led by Ayatollah Khomeini emerges. Saddam Hussein takes over as president of Iraq.

Saddam accuses several Baghdadi Jews of espionage and executes them. He then executes leading religious scholar Sayyid Muhammad Baqir al-Sadr and his sister Bint al-Huda. He liquidates many members of the Ba'ath Party disloyal to him.

1980   The Iran–Iraq war is provoked by Saddam Hussein.

Entire communities, predominantly Shi'a, are forcibly deported from Iraq.

1982   Israel invades Lebanon.

1988   Hadi dies and is buried in Damascus.

Saddam Hussein launches the genocidal *al-Anfal* campaign against the Kurds. Civilians in Halabja are gassed with chemical weapons. As many as 182,000 people are killed.

The Iran–Iraq war ends in stalemate, with over one million dead.

1989   Bibi dies and is buried next to Hadi, despite her wish to be in interred in Najaf.

Ahmad is in Jordan when martial law is declared. Under threat of handover to Saddam, Ahmad flees Jordan and enters fully into opposition politics.

Ahmad and his family move to London.

1990   Saddam invades Kuwait. The First Gulf War, led by U.S.-coalition forces, forces Saddam to withdraw.

1991   United Nations sanctions are imposed on Iraq.

1992   The Iraqi National Congress, an umbrella group for all forces opposed to Saddam's regime, holds its founding

conference. Ahmad plays a leading role. He later moves to
Iraqi Kurdistan, to a U.N. no-fly zone, where the INC
holds its first conference on Iraqi soil.

1994   Tamara first visits Iraq.

1996   Saddam Hussein's forces attack Arbil, a town in the no-fly
zone, kill INC members and demolish their set up. Many
flee to the Turkish border.

1998   Rushdi dies and is buried in London.

Ahmad plays leading role in lobby for Iraq Liberation Act
that is passed by U.S. Congress.

2001   Najla dies and is buried next to Hadi and Bibi are now in
Damascus.

Al-Qaeda attack of the World Trade Centre in New York
City. The U.S. declares war on the Taliban in Afghanistan.

2003   In January, the INC crosses on foot into Iraqi Kurdistan.

In March, U.S. coalition forces attack Iraq to topple Saddam
Hussein's regime.

The Iraqi Museum and National Archives are looted.

Ahmad, Tamara and the INC group arrive in Baghdad on
April 15.

In May, U.N. Security Coucil Resolution 1483 declares a
U.S.-coalition occupation of Iraq. The Iraqi Governing
Council is appointed in July. Ahmad is one of nine rotating
members.

Saddam Hussein is captured in a hole in the ground by U.S.
forces.

A Special Tribunal is established to try Saddam and senior
Ba'ath government members for crimes against humanity.

2004   Transitional Iraqi government is established.

Bibi's remains are transferred to Najaf.

2005   First nationwide elections are held in Iraq.

Ahmad Chalabi becomes Deputy Prime Minister.

2006   Saddam Hussein is executed.

2007   Hassan visits Baghdad for the first time since 1969.

Prologue

THE KITCHEN WAS bare, an abandoned room. The sole trace of its former occupants was a squat, white bone-china teapot. I reached for it, turning it over in my hands. On its underside were stamped the words 'State of India'. Alone in this silent space, the teapot spoke to me of a bygone era that had come to an abrupt end.

It was 19 April 2003, ten days after the fall of Baghdad to the US-led coalition forces, and the city, depleted and derelict, was grappling with a new reality. The heat of the day was intolerable, and I could feel my very eyeballs become coated in perspiration, a strange and unwelcome sensation. This was my first ever visit to Baghdad, my father's home, his parents' and grandparents' before him, and theoretically mine as well. I had arrived in the capital after a long car journey from the south in the company of my father – Ahmad Chalabi, a leading opposition figure to Saddam Hussein's fallen regime.

Everybody asks me about my father. He has been labelled a maverick, a charlatan, a genius. He has been named as the source of supposedly faulty intelligence that led America into the war in Iraq. He has been called a triple agent for the US, Iran and Israel. But this is my story. He has his own tale to tell, although I acknowledge that my father has played a pivotal role in shaping my relationship to his country, Iraq. As with everything in the Middle East, nothing makes sense until you understand the past, and the past is never straightforward.

During this, my first visit to Baghdad, whole convoys and fresh hordes were descending on the capital: the streets were busy with an assortment of opposition leaders, formerly exiled professionals, gold diggers and prospectors, sceptical foreign journalists – and ordinary Iraqis: doctors, lawyers, carpenters and shopkeepers who were returning home. For many, their homecoming was clearly a source of mixed emotions. For my part, as I entered the city with a large group of Iraqis who had been working for the opposition in exile, I swiftly understood that my life here would not be governed by a familiar set of values based on logic, chronology and order.

All of my companions, including my father, had their own personal memories of Baghdad. Like little children, they sparked with enthusiasm and anticipation when we entered the city in which they had been born. Many kissed the ground in tears before rising hastily, anxious to find their relatives and loved ones. I had none to find here. I stood by, silently searching their faces for an emotion I could recognize. None came. I felt cold and detached. This place was as foreign to me as any other, and I had no memories to draw upon to make me feel otherwise. What came instead was an image of Beirut, my birthplace. I remembered clearly the feelings of comfort, safety and warmth I always had deep inside whenever I was on a plane coming in to land in Beirut, the sea shimmering against the horizon. As much as I wanted to push that image away and connect with the ground beneath my feet in Baghdad, I couldn't.

It quickly became clear on our arrival that the promised 'liberation' had not happened. The sense of excitement and expectation with which I had travelled was replaced by a deep foreboding as I entered a shattered world. I went to my grandparents' house in Baghdad. Forty-five years had passed since they had been forced to flee the country. A big, solid, four-storey home, it was designed in the Bauhaus style and built in the late 1940s. The clean lines of the windows, the large rooms and elegant staircases were all suggestive of that era's faith in a better future. The place smelt the same as my grandparents' subsequent homes in Britain, infused with an aroma of rice and something indefinable. In London, they had recreated what they could of all that was soothing and familiar to them, building altars to their old life through the objects that had followed them into exile – their photographs, silver and precious carpets. However, they had merely been repeating a process they had already been through during an earlier period of forced expatriation, in Beirut, before the Lebanese Civil War drove them on once more.

I knew this house from the stories of other relatives, stories which had been told to me over and over again, but I could never have imagined the sense of emptiness that echoed down the long corridors and through the airy rooms. I tried to remember the rhythms of my grandmother's deep voice as she spoke of her former home when I was a little girl: 'You can't imagine the wonderful life we had in Baghdad, Tamara. I was like a queen ...'

A life-size stone statue of a deer stood in the withered garden outside the house. I knew that my grandfather Hadi had loved that deer as much as his father before him. Someone had beheaded it. My first impression was that the deer looked almost offensive among the unkempt grounds, as it suggested a more carefree time when the people and the country had been very different. It was now a dirty ivory colour, yet there remained a certain sensuality about it as it stood proud, the fluidity of its hind muscles elegantly carved. Even the amputated head lying on the ground was playful. Its large dark eyes were well defined and penetrating, their gaze frozen in time.

My journey to Iraq had really begun in my head many years earlier, in my grandparents' house in Beirut. It was 1981. I was seven years old. A man's voice, sonorous and beautiful, cut across a crowded room, singing about a land I did not know.

> A man fired an arrow that slayed the child.
> Oh my child, they killed a child
> Woe is me, woe is me ...

Although the singer was tucked away in a corner, his voice held the room captive. I could not understand why the audience wept as he sang about a thirsty child killed in his father's arms. I had never heard anything like it before. It disturbed my sense of the established routine and quiet of my grandparents' house.

I crawled through the legs of the grieving adults towards the familiar figure of my uncle Hassan. He sat listening intently, inscrutable in the dark glasses he wore to mask his blindness. I squeezed myself in next to him, watching as he tapped his knee with the palm of his hand in time to the song. I asked him why everyone was crying. He told me that it was in memory of Imam Hussein.

'Did he die today?' I asked.

'No, no, Tamoura,' he said fondly, calling me by the nickname he had given me. 'He died a long time ago, before any of us were born.'

'So why are you still crying?'

He explained that the singer was commemorating the Battle of Karbala, when Imam Hussein, the Prophet's grandson, was confronted by Caliph Yazid's forces of 4,000 men. A very long time ago Hussein had gone to war, taking his family along, and a small army of only seventy-two men, many of whom also went into battle with their women and children. When the armies clashed on the banks of the Euphrates River, in the month of Muharam, Hussein was defeated. He, his infant son and his men were slain and the women and children taken into captivity.

My uncle smiled sadly. He said that time did not lessen the sense of tragedy of an act that had the power to haunt people forever. He told me that Hussein had been killed by an evil man for the sake of *haqq* – truth and justice.

'But if it was so long ago, then why are you still crying?' I persisted.

Hassan told me that during the first ten days of Muharam, which were called Ashura, this event and its consequences were remembered. My grandfather Hadi used to host a recital in Baghdad on the last day of Ashura, and hundreds of people would go to his home to commemorate it. He added that Ashura was especially painful for our family, because it reminded us that we had been deprived of our own country.

'We are foreigners everywhere, and we have lost so much,' he said. He touched me lightly on the shoulder. 'You should know these things. They are part of your history, of who you are.' I hated what he said. Surely I belonged exactly where I was? My uncle sensed my discomfort. 'Do you deny your roots?' he asked, smiling. I didn't understand what he meant; he explained that he, my father, my grandfather and grandmother had once had another country, but that they had lost it. Their homeland was my home as well. I scowled. Lebanon was my country and my mother's country, Beirut the city where I had been born. I was not a foreigner here.

A slice of chocolate cake soon made me forget what my uncle had said, but on some level I dimly perceived that the grievance captured in the words of the song was the same as that which made my father's family weep in their exile. They were waiting to return to their homeland. Their lost country maintained a hold over them, the legacy of an inheritance centuries old.

The earliest indications of a settled civilization in the world are found in the region that is known today as Iraq. Between the fifth and fourth millennia BCE, lower Tigris and Euphrates basin cities such as Ur, Uruk and Larsa emerged and stratified societies developed within them. Mesopotamia – as the Greeks referred to the region between the two rivers – covered roughly the central southern part of what is now Iraq. Mesopotamia was also the term used to describe the provinces of the Ottoman Empire that belonged to this region. The ancient history of Mesopotamia is now lost to us, but it was mythologized by the Sumerians in epics such as the story of Gilgamesh, which was first written down in around 2000 BCE and which is a story of kingship and heroism that has informed and inspired people ever since.

The region that corresponds to the north of modern-day Iraq was the birthplace of the world's first empire. The Assyrians, descendants of the Akkadians who settled in the land of Sumer, engaged in what amounted to a conquest of the known world of their time – from Persia to Egypt. The Sumerians, Akkadians, Assyrians and later the Babylonians created what are, in effect, the foundations of civilization today. Our seven-day week, sixty-minute hour and much of our understanding of the constellation of the skies are the direct legacies of this defining period in human history. The mythologies of a large cast of gods and goddesses survived from this period too: Anu, the heaven-god of Mesopotamia, was the equivalent of Greek Zeus, while Ishtar or Inana was the goddess of love, war and fertility, and the precursor of Egyptian Isis.

The fortunes of Mesopotamia were largely dictated by its geography, in particular its position in relation to the frontier lands of the

Graeco-Roman and Iranian worlds. The territory changed hands as part of the ebb and flow of the respective powers of these historical entities. Alexander the Great and Darius the Persian fought over the land, while many prophets passed through it, including Abraham and Mani, the founder of the ancient but now extinct religion of Manichaeism.

The Muslim conquest of the region in the seventh century CE reconfigured the coordinates of Mesopotamia, and the Islamic empire transformed this former frontier land into the centre of a global empire. The region became central in shaping the ensuing Islamic civilization. In the ninth century, the Caliph al-Mansur ordered the building of a round city with four gates, which grew into a dazzling capital: Baghdad. Baghdad was not only the capital of the Abbasids' Islamic empire, but of a civilization. The *Tales from the Thousand and One Nights*, many of which speak of Baghdad, represent a vivid example of the city's fusion of cultures, mythologies and styles. The city also became an important trading centre on the Silk Road.

By the thirteenth century, Mesopotamia was a frontier territory once again after the Mongols' invasion that left its cities and sophisticated irrigation system devastated. In 1534, the region was captured by Ottoman Turks, but from 1623 to 1638 it lay in Iranian hands. My father's family originally came to Mesopotamia with the army of the Ottoman ruler Murad IV, Sultan of Sultans and God's Shadow Upon Earth. Murad was a warrior prince, famed for his prodigious strength and the last Ottoman Sultan to command an army on the battlefield. His campaign against Persia led to the invasion of Azerbaijan and Armenia. And, in the last decisive feat of Imperial Ottoman arms, Murad recaptured Baghdad from the Persian Shah Abbas I in 1638. The city remained under Ottoman rule for nearly four hundred years.

The three Ottoman *wilayets*, or provinces of Mesopotamia, that were referred to as the *pachalik* of Baghdad included Mosul in the north (which comprised part of the high Zagros mountain range extending from Turkey to Iran), the Kurdish regions, Baghdad itself and Basra in the south, perched on the Persian Gulf. These were subject to various different forms of administrative rule after their conquest by the Ottomans, whose central government was based in Constantinople in Turkey. Most usually, the *pachalik* was administered through indirect rule, which meant that local families or tribes

controlled the areas but paid taxes to the central Ottoman government. The system was changed in the first half of the nineteenth
century, when the authorities in Istanbul decided to impose direct rule
and sent an army along with a *wali*, or governor, to re-establish their
authority over Baghdad and to collect taxes in the name of the Sultan.
This diminished the power of many of the local leaders, especially
amongst the tribes, who remained resentful of the central government
in Istanbul.

By the early twentieth century, the Ottoman Empire was contracting. It had lost control of the Balkans and earlier of Greece, and was
gradually whittled down to half the size it had been in the sixteenth
century. As in the rest of the Empire, there was a multi-ethnic and
multi-religious population in the area of Mesopotamia that became
Iraq, consisting of Arabs, Kurds, Turks, Persians, Lurs, Sunnis, Shi'a,
Christians, Jews, Mandaeans and Yezidis, among others. The Iraqi
dialect of Arabic had strong Turkish and Persian influences. The blend
of cultures made for a rich, diverse but highly complicated society.

By the time they were exiled in 1958, as a consequence of a revolution that overthrew the monarchy, my family had become firmly
entrenched in Iraq's political life and society. Over the course of three
centuries they had transformed themselves from warriors into administrators and the confidants of the ruling family. They had arrived as
Sunni Muslims, but they left as Shi'a Muslims. They became the administrative rulers of Kazimiya, where they lived, which lay across the
Tigris to the north of Baghdad, but which is now a part of the city
itself. With the Ottoman reforms, the family's administrative role in the
town came to an end by 1865. However, they retained their high standing in society.

Across the region, the Sunnis were dominant, with the exception of
shrine cities such as Kazimiya. Generally excluded from political
power, the increasingly disenfranchised Shi'a populations of these areas
immersed themselves in learning and religion, criticizing their Sunni
overlords from the high ground of their religious authority. As Shi'as
who were deeply involved in politics, my family was caught between
two worlds. They were both insiders and outsiders.

* * *

The rudiments of my family history, with its tale of loss and privilege, were relayed to me principally by my uncle Hassan over the course of a few years in Beirut and London. The story whose seed he had planted in my seven-year-old head gave Iraq a status that grew inside me as I grew, and slowly came to embody my sense of the future: I created the country in my mind long before I ever saw it. Its importance was heightened by the impact Iraq had on my family once my father entered the world of politics in opposition to Saddam Hussein's regime in the early nineties, adding a layer of gravity, urgency and uncertainty to our daily life. Iraq came to dominate my thoughts, and I poured my imagination into this mythical place.

I first found a doorway to my Iraqi inheritance through learning about Iraq's culture and history. I imagined a place of scholars and antiquities, music and poetry, a multicultural haven of different peoples – Arabs, Kurds, Turks and Persians – and languages mixing together peacefully in a green and lush land by the riverbanks. This vision defied all the horrors of the country that I read about in the news.

But it was really anger that triggered me to write this book. My anger grew out of my experiences in Iraq in the aftermath of the war in 2003, with the US occupation of the country and the new political powers that were in place. I was angry at what I perceived initially as a country hurled back to the Middle Ages through misrule, neglect and sanctions, and a beaten people who had lost their voice long ago. I was also angry about what I saw as the expropriation of those people's silent voices, and of Iraq as a land by the US civil administration and the international press to serve their own agendas, political and otherwise. They became the designated spokespeople for an Iraq they barely knew and didn't care about, in the shadow of a greater preoccupation with the role of America in the region. They reduced Iraq to a desert of tanks, screaming women and barefoot children. The country's ancient history and cultural output over millennia meant nothing to them. I tried to understand the silence of the Iraqis themselves – perhaps it was the consequence of enduring fear, or a habit developed as the result of decades of oppression; perhaps it was their unfamiliarity with the latest means of communication owing to those long years of sanctions, I didn't know. One of Iraq's burdens has always been the way it is presented to the outside world as patchy, Manichaean, extreme. It is a nation that is portrayed either through its politics, most notoriously through Saddam and his regime, or through its ancient and glorious history, but never through its *people*.

The Iraq that I witnessed in person for the first time challenged all my preconceptions. It continues to do so, throwing back at me contradictions and tangents just when I think I am beginning to understand it, raising as many questions as it provides answers. It makes me wonder why there should be such a strong attachment to the country in my family. What does this attachment suggest? Does it represent a refusal to move on, to grow and embrace the world?

In the wake of what I saw for myself in their homeland, my family and their stories made me wonder anew about my own origins. Writing about their experiences challenged my notions of language, as I tried to render an Iraqi Arabic with all its idiosyncratic nuances into English. Most of all, it made me wonder about the very concept of Iraq: as a modern state, an ancient land, a nation, a word, a song, a river, a grave, a shrine, a statue of a deer.

In writing this book, I have been fortunate to have had access to a wealth of material: oral histories, archive materials, newspapers, buildings, relics, memoirs, music, interviews and photographs. This book is my attempt to make the unruly disciplined, to assemble the disordered, unorganized parts of the past into a cohesive narrative. As Iraq has an ancient oral tradition, and a great deal of this story was transmitted to me orally, I have tried to respect those elements, and to remain faithful to and respectful of the memories my family have entrusted to me. The timescale of memory is not the same as the timescale of history. Major periods of history can be summarized while minor periods can be expanded. This was certainly true of my family, who when speaking to me dwelled on their happy childhoods in Iraq, but for whom the revolution and many of the years following it passed in a blur. My family's stories of Iraq are more personal and intimate than a dispassionate and neatly constructed history. They show the country through the lives of people who have loved it.

BOOK ONE

Fallen
Pomegranates

DECEMBER 2007

I am walking through Kazimiya's alleyways, exploring the crumbling houses with Fatima, a friend from the town, and looking for the old family home. We are following the directions of an elderly historian who is too frail to show us the way in person. He has directed us verbally: turn left by the old train station, right at the donkey stables and left again by the old water pump. Because Kazimiya is a shrine town, and therefore quite conservative, I am wearing an *abaya* – a long over-garment – which keeps slipping off my head.

Narrow channels of water flow through the middle of the cobbled streets we walk along. Children play and old men sit in their shop fronts, watching us and muttering to each other as they wonder what these strangers are looking for. I wince, thinking about the century that has passed since my grandmother walked these same streets as a little girl, a daughter of this town, and of the many waves of people who have passed through this frontier land, contested between the Ottomans and Persia over many centuries, and later between the British and now the Americans. My grandmother was born an Ottoman subject, just like a native of Istanbul or Izmir, and here I am trying to find remnants of that time and place. She certainly wouldn't have struggled with the *abaya* as I do now, nor needed directions to find the main square.

We head towards the side gate of the shrine, Bab al-Murad, said to have been designed by the angel Gabriel, where my grandmother and many others once gave offerings to the poor in gratitude for prayers answered by the seventh Imam. I stop and do the same as I wait to go inside to make my wish.

I

Duty Calls

A Busy Day for Abdul Hussein

(1913)

ABDUL HUSSEIN CHALABI rose early, performed his ablutions, uttered his prayers unto Allah and the Prophet, then sat down to a large breakfast of his favourite food in all the world: fresh *gaymar*, cream of buffalo milk, velvety-smooth in texture, spread over just-baked bread and crowned with amber honey from the Kurdish mountains.

He was still tired. He had slept poorly and his troubled dreams had been of his dead sister. In them, she had stood silently before him, her eyes burning with reproach. He had tried his best to appease her, to explain that the family's decision had really been for the best. She had opened her mouth as if to speak, and then he had woken with a start. Alone in his bedroom, he had taken a few minutes to recollect himself before rising.

Feeling better for having eaten, he adjourned and sat with his head thrown back while his butler shaved his plump face. He could hear his mother, Khadja, barking orders at her servant across the long corridor that led to her quarters. Once shaved, he dressed as always after breakfast – never before – so that the belt of his cloak would not cramp his enjoyment of the best meal of the day. At the age of thirty-seven, Abdul Hussein remained very particular about his appearance, forever sending the servants into panic attacks with his complaints of poorly-ironed shirts and badly-polished boots.

He wore the typical attire of a sophisticated urbanite: a traditional robe tailored in Baghdad from *sayah*, a delicate striped cotton material

Oil portrait of Abdul Hussein Chalabi.

bought in Damascus, over white drawstring trousers. On his head he wore a fez, decreed by the Sultan in Istanbul to be the appropriate headgear of the modern Ottoman Empire. Abdul Hussein not only embraced this symbol of modernity; he believed that it suited his full face rather better than the old-fashioned *keshida* still sported by his eldest son, Hadi. A cloth wrapped around a conical hat, the *keshida* was also much more cumbersome than this new headgear. As a final indulgence to vanity Abdul Hussein smoothed the hairs of his moustache with a small bone comb he had purchased in Istanbul.

As for any powerful and influential man, his day was ruled by a rigorous schedule. He barely had time to browse the morning papers before a servant came to inform him that at least ten people awaited his presence in the *dawakhana*, the formal drawing room in the men's quarters of the large yet overrun house.

As a mark of his status it was Abdul Hussein's lot to sit and receive men all morning in the *dawakhana*, a ritual chore inherited from his father and from the grandfather he had never known. Men came to him in need of services, favours and assistance. They presented him with their problems concerning their lands, the government, the tribes, the mullahs, the weather, even – on occasions – God Himself. Abdul Hussein would sit in a wooden armchair and listen carefully as the *chaiqahwa*, the tea-coffee boy, made his rounds of the assembled visitors.

Today, his enthusiasm for the job in hand was at a particularly low ebb. He was still unsettled by reports of the Ottomans' latest defeat in the Balkan war and the subsequent loss of the majority of the Empire's European provinces. At a meeting in Baghdad the previous day, the Governor had broken the bad news to the *Mejlis-i-idare-i-Vilayet*, the advisory council for the Baghdad Vilayet, of which Abdul Hussein was a member. The prognosis was dispiriting; the Ottoman Empire was in decline.

A servant interrupted his thoughts with a message from the house of his sister Munira. She was married to Agha Muhammad Nawab, a wealthy Indian Shi'a notable who, Abdul Hussein learned, had just returned from a long trip to India. Abdul Hussein read in the note that Munira had recently taken delivery of a new arrival, which, the Nawab promised, would interest him greatly. He sent his sister a reply to let her know to expect him for lunch. Today's challenge would be to discharge his duties as swiftly and shrewdly as his wits would allow him so that he would be free to call on the Nawab in the afternoon.

The *dawakhana* began with the usual exchange of greetings. Ibrahim, the wiry manager of one of Abdul Hussein's estates, was waiting to give him his weekly report on the progress of the crop in one of his citrus and pomegranate orchards, which lay north by the river. Although Abdul Hussein had visited the land only a few days earlier, he could always rely on Ibrahim to show up with something fresh to complain about – any excuse to visit the *dawakhana*. Addressing Abdul Hussein as *hadji* in reference to his recent pilgrimage to Mecca, Ibrahim began to explain the reason for his presence.

He complained that the Bedouins were cutting the telegraph lines again – he had even caught one them using the lines to tie goods to his donkey – and that his men were having to waste their time throwing the nomads off the land, while the authorities did nothing to tackle them.

Abdul Hussein knew his manager was right – that over the past few decades the area's administrators had fallen into slipshod ways; the vandalism of the telegraph lines was only part of an ongoing problem which the Ottoman Governor had not been able to resolve. Whenever there was a lull in security, the desert tribes attacked the towns. Lying a few miles to the north-west of Baghdad, across the River Tigris, Abdul Hussein's home town of Kazimiya's proximity to the open desert to the north made it vulnerable even though, like Baghdad, it had walls and gates that were meant to protect it.

Abdul Hussein felt frustrated that he could not do more himself to maintain public order. But the grand old days when his family had been Kazimiya's rulers had passed. The family surname, Chalabi, was an honorific title that came from the Turkish *Çelebi*, a term which had several meanings, amongst them 'sage', 'gentleman' and even 'prince'. It had been bestowed on the family when they had administered the region for the Ottomans. As direct rule had been imposed on these parts by Istanbul some decades earlier, the Chalabis no longer performed those duties, although they remained at the heart of Kazimiya's political and social life.

Ibrahim's father had worked for Abdul Hussein's powerful father Ali Chalabi, and Ibrahim had grown up hearing stories of the latter's iron fist and courage. Ali had been feared in Kazimiya, even hated by some, but he was admired by Ibrahim: this much Abdul Hussein knew. And so Abdul Hussein could hardly object when Ibrahim decided to share with the *dawakhana* a story he had heard concerning Ali's ingenuity.

Abdul Hussein's younger brother Abdul Ghani joined them as Ibrahim began to tell their guests how, one day many years ago, news reached Kazimiya of a plague of locusts that was approaching the city from the north. The sight was terrifying: a cloud of dark green insects rolling towards them at startling speed. No force on earth could push them back, making the townsfolk panic as they prepared to ride out the attack behind locked doors. Many had already resigned themselves to losing their crops, and with them their annual profit, but not Ali *effendi*. That year he had decided to plant a new crop of tomatoes. Determined to save them, he summoned the farmers to discuss what could be done. Then he threw his camel-hair cloak, his *abaya*, to one side and paced up and down his land for two days in his muddy boots.

Ali Chalabi, seated centre, holding his youngest daughter,
surrounded by family and friends.

His guards followed him, their rifles on their backs. Shaking his head,
Ibrahim said, 'None of his employees had ever seen him behave like
this.'

The rumour spread around Kazimiya that the great Ali Chalabi had
gone mad. Witnesses described how he would stop in his tracks,
scratch his dark beard, look up to the sky, turn around and resume his
pacing. By the second day, his guards were wilting in the heat and gave
up marching after him, yet Ali was so distracted he didn't even notice.
Suddenly, they heard him shout, 'I've got it! I've got it!'

9

He ordered his men to go to all the markets and buy every single mud pot they could find. 'And don't you dare tell anyone what you're buying the pots for!' he added. Intimidated, his men scoured every market in Baghdad. As soon as one returned with a batch of pots, Ali would send him off to buy more. When he had laid his hands on every pot in the area, he ordered the farmers to take them and cover every single tomato plant, hiding them from the locusts and making his fields look like a pockmarked sheet of baked mud. No one else in Kazimiya had thought of the idea, and when the locusts arrived the next day all the crops were ruined except for Ali's tomatoes. He made a fortune.

'*Allah Yirahamah*, may he rest in peace. We need men like him today to guide us through these changing times,' Ibrahim concluded. Abdul Hussein had never approved of his father's severe nature, but he knew what Ibrahim meant. The teetering Ottoman government had been radically transformed in 1908, when a group of nationalist Turkish army generals – the 'Young Turks' – had seized the reins of power from the Sultan in Istanbul, limiting his role and facilitating a new constitutional era. The events of 1908 had at first brought with them a new energy, promising freedom and equality for the many multi-ethnic communities of the Empire. But gradually it became clear that the Young Turks were promoting a European-style nationalism, with Turkishness as its main identity. The Arab people of the Ottoman Empire had begun to feel increasingly marginalized and disadvantaged as swathes of secular modernity swept through the Empire to the west of them.

Abdul Hussein had entertained certain hopes for the modernizing projects proposed by Istanbul. On paper, the proposed German-engineered Berlin-to-Baghdad railway had been more exciting than anything the locals had dreamed of ... but now, what of it? The Germans were still viewed positively as an advanced industrial people who had come to help develop Mesopotamia, yet very little track had actually been laid since the project's inception a year ago. Many pieces of machinery already lay abandoned, surrounded by rubble, collecting dust or rusting. To Abdul Hussein it was a source of bewilderment that even deepest Anatolia had already been linked up to the rest of the world by rail. *Why not us?* he wondered.

He could see that new ideas did not grow as freely or as quickly in Baghdad as in Istanbul, for all the new cafés, newspapers and government schools that were now springing up. Kazimiya was even slower to embrace change, partly because it was a Shi'a shrine town and therefore more religious in outlook. Abdul Hussein felt as many did that this backwardness was enforced from above, as a consequence of the town's predominantly Shi'a character: under an Ottoman system dominated by the Sunnis the Shi'a were never going to receive their proper due. Politically they were weak, and everybody knew it.

He sensed a haughtiness and disregard among Ottoman officials when it came to his people and this land. Midhat Pasha had been the only governor to do anything for Baghdad, but he had left the area in 1873, three years before Abdul Hussein was born. The new constitutional reforms, with their accompanying bureaucratic language, threatened to alienate Abdul Hussein from his own heritage. And now he had to shed his claim to it, because in the eyes of central government he was an Arab from Kazimiya. Yet he considered himself every bit as Ottoman as any Istanbuli, whether they liked it or not.

Next, an anxious young man introduced himself as coming from Baghdad and explained that he was visiting the *dawakhana* with a common acquaintance. Clearly upset and frustrated, he said that he had been attacked earlier that morning by some robbers in Agarguf, an archaeological site located in the desert outside Kazimiya. Like Ibrahim, he blamed the government, and complained to the *dawakhana* about poor security, poor governance ... Abdul Hussein soothed the man, and called one of the servants to attend to him and, when he had rested, to hail a *rabbil*, a carriage, for him from the station down the road.

No sooner had Abdul Hussein turned around than another man raised his voice to air his grievances. His brother had broken his shoulder in an accident the previous week, and he wanted to claim compensation from the *trammai* company, in which Abdul Hussein and his family were major shareholders. The tram tracks were now more than thirty years old, and – like all else in the locality – suffering from government neglect. 'May God heal your brother!' he reassured the claimant. 'We will give him compensation, don't worry.'

Over the next couple of hours Abdul Hussein listened carefully to a torrent of requests and concerns, at times instructing his clerk to note down information for him to follow up. His visitors were more than

The Baghdad–Kazimiya tram, circa 1910.

living up to the locals' reputation for cool nerves and slow conversation. The *Kazmawis*, the townsfolk of Kazimiya, were not nicknamed 'cucumbers' for nothing.

Finally his steward Sattar burst into the room to tell him that he was awaited at the shrine, where he had promised to be by noon. Abdul Hussein apologized to the remaining men in the *dawakhana*. All nodded knowingly; when the shrine called, everyone heeded. The men rose in unison to say their goodbyes.

Abdul Hussein went to the shrine with Sattar on foot. The ten-minute journey would have been difficult to negotiate in his preferred mode of transport, his Landau carriage. Instead, the two men manoeuvred their way through dark unpaved alleyways until they reached the first fruit and vegetable stalls on the fringes of the main square. An old woman was selling *baklava*, struggling to bat away the flies that buzzed around her tray of wares. The overpowering smell of the market assailed them, the stench of the dirt on the ground mixing with that of the running water in the open culverts.

There were a good many foreigners in the vicinity, and a medley of languages filled the air. Many Persians were milling around, others sitting on the ground to sell their goods, mostly foodstuffs. The locals relied on the steady influx of these visitors, who sold important supplies of Persian specialities, particularly the much prized saffron. The market was also busy with Indians dressed in their *salwar kamizes*, Afghans in heavy coats and curly-moustachioed Iranians. There were as many types and styles of hat as there were people – the *charawiya*, the *yashmak*, the *arakchinn*, the *keshida*, the *saydiya*, the *fina*, the *'igal*. There were even a few travellers from as far away as Rangoon. The horse-drawn tram had just made its final stop at Kazimiya station, which lay close to the shrine, and people were pouring out of the double-decker carriage, old women complaining loudly as they were jostled by young boys.

In Kazimiya, the courtyard of the shrine was the first port of call for anyone wanting to find out the business of the day, from crop prices to political intrigues, to the state of the river. It was, therefore, an indispensable meeting place for any influential man in the town. As with the other shrine cities of Mesopotamia, there was a strong Persian influence in Kazimiya, and the beauty of the shrine owed much to the piety of the Iranian and Indian Shi'a, whose financial and artistic contributions had flowed into it through the ages. Unlike other Shi'a shrines in

The shrine of Imam Musa al-Kazim and Imam Muhammad Jawad in Kazimiya was renowned for its two domes and four minarets.

Mesopotamia, Kazimiya's boasted four golden minarets on the outer side of the larger square complex that contained it, and two golden domes. They gave a dramatic dressing to the building, dominating the skyline.

The shrine housed the bodies of Imam Musa Ibn Ja'afar, the seventh Imam, and of his grandson Muhammad Jawad, the ninth Imam, who were direct descendants of the Prophet through his daughter Fatima. The seventh Imam had lived during the golden age of Baghdad, before being poisoned by the Caliph, Harun al-Rashid, the notorious figure mentioned in the *Thousand and One Nights*.

Abdul Hussein and Sattar entered the shrine through the silver-gilded main gate, the Bab al-Qibla, which opened onto a vast white square courtyard. Immediately they veered right under the arcaded, turquoise-and-yellow-tiled gallery. They were recognized by one of the many children who virtually lived at the shrine, and the boys dashed over to them, tugging at their robes. It was well known that Abdul Hussein always carried *numihilu*, lemon-flavoured sweets. He didn't disappoint them, but reached deep into his pockets and gave each child a sweet. They tended to know better than his adult friends inside what the news of the day was, and they never hesitated to repeat what they had heard that morning.

In turn, Abdul Hussein took pleasure in gently scolding them. 'You again? I thought I told you to go to school, *ya razil*, you rascal! What will loafing about here with your friends do for you when you grow up?' He would hold up his hands in mock horror. '*Yalla*, don't come back to me for a sweet if you don't go to school!' Part of him knew such chastisement was futile. These children had no compass in life. It wasn't their fault – their parents were no better – and there weren't enough schools in the locality to take on all these children.

Leaving the clamouring children in their wake, Abdul Hussein and Sattar crossed the courtyard. The shrine was half-empty. It was late morning, and the seminary students who occupied several of the alcoves in the arcade galleries were sitting, legs crossed, listening to their teachers. Their black and white turbans, indicating their lineage, or lack thereof, to the Prophet, moved up and down as they looked from their books to their teacher and back again.

Abdul Hussein's friend the Kelidar, the shrine's hereditary overseer, was standing with a group of men near the main entrance to the tomb room. Several old ladies in thin black *abayas* were seated nearby on the tiled floor. They leaned against the outer wall of the tomb room, chatting. Against the yellow brick, the patterned tiles shone turquoise, white and navy in the sun.

Before he could join the Kelidar Abdul Hussein felt his attention drawn to a well-dressed older man sitting alone in a corner, weeping. Puzzled by this distressing sight, he walked towards the man. '*Assalamu alaikum* – peace be upon you, my friend,' he said courteously.

'*Wa alaikum assalam* – and upon you,' the man replied, brushing tears from his eyes with the heel of his hand.

'*Khair*, what is the matter with you? Why are you crying?'

'What can I tell you, *ammi*? Life has dealt me a cruel blow,' the man said, pulling himself upright. He cleared his throat and explained: 'God blessed me with a good fortune and I decided to divide it between my three sons. And now their mother has died and they don't want to see me. I go to my eldest son and he barely offers me tea, the second one is always travelling and the third one is too scared of his wife, who doesn't like me.' He shook his head sadly. 'I've always been a good Muslim, praying, fasting and giving alms – and I can't believe this has happened to me. I've been left all alone. What kind of children are these?' he asked in despair.

Abdul Hussein nodded gravely. 'Don't worry, my friend; there is a solution for everything.' He promised to discuss the matter further with him once he had concluded his business with the Kelidar. Politely, he took his leave and crossed the courtyard.

'*Hadji*, good that God brought you!' the Kelidar exclaimed. The two men greeted each other warmly. 'And how is the child?' the Kelidar asked. Earlier that week Abdul Hussein's son Hadi had given refuge to a little lost soul who had been haunting the shrine. The small boy had been found sitting in a corner of the shrine, crying and refusing all offers of help, even turning away a glass of water. But Hadi had broken through the boy's misery, established that his name was Ni'mati, and offered him a place to stay under his father's roof. Abdul Hussein assured the Kelidar that the child was settling in well.

'That's wonderful news, I can't thank you enough!' The Kelidar then proceeded to business. He explained that the town was organizing a welcoming committee to accept the gift of some new carpets which were arriving from Iran the following week, having been purchased through the Oudh Bequest. The result of a complex diplomatic agreement, the Oudh Bequest involved the political authorities of the Indian Raj and the Shi'a religious powers of Mesopotamia, and benefited the shrines in the region with regular improvements and maintenance. However, the mayor of Kazimiya had fallen out with the Nawab family, the Indian Shi'a notables who were in charge of administrating the bequest. One of them, Agha Muhammad Nawab, was Abdul Hussein's brother-in-law, and the Kelidar hoped that Abdul Hussein could smooth the way for the ceremony.

For Abdul Hussein the request was a godsend; it meant that he could safely ignore the many other matters that the Kelidar hoped to raise with him, on the pretext of hastening to undertake this vital errand. He brushed his moustache with his thumb and index fingers, and said, 'Zain – fine. I'll undertake this without further delay.' He excused himself and made his exit from the shrine, sending Sattar home ahead of him to prepare the landau for the short journey to the Nawab's house. Following on Sattar's heels as fast as he could, he huffed his way through the heat to the stables next to the house. 'Bring me some water, quickly!' he called out into the courtyard.

His son Hadi's new friend, Ni'mati, came out carrying an engraved copper pitcher of water and a matching copper cup on a tray. Abdul Hussein smiled at the boy and rested his fez on one knee while he mopped the sweat from his brow. Then he drank deeply while his carriage was readied. Refreshed, he took his place on the left-hand side of the carriage seat. Accompanied by the steadfast Sattar, who never left his side outside the house, Abdul Hussein was soon on his way.

The palatial home of Agha Muhammad Nawab, Abdul Hussein's brother-in-law, lay outside Kazimiya, south towards Baghdad. Abdul Hussein liked to take the picturesque route to it that ran closest to the river. There was one field in particular that he adored because of its wheat and reed sheafs, which stood nearly as tall as him. Their golden hue when the sun shone upon them was one of his favourite colours, and a momentary sadness always surfaced in him when the time came to cut them down.

The carriage drew up in front of the house and Abdul Hussein dismounted. Standing placidly in the waters of the rectangular pool was the new addition to the estate: a life-size stone statue of a deer. Abdul Hussein clapped his hands and repeated '*Ayaba, ayaba*' in admiration as he walked back and forth around the pool, studying the stag from all angles.

Some of the garden staff and stable boys drifted over to watch, amused by his uncustomary excitement. They explained to him that the deer had arrived yesterday as a surprise gift for their master's wife. Praising it, Abdul Hussein supposed that the Nawab had bought the statue to remind him of Hyderabad; there, deer hunting was a noble sport often portrayed in exquisite Mughal miniatures. In giving Munira the stag as a gift, the Nawab was also bringing a little of his home country to Baghdad.

A gardener told him that the stone deer had made the journey in a large wooden crate, travelling from Bombay to Basra by boat; then by steamer to Baghdad, with many stops along the way, until it had been conveyed by a small vessel to the jetty at the bottom of the Nawab's garden, and finally installed in the middle of the pool.

The deer statue through the gates of the Deer Palace, circa 1925.

As absorbed as he was in the deer, Abdul Hussein was nevertheless looking forward to lunch, although not particularly to the company of his sister Munira, whose long face had tested his patience of late. Yet he hoped that she had prepared the food herself, as she sometimes did, because her culinary skills were superior to those of any of her servants. Her *turshi* pickles were legendary and, like a good bottle of wine, only improved the longer she kept them. Many years ago she had pickled an exceptionally fine batch of cucumbers, storing them in jars, one of which remained in Abdul Hussein's pantry, where it was coveted by all.

When Abdul Hussein crossed the threshold he learned from a servant that the ageing Agha Muhammad Nawab was already taking his afternoon nap, recovering from his long trip. The dining room was empty, except for a servant girl who was laying out the dishes on the table. Munira was nowhere to be seen, but Abdul Hussein could hear her voice issuing faintly from the kitchen, where she was supervising some final touches. That was a good sign, he thought; she must have done the cooking.

As he sat down, Abdul Hussein automatically reached for the pickles, stuffing one into his mouth. He had barely begun crunching on it when his sister appeared and sat down in silence across the table from him.

Greeting her, he said, 'What a wonderful deer the Nawab has brought for you! Your husband has really outdone himself this time!'

'Yes,' Munira replied dully. Her eyes with their deep shadows remained fixed on her hands.

Abdul Hussein held his plate out to be served another favourite dish: aromatic basmati rice infused with dried lime and saffron-flavoured chicken. He tried another tack. 'Have you heard about the little boy Hadi found by the shrine, crying?' Munira raised her head and glared across the steaming dishes at him. This time Abdul Hussein avoided her gaze.

He knew in his heart that she wouldn't want to hear about children; that she blamed him for her childless marriage to the man she held responsible for their sister's death. Their sister Burhan, the Nawab's first wife, had been subjected to all the same rumours of barrenness and inadequacy that taunted her now. True, Abdul Hussein had thought that the match would be a good idea; the Nawab was a rich and influential man who could give Munira a good life. When Burhan had died and Munira had been married to him, nobody had known for certain that he wouldn't be able to sire a family.

'He doesn't speak a word of Arabic,' continued Abdul Hussein. 'Probably from Hamadan or somewhere around there.' When Munira sniffed unsympathetically, he snapped, 'Not every fifteen-year-old would have done what Hadi did, and brought him home. You could at least be proud of your nephew!'

Munira glowered, but Abdul Hussein persevered: 'Anyway, he seems to have taken well to the horses in the stables. We'll sort him out. His name is Ni'mati.'

Munira leaned forward, her face covered with one hand, her elbow propped on the table-top. Abdul Hussein abandoned any further effort to enjoy his lunch. 'What is wrong now, sister? Speak!' he exclaimed. 'You have everything you could possibly want – this palace, a respected and rich husband. A beautiful garden, that wonderful deer. All these servants … What on earth is *wrong*?'

'This deer will be a curse,' Munira said sullenly. 'There has already been a crowd outside, staring at it. And they've started calling our house "the Deer Palace".'

'What rubbish!' Abdul Hussein erupted. 'Let the people talk – they talk anyway, and now at least they'll have something pleasant to gossip about.' He rose to his feet, threw his napkin on the table and curtly took his leave of her.

Munira remained in her seat. Her fingers clenched her water glass, which she suddenly hurled at the wall. She watched as the liquid dribbled down to the floor. The servants could clean up the broken shards later.

Abdul Hussein called for Sattar and the carriage driver, but they were nowhere to be found. He was so angry that he forgot to leave the Nawab a message about the important business of the shrine and the carpets. He tucked his fez under his arm, squashing it with the sheer force of his irritation, and started to march out of the garden, eyes fixed on the ground. Belatedly, he shouted back at a gardener, 'Tell those imbeciles to follow me now!'

He crossed the grounds that faced the newly-named Deer Palace and walked down towards the riverbank, where he waved at a boatman to bring his *guffa* over. Guided by the expertise of such boatmen, round-bottomed *guffas* had been whirling their way down the Tigris for thousands of years. As the little boat transported Abdul Hussein towards Baghdad, the soft breeze calmed his heated temper a little. He could see boys flying homemade kites from the rooftops that lined the river. On the other side, he spotted one of the steamers of the British Lynch company heading south to Basra. It was time for his siesta, but he was still too agitated to rest.

He decided to cross to the eastern bank near the old city, where he could sit in one of the cafés near Maidan Square and smoke a *nargilleh* – a water pipe. Many cafés had sprung up there in the last few years, havens of music and liveliness, and Abdul Hussein was sure a visit to one would lighten his mood. But as the small boat neared the bank, he remembered the weeping old man back at the shrine, saddened by

A *guffa* on the Tigris in Baghdad,
circa 1914.

the behaviour of his three errant sons ... An idea came to him, and he told the boatman to turn around and take him to the pontoon bridge at Kazimiya.

The pontoon bridge consisted of wooden boats tied together. As Abdul Hussein approached it, it gently rocked from side to side. Observed from a distance, the crowds of women in their black *abayas* who were crossing the bridge formed a single swaying mass.

Disembarking nearby, Abdul Hussein paid the boatman and made his way back home. There he found Sattar, and asked him to send one of the servants to fetch a builder and his tools. He ordered another member of his staff to find a huge metal cooking pot. The boy returned with the household's largest pot, which could hold enough rice for fifty people. The boy must have thought his master had gone mad when he told him to fill it with soil and then cover it carefully so its contents were not visible.

Next, Abdul Hussein sent Sattar to visit the weeping man's eldest son and invite him and his brothers to come to his house straight away. Surprised, the young men returned with Sattar to find the large covered pot waiting for them in Abdul Hussein's courtyard. Abdul Hussein grinned at his guests and gestured to the pot: 'Your father has left this pot and its contents with me in safekeeping for you. I'm instructed to give it to you once he has passed away.'

Presumably concluding that their father had even more money than they had imagined, the three brothers obediently followed Sattar and the other two servants into Abdul Hussein's house. There, Abdul Hussein introduced them to the builder he had summoned, and explained, 'I am going to store the pot here in this corner of my stables. This builder will construct a small box to cover it so that no one can tamper with the contents until it's time. I want you to witness his work now.'

Some weeks later the old man came to visit Abdul Hussein, looking very much happier. 'I don't know what you've done,' he exclaimed, 'but my boys have come back to me! They're completely changed, and now each one takes his turn to look after me. I'm so relieved.'

'That *is* good news,' Abdul Hussein said warmly. 'I simply reminded them of the Holy Book's recommendation that we care for our parents.'

'*Allah yikhalik* – may God protect you. They seem to have heeded your advice. I don't know how to thank you.'

Abdul Hussein smiled. He was quite sure those sons deserved the eventual disappointment of discovering that the pot was filled with mud. The important principle, as always, was that until then good order and harmonious relations be restored within the family.

2

Stacking Rifles

Hadi and the War

(1914–1916)

BY EARLY NOVEMBER 1914, the frivolities of stone deer and spurious pots of gold were far from Abdul Hussein's mind. Turkey had officially entered the Great War on Germany's side.

The Germans had been consolidating their relationships with countries in the Middle East for several decades through a variety of measures, such as assisting in reforms within the Turkish military and helping to build railways. The ambitious Berlin–Baghdad railway project had been drawn up to give the Kaiser direct access to Mesopotamia's oil fields. Given the strong German presence in the Ottoman Empire, there had been no real choice of sides for the Sultan to take once the war broke out.

A few days after this ominous development, the menfolk of the town convened in Abdul Hussein's *dawakhana* in the late afternoon. The mood was bleak. They sat puffing away on their cigarettes, drinking *istikan* after *istikan* of tea nervously and noisily. Some slumped back in their chairs; others were hunched forward, chins propped glumly in their hands. Indeed, their heads looked so heavy that their various headdresses – fezes, *yashmaks* with *i'gal* cord which held the cloth on the head, *charawiya* caps – seemed to be falling off or else tilting sideways. All the assembled men, Abdul Hussein included, had but one thing on their minds: the lives of their sons.

The military had begun to enlist young Muslim men to serve at the front. During past military campaigns there had been a systematic procedure of conscription according to ages and professions, but this time the Turkish Sixth Army division, headquartered in Baghdad, had simply sent out sorties with instructions to bring back all able-looking men.

There were reports of boys as young as fifteen – a year younger than Abdul Hussein's eldest son Hadi – being dragged screaming from their homes by roving patrols, even as their mothers pleaded with the soldiers. Men took to hiding, and locals helped each other to avoid conscription. At the sound of the first drumbeat in the town square, and the rallying cry of 'Safarbarlik var, safarbarlik var', many ran out of their shops and homes, some disguised in women's *abayas*, some fleeing the city to seek refuge among the tribes. And so the army adopted a wilier and yet more pitiless strategy, arresting the next of kin in order to put pressure on dodgers and deserters. Some deserters were even hanged, *pour encourager les autres*.

Non-Muslims such as Christians and Jews were exempt from conscription as was traditional in the Ottoman Empire, where all wars were fought in the name of Islam. Instead, a hefty exemption tax of thirty Ottoman gold pounds was imposed on them. Military doctors and conscription officers profited by taking bribes and accepting favours from the *effendis,* the urban elite, to send their sons to local posts instead of to the front.

Abdul Hussein's fears for Hadi were exacerbated by his worries over the wider political situation. The Ottoman Sultan had called on all his Muslim and Arab subjects to fight the British infidels who were attacking the realm. The pronouncement was clear: they should mobilize as Muslims in this war, a sentiment that resonated deeply in all of them. Yet this feeling was clouded by disquiet. Did the state really represent them any more?

For all his disappointment at the slow pace of reform, Abdul Hussein had a very deep attachment to his Ottoman world, yet much about Turkey's involvement in the war seemed illogical to him. As the men in the *dawakhana* weighed the war in the balance, tempers became increasingly frayed and voices were raised. He endeavoured to be the voice of reason.

His brother-in-law Abdul Hussein al-Uzri was among those to fan the flames by dismissing the war as a European conflict that was irrelevant to their lives in Mesopotamia. Married to Abdul Hussein's sister Amira, Uzri was a poet and the editor of a local newspaper, *al-Misbah*, in which he freely aired his views. His comments provoked a furious response from Abdul Ghani, Abdul Hussein's younger brother, who argued that war in Europe concerned them greatly; that the Germans needed the Sultan to be on their side in order to buffer the Ottoman lands from the encroachment of the British and Russians. Moreover, he said, it was their basic religious duty to fight *el-Ingiliz*, the English, who were coming to occupy their land. He turned to Abdul Hussein for support.

Abdul Hussein chose his words carefully. 'If they are attacking Islam we must defend our faith – after all, our Sultan is the head of the Caliphate – but we must be clearer on the premise of war.'

'I say we must fight, we must defend our faith! They will attack our land, they will control our holy sites,' Abdul Ghani insisted.

'What a fool you are, Chalabi!' Uzri countered. 'They're not interested in our holy sites; they want to protect the oil fields in the south, in Persia; that's what they want – they don't care about our Imams. *This war is not our war; it doesn't serve anything!* Istanbul doesn't care about us, the Ottomans simply want to sacrifice us for their vanity,' he shouted excitedly, throwing his fez on the ground with force.

Abdul Ghani folded his arms stubbornly. 'I disagree. This is about our faith, and we have to defend ourselves. What do we have if not our religion?'

Abdul Hussein watched in dismay as the dispute grew more heated. Several of his guests were of the view that the Ottomans were ill-prepared for war, whatever God's will for the outcome might be. One man glumly volunteered that the British war machine would crush any Ottoman opposition. Another spoke in favour of adopting Iranian papers, as Iranians were exempt from conscription. Finally, one asked Abdul Hussein to tell them his opinion.

Abdul Hussein reflected for a moment before replying. Should he tell them what he really thought? That this was the beginning of the end? He cleared his throat. 'Those new men in Istanbul have changed things so dramatically,' he said. 'They want to be Turkish now, not Ottoman or Muslim. We're an afterthought for them.' He was referring to the fact that the Ottoman Empire had always been a multi-ethnic

Muslim territory even though its rulers were Turkish, but now the Young Turks were placing their own nationality centre stage. He shook his head. 'This is a bad war, and I don't know why the Sultan has agreed to be part of it. But he is our Caliph and he has declared *jihad*, holy war.'

When Basra fell to the British a month later, the Sultan called for *jihad* across the Empire. All of Baghdad's mosques rallied men to join the fight, and Kazimiya's men were roused to action by the fiery Friday prayer speeches of the mullahs at the shrine.

Hadi was only sixteen years old. After many sleepless nights, his father, desperate to save him from conscription, used his influence to secure him a post under a Turkish general who had taken up residence with his retinue in the Deer Palace. Abdul Hussein's brother-in-law Agha Muhammad Nawab had died of old age in the summer, and Munira, now widowed, had wasted no time in moving to the Nawab's other house in Kazimiya to be nearer to her mother and family. She had inherited property as well as a considerable fortune from her late husband, and the move also meant she could be closer to her farms. Curiously, Munira seemed much happier than when Abdul Hussein had visited her for that miserable lunch a year ago. These days she listened attentively to what he had to say, and took a real interest when he spoke of his concerns for Hadi's safety and the problems with the education of his other children in these difficult times. Their sister Burhan no longer seemed to cast a shade over their conversations.

One evening Abdul Hussein returned home with the news that his son was to report for duty at the Deer Palace the following morning. Despite his initial disappointment at the lowly and loosely defined post assigned to him, Hadi approached his job with enthusiasm. Every day he rode out very early in the morning, often accompanied by Ni'mati, who was as sinewy and dark as Hadi was robust and fair, to collect fresh fruits and vegetables from his father's lands for the officers. Knowing his father's fondness for *gaymar*, Hadi purchased this for the men's breakfast from a woman who lived in one of the reed huts by the riverfront further up from Kazimiya. The woman kept buffalo, and a few clay pots would arrive daily, filled with their lightweight, fluffy, extra-white cream, which was devoured by the officers.

A young Hadi, wearing the typical *Keshida*, standing behind
male realtives, circa 1912.

In addition to ensuring that the Deer Palace was well stocked with
fresh produce, Hadi soon became a messenger, delivering letters to and
from the Military Headquarters in the Citadel in Baghdad. This gave
him the opportunity to discover the city itself. Ni'mati, spared
conscription like many of the Iranian household staff because he wasn't
an Ottoman subject, was often his companion on these errands too.
Abdul Hussein's steward Sattar, on the other hand, had fled north to
hide among his relatives in the Kurdish mountains as soon as the forced
conscriptions started. Two stable boys had also disappeared overnight
without warning.

Usually dressed like his father in traditional attire, Hadi replaced
his civilian clothes with a basic military uniform and carried a satchel
for the post. He cut a pleasing figure in his new outfit, and was generally considered a charming young man. Raised and educated in the
town, he was very attached to Kazimiya, and was known for his active
involvement in community events. Every year he helped to organize the

Ashura processions, and he was a talented horseman who enjoyed displaying his skills at both Ashura and the Eid holiday that marked the end of Ramadan, when he would parade through town on horseback, sporting a sword and shield.

For all his popularity, Hadi was humble by nature and earnest in his enthusiasm and concern for people. His honest round face appeared all the brighter for the dark fez he had begun to sport, again in the manner of his father. He had great respect for Abdul Hussein, but he was more impulsive than his father. He already had a keen eye for the ladies, but was skilful at concealing it most of the time.

Now a part of him wished that he could indulge his adolescent dreams by putting on a proper soldier's uniform and fighting in a battle. But he was also aware of the harshness of the army, and the cruelty with which soldiers were often treated. He was unsettled by the way in which civilians were sometimes pushed around by low-ranking soldiers, who were simply replicating what their superiors did to them. And he was disgusted by the sense of entitlement many of the officers displayed, showing no respect for people's property. Some of them even raided shops for personal profit. Most of all, he hated the Turks' disdain for the Arabs. He couldn't understand why this should be so, since the Turks and Arabs shared a religion and a king.

Many decades later, Hadi would remember his humble role during the Great War, which he felt had imbued in him both his curiosity and his resourcefulness, which allowed him to thrive amidst chaos.

Employed in the offices of the Commander-in-Chief of the 6th Army, Nur al-Din Beg, Hadi learned how to make himself virtually invisible in a room, all the better to observe proceedings and acquire knowledge. Besides the idiosyncrasies of the generals he encountered, particularly their asperity and ill humour, he was fascinated by the organization of the military. He learned to appreciate the importance of timekeeping and personal accountability. While general morale was low, as the citizens of Baghdad and Kazimiya continued to feel that the conflict in their region was of marginal interest to the Ottoman government, it was still a war, and lives were at risk.

* * *

As more family friends and acquaintances became aware of Hadi's new position, he was increasingly entrusted with letters to pass on to Nur al-Din Beg's office. Initially this caused him concern: the proprieties of rank meant he was in no position simply to place such letters directly into the hands of his superiors, and he worried about how best to deliver them without breaking with protocol. He was uncomfortably aware that a good many of them were requests for compensation, usually relating to goods taken by the army without payment. However, some contained military intelligence about battles being waged on the front to the south of Baghdad.

It was Hadi's good fortune that the first time he dared to hand over a letter it contained useful intelligence for the Commander rather than a simple grievance. Encouraged, he handed over more; some proved useful for gauging the mood of the civilian population, while others offered detailed information about the tribes further south, the morale of the enemy troops and even the weather, including the hazards of sandstorms and dust clouds.

Hadi began to feel personally involved in the war effort, and studied the generals carefully whenever he got the opportunity. He observed the tensions between them and the ways in which they organized their staff, and was horrified when he saw soldiers being flogged for their misdemeanours, or when he was forced to be present at the execution of deserters.

Closer to home, Hadi's maternal uncle, the poet and newspaper editor Abdul Hussein al-Uzri, was rounded up from his house in Kazimiya in the spring of 1915 on the orders of Nur al-Din. Uzri had been outspoken in his editorials, criticizing the Ottoman position and calling for self-determination and Arab independence. As a punishment, he was imprisoned in Kayseri, ancient Caesaria, in the Anatolian heartland. With him were several Baghdadi men of letters, including the notable Pere Anastate al-Karmali, a Jesuit scholar who has contributed substantially to Iraq's literary heritage.

Although he was at the Citadel that day, Hadi only learned about what had happened when he came home in the evening to find the house filled with a cacophony of raised voices. He panicked, thinking at first that someone had died. His young cousins were huddled together, holding on to their mother, Abdul Hussein's sister Amira, who was sobbing loudly. Hadi had never seen his aunt like this. He knew her as a tough woman who was cowed only by her mother Khadja; not even Uzri's fiery temper could intimidate her.

The heavy shadow of terror fell over the household as the family fretted over the possibility that further retributions might come their way. In these times of war, even Abdul Hussein's good relations with the authorities could not be counted on to protect them.

As the days passed, Amira cried less and shouted more, becoming short-tempered with everyone, especially the servants, who tried to stay out of her way. Hadi's mother Jamila, on the other hand, simply lost her appetite. She would only drink tea and nibble on bread, like the fragile little bird she resembled. She slept very poorly, yet she made sure to be up early each morning to see Hadi before he left for work, anxious that it might be the last time she saw him if he was also taken away.

Hadi, however, equipped with his enthusiasm for tackling every challenge that came his way, continued to make himself indispensable at the Citadel. When he was not at work, he preferred to stay away from his father's *dawakhana*, where the endless complaints about the authorities bored him. Instead, he wandered among the bazaars near Headquarters in Qishla. From the stallholders and café owners he gained an insight into the soul of the country: what people bought, what they wanted, what they required. He found that the mechanics of the market interested him, and his eyes were opened to the world of commerce. The war had depleted the bazaars, but even as a young man Hadi smelt the endless opportunities that might lie ahead.

He was chatting to a pomegranate-juice seller late one afternoon when the sky suddenly seemed to rip in two above them.

'*Ya Allah*, what is it!' yelled the juice seller, instinctively ducking. Other men nearby had pressed themselves into doorways, or against walls, their eyes wide with fear.

Looking up into the blue overhead, Hadi spotted a trail of white, then the sun glinting on the wings of a flying machine. His heart was beating furiously, but he could hardly contain his excitement when he realized what he was looking at: an aeroplane! He had heard his father talk of such things, and now he had seen one. He rushed home to tell

Muhammad Ali and Ibrahim
in a studio shot.

his younger brothers, Abdul Rasul and Muhammad Ali. The future
was coming to Baghdad.

New technology was not the only thing to arrive in the city. After a
defeat at Shuyaba in the winter of 1915, the Turkish army reshuffled
some of its military leaders in Baghdad. One of the newcomers was
Field Marshal Colmar Freiherr von der Goltz, on 'loan' from Germany
to the Ottoman army. With a long history of military service, beginning as a Prussian officer before German unification, Goltz had
contributed to the modernization of the Ottoman army in the 1890s.
His arrival in Baghdad on 15 December 1915, in the company of thirty
German officers, caused some consternation as he made strategic decisions from the outset without consulting the leading Turkish
commander. Nonetheless, when he arrived he was ceremoniously
welcomed in the streets by crowds of school children.

Hadi was curious about the Germans' motives in posting these officers to Baghdad. He wondered why they were here, fighting with the Sultan in Mesopotamia, rather than fighting the British in their own country. Although he listened carefully to his father's explanations, and to Uncle Abdul Ghani's arguments about fighting the infidel, he wasn't sure if he understood why this war was being fought at all. And if what his uncle said was true, then surely the Ottomans shouldn't be fighting *alongside* the infidel here, in Baghdad.

He hadn't witnessed many encounters between men of the East and Europeans, and he found the interaction between Goltz and his Ottoman colleagues absorbing. When irritated, Goltz would take off his round spectacles and wave them around, while his face turned very red, and he would lift his hands up to his hair and down again in rigid, mechanical fashion like a wind-up toy. He was especially impatient with his local staff, reprimanding them in his pidgin Turkish for the slightest mistake. Hadi once witnessed the flogging of a tea boy who had accidentally dropped a glass on one of Goltz's documents.

One day, while walking back to Headquarters, Hadi spotted Goltz patting his handsome pair of Turkish Kangal sheepdogs. The way he fussed over them, murmuring to them and affectionately stroking them, was in complete contrast to the way he treated people.

Whatever his personal idiosyncrasies, Goltz earned his military reputation. He was regarded as a hero by many for his successful planning of the famous siege of Kut in Mesopotamia in 1916, which inflicted a humiliating defeat upon the British. All the same, having seen how he treated his staff, Hadi wasn't sure if Goltz really cared about the fate of the Arab people.

The fighting continued unabated as the British pushed north towards Baghdad. The sound of weeping became a constant in the streets as women feared for the safety of their conscripted sons, husbands and fathers. There were shocking reports of children dying of starvation, of women selling themselves in order to survive, and of harsh reprisals by the Ottoman military authorities. Hadi knew the last of these to be true, as he had seen for himself the bodies of army deserters left to rot on poles in several of Baghdad's squares.

One morning, a woman approached him as he stood outside the Citadel talking to a friend. In spite of her youthful voice, she looked old. She was haggard with worry, and had barely started talking when her tears welled up. Both her sons had been conscripted a year earlier, she explained, and she hadn't even seen them go as they had been forcibly carted away from their shop in the *soug al saffafir*, the metal market, where they were coppersmiths. She begged Hadi to find out where they were, as no one had responded to her many pleas. He wrote down their names and told her he would do his best.

Hadi approached some of his colleagues, who simply shrugged their shoulders and said that it was probably lucky the boys' mother didn't know their fate, as they had most likely perished on the Eastern Front in Russia. Unable to give the woman the news she wanted, instead he started to give her food secretly, which he could arrange fairly easily as he was delivering supplies to the officer at the Deer Palace. She took the food gratefully, especially as the price of staples such as sugar and wheat had risen drastically in recent months. Yet the look of hollowness in her eyes never left her as she waited for her sons to return.

The horror of the war was never far away. A constant flow of wounded soldiers streamed into Baghdad; the bodies of the dead lay in flimsy open coffins, attracting swarms of flies. There were never enough doctors or medical supplies, and many of the wounded died unattended. These sights terrified the local population, who could only assume that their conscripted loved ones suffered similar fates on more distant fronts.

The Ottoman military casualties on the Mesopotamian front amounted to approximately 38,000 lives out of an estimated total of 305,085 lost Empire-wide. Civilian casualties were even higher. There were increasing numbers of destitute women begging on the streets, many with infant children, who had escaped from the ravaged villages south of Baghdad where the fighting continued, or whose menfolk had been taken to the front, leaving them to fend for themselves. Some were even imprisoned by the authorities for their husbands' desertions. The plight of these women moved many, including a leading poet, Ma'ruf Rusafi, who wrote:

He died, the one that gave her safety and happiness
And fate, after his absence, lumbered her with poverty ...
Walking, she carried her infant on a tear-covered breast;
His swaddle from rags, repelling any onlooker.
No man, but me did I hear her
Pleading with her God, her suffering life ...

3

All That is Good Will Happen

A Marriage Prospect

(1916)

THE INSANITY ON the streets outside afflicted Hadi's grandmother, Khadja. Gossip abounded about the general state of moral turpitude in Baghdad, now that the city streets were awash with refugees, and Khadja was concerned by the long hours her grandson spent at Military Headquarters, adrift in a sea of corruption. At home, she had caught him stealing interested glances at several young women who had come with their mothers to visit her; she also suspected him of flirting with their pretty new maid. She concluded that it was time for Hadi, now aged eighteen, to marry.

Ensconced within her own quarters in the large Chalabi house, Khadja had outlived her husband, Ali Chalabi, and her robustness and energy were boundless. She had a reputation as a domestic tyrant who never had to repeat her decrees more than once. A fair-skinned woman with small eyes and a delicate physique, she occupied herself by matchmaking and initiating divorces between couples, applying equal effort to both activities.

Summoning her three eldest daughters, Munira, Amira and Shaouna, and her son Abdul Hussein, Khadja delivered her verdict with respect to Hadi. Hadi's mother, Jamila, was excluded from the meeting on the basis that she was an outsider. Although she had been married to Abdul Hussein for many years, she was still disliked by her sisters-in-law because they had originally wanted another wife for their brother, believing he was too good for her.

An oil portrait of Jamila,
Abdul Hussein's wife.

Khadja was so feared that no one else dared approach her *kursidar*,
her private sitting room, unless invited, except for the servant who
brought them the tea at the beginning of the meeting. Resplendent on
her satin-covered seat, Khadja smoked her *nargilleh* and ran through
the list of potential brides for her grandson.

The name Bibi Begum was mentioned a few times. Although she
was personally unknown to the family, the girl was the niece of the
wife of a distant cousin of theirs and the daughter of Sayyid Hassan
al-Bassam, a respected merchant who had died five years earlier. They
also knew her mother Rumia well. Rumia was a highly regarded, God-
fearing woman, famed for her culinary talents and her lineage – her
mother was a granddaughter of the Persian Qajar Shah, Fath Ali Shah.
She came from a well travelled and erudite family, the Postforoush
from Azerbaijan, who had settled in Kazimiya several generations
earlier.

For hours the three Chalabi daughters discussed the advantages and
disadvantages of such a union. Munira preferred another family, the
Qotobs, whose daughters she thought much prettier. Amira disagreed,

considering them too haughty. But finally and inevitably they agreed with their mother, settling on Bibi.

When he came home later that day, Hadi was informed of their decision. He knew that he was expected to get married; it was a part of life. In addition to being a religious duty, marriage was a rite of passage that everyone went through. Love, if it came at all, was expected to come after marriage. It would have been impossible for it to come before, as there was almost no opportunity for a young man such as Hadi to meet a suitable bride outside the family in any respectable setting. He accepted the decision with a combination of excitement and trepidation, trying to imagine what Bibi looked like from the description that was given to him. But none of the Chalabi women – not even Khadja – had an inkling of the true nature of the girl's personality and temperament.

Sixteen-year-old Bibi had recently had an argument with her mother, Rumia. Her grandfather Sayyid Nassir had summoned her to his sitting room, where he and Rumia were drinking tea. A willowy and cultured woman, Rumia sat quietly out of respect for her elderly father-in-law while he informed Bibi that he had been approached about the prospect of marrying her to a distant cousin who was a mullah. He explained that the cousin was moving to Persia, where she would join him if the match went ahead.

Incensed, Bibi declared brazenly, 'I don't want him!' The force of her response silenced the room.

Rumia covered her eyes in despair, fearful of the damage her daughter's character would inflict on her reputation. After a moment she looked up and pleaded in a small voice, 'Bibi, be reasonable, what is it that you expect? He's a good man. Don't become blind with your empty dreams – life requires sacrifice.' Rumia knew that her daughter wanted to live comfortably, to mix in good company and travel the world in style. Apparently all those Persian love poems she had learned when younger had gone to her head.

Bibi was adamant. 'Yes, mother, you remind me of that every day, and I can see it all around us,' she said firmly. 'But why should I sacrifice myself to this man? *I don't want to be a mullah's wife; I don't want him* – I won't discuss it!'

Bibi retreated to her room, where she kept a hidden stash of hand-rolled cigarettes that she had discreetly stolen from Rumia over time. She lit one up and sat with her back against the door so that no one could come in. She had been a smoker since the age of twelve, and now she played an old game: with every puff, she followed the rising smoke, seeing what shape it suggested and interpreting this in relation to one of her many wishes. If the puff of smoke retained the same shape for a count of five, it meant her wish would come true. Now she wished for a suitable man – handsome, intelligent and well-to-do.

Bibi had been her late father Sayyid Hassan's favourite, and nothing could ever compensate her for his loss. Even as a young girl, precocious and with a sharp turn of phrase, she had commanded and demanded attention from all, particularly from her father. Her name, Bibi Begum, was partly a testament to her father's travels. Not content with one title to call his daughter, he put two nouns together: 'Lady Madam' in Urdu and Hindi. However, in picking the name Bibi, he also stripped her of the title she would earn when she eventually became a grandmother, as 'Bibi' was the name grandmothers were given in Mesopotamia. Long before she became a grandmother, Bibi's name imbued her with matriarchal qualities. It seemed to give her the foundation upon which to build her life.

As a little girl, Bibi had always been excited as she waited for her father to come home from his latest travels in the East; from Persia and India where he bought goods for his wholesale provisions business in Baghdad. Everyone else in the household would rush around, preparing for their master's return, except for Bibi, who would flit in and out of the courtyard to check whether he had arrived.

She always wanted to be the first to greet him, ahead of her mother and her two brothers. Once she had spotted him from a distance, walking along the alleyway with several men. Trailing behind them came a cart piled high with cases. Bibi couldn't contain her excitement a moment longer and ran over the cobbles to her father, flinging her arms around him as he reached down to pick her up.

'My, my, you've grown, my *khatuna*, my darling,' he chuckled as he kissed her warmly on her flushed cheeks.

'Did you bring me back lots of presents?'

'You naughty girl, is this the first thing you ask your father after such a long trip?' Sayyid Hassan laughed.

'Well, did you?' Bibi insisted.

'Of course I did,' Sayyid Hassan replied with a smile.

'And do you have lots of stories to tell me?'

'Lots and lots. Let's go inside.'

No one embraced her like that any more, and no one gave her the sorts of gifts her father had once lavished upon her. He had had a good eye, always returning from his travels with lovely objects for the house and beautiful jewels for his wife and daughter.

Sayyid Hassan had taken a deep interest in Bibi's education. She appeared to have inherited from her mother's cultivated and bookish family a talent for poetry and learned conversation, and he saw her interest perk up whenever her maternal uncles visited, when she would hang upon their every word. However, there were no girls' schools in Kazimiya and only a handful in Baghdad, to which few of the local townsfolk ventured.

As a solution to this dilemma, two male teachers were hired to teach Bibi. One was a sheikh who taught her to read the Quran and gave her lessons about Islam; the other taught her literature and poetry. Bibi proved to have a knack for memorizing and reciting verse and songs that she retained all her life. Her language teacher was an Iranian resident of Kazimiya, and he included Persian poems, which were both tender and spiritual, in Bibi's curriculum.

Bibi was also encouraged to pray, having watched her parents do this every day since her infancy. As a little girl, one of her favourite things was to recite prayers to her father, thus commanding his full attention.

With her mother, things were different. Religiosity was perpetually in the air around Rumia and she seemed obsessed with doing good, a trait that Bibi absorbed unconsciously, even though she felt overwhelmed by her mother's devotion to God, and became ever more critical of her ascetic ways. Since Rumia was busy with the household chores and managing her staff, she had little time in which to give Bibi the undivided attention she craved.

When Sayyid Hassan died unexpectedly after an attack of pneumonia, Bibi's life crumbled. She was only eleven years old; she had been inconsolable as she watched him losing his life force breath by breath. Even now, she still had occasional nightmares about the men who had removed his body from the house, and the wailing of the women during his funeral rites.

With his death, life had changed drastically for the lively household; it had shut itself away from the outside world, and Rumia too had seemed to fold into herself. Only Saeeda, Rumia's young maid and confidante, had been able to lull Bibi to sleep, as she had become terrified of death, which became a lasting obsession for her. Saeeda soothed her, putting her strong arms around her and rocking her, recounting her own tragedy.

Saeeda had lost her family when she had been kidnapped as a little girl in Sudan. She had been sold into slavery several times over before arriving in Mesopotamia at the age of ten, whereupon she had been freed by a renowned Persian sage known as Al-Qotob, the Pivot. She had become a servant in the household of the Pivot, and as his family had close ties with Rumia's household, she had eventually found her way into Rumia's employ. Her story gripped Bibi's imagination and demanded her sympathy, shaking her out of herself. She tried to put herself in Saeeda's shoes, but the experience was too painful. The two girls, at least, had a common bond: both had suffered a terrible loss at a tender age.

Khadja sent word that she would like to pay Rumia a visit. Although she had approved the choice of Hadi's bride in theory, she wanted to make sure there were no hidden flaws in the girl. 'You can never trust anyone else's eyes,' she told herself.

Rumia realized what the honour of Khadja's visit implied: there was interest in Bibi from the Chalabi household. She calculated that the prospective groom must be Abdul Hussein's oldest son, Hadi, as his other boys were little more than children. She let Bibi know that they were expecting important visitors.

On the given day, Khadja set off on foot with her three daughters in tow. They were accompanied by a servant girl each, as well as by Jamila, the groom's mother, who walked meekly behind. Jamila dreaded spending any time with her female in-laws, especially during social visits, when she could never get a word in edgeways.

Clutching their *abayas*, Khadja and her daughters muttered to each other in irritation. Jamila was certain they were talking about her. She was very unsure about this afternoon. Her son was too young to get married, she felt, but she was unable to express her opinion freely, especially to her mother-in-law.

Naturally apprehensive about the visit of the Chalabi women, Bibi prayed that her mother's quiet manner wouldn't turn the day into a disaster. But, to her immense relief, Rumia presented the visitors with a feast of a tea, dazzling them with her culinary talents and her natural elegance. She had used her skills in the marketplace to obtain goods that were now in short supply, owing to the military's requisitioning of fresh produce. Saeeda had also proved invaluable, scouring the main square for ingredients. Back home in her kitchen, Rumia had prepared a variety of sweet pastries and stuffed bread, which she served on the delicate silverware her late husband had bought during his travels many years earlier.

A woman in an *abaya*
walks by the river.

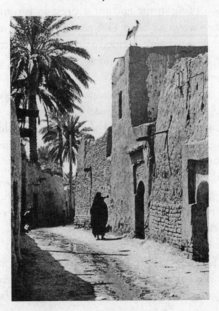

Khadja and her middle daughter, Amira, did most of the talking. Amira took after her mother, and rumour had it that she wore the trousers in her marriage. During the years that her husband Uzri was in prison, time had spun her misery into anger, which threaded its way through everything she said. An eccentric woman, she had a great dislike of the cold, and took to her bed in November each year, rising only in April once the weather had warmed up. During that time she received visitors in bed, had her meals there and continued to run the household from her bedroom. Her husband's incarceration had compromised even this ritual, and she commiserated with Rumia on the hardships of living without a spouse.

The Chalabi women were impressed by Rumia's elegant presentation, and the delicacies they ate sugared their moods. They exchanged glances with each other. Khadja even let slip a smile when she saw the trays of sweets, which were covered with delicately embroidered doilies and decorated with freshly cut rosebuds.

Custom dictated that Bibi remain absent from the room until near the end of the women's visit, when she could come down to greet them. Having worried all morning about her appearance and fussed over what to wear – opting in the end for a dress in turquoise, that most regal of colours – after an hour Bibi couldn't wait a moment longer. She knew her fate was being decided as the guests chatted away in the drawing room, and decided to stand behind the door, eavesdropping, until she was sent for.

She became engrossed when Khadja spoke effusively of her grandson Hadi, his good looks and his excellent prospects. Although there had not yet been an official proposal of marriage, the old woman could not help letting Rumia know of the great honour she, Khadja, was bestowing on her with this association.

The door Bibi was leaning on was suddenly pulled open, as a maid came out with a tray of empty tea glasses. Unbalanced, Bibi stumbled into the room. Rumia closed her eyes and covered her mouth in shock and embarrassment, terrified that this display of bad manners might spoil her daughter's chances of marriage.

The visiting party stared at each other in surprised disapproval, until Bibi's aunt Fahima chirped up, 'Ah, you're here, my dear! I was about to come and get you. Come and meet Khadja *Khanum* [lady or madam] and Jamila *Khanum*, and Hadi's aunts.' The three aunts pursed their lips in unison, not amused to be introduced after the wretched Jamila.

Bibi desperately willed herself to look demure, and focused very hard on the flower-patterned carpets covering the floor. Her cheeks glowed from embarrassment, although she was by no means a naturally timid girl. But this occasion was different: she had heard of Khadja's viciousness, and knew that the matriarch had the power to make or break this union. So the best course of action for her was suddenly to become shy. As she greeted each of the Chalabi women she was aware of their scrutinizing eyes roaming over every inch of her, and she silently recited a short verse from the Quran in the hope that it would temporarily blind them when they reached her neck, lest they see how short it was. If there was one thing she really envied her mother, it was her long, thin neck – that, and her height. Rumia was tall and graceful; Bibi was not. As all these thoughts tumbled through her mind, Bibi was so nervous that she nearly forgot to breathe.

When she finally approached Hadi's mother, Bibi felt some relief. Jamila was warmer to her than the other Chalabi women had been, and asked her to sit next to her, complimenting her on her silk dress. Bibi missed the dark looks exchanged between Khadja and her daughters. However, the tension was broken when another tray of walnut pastries appeared.

Walking back home, Shaouna declared to the others that Bibi was going to be quite a handful. She turned to Jamila and said snidely, 'You won't be able to control her. She'll walk all over you.' Jamila didn't reply, but her sisters-in-law started to laugh. Jamila could feel herself burning from all the anger she had stored up inside from the day she had married; she had never yet heard a kind word from any of these women. Her silence irritated them, but so did her words on those rare occasions when she had dared to answer back. Today, she kept her own counsel; she liked Bibi.

After the visitors had left, Bibi asked her mother what she thought of the Chalabis. Rumia was not very forthcoming; she told her that they seemed decent enough, but that it didn't do to rush into such an important thing as marriage. Bibi was upset by what she took to be her unenthusiastic response. Her mind ran away with her as she imagined what her life would be like as an unmarried spinster stuck in her mother's house.

The prospect was very bleak. She would always be at the mercy of her grandfather, her uncle or her brothers, and, worse, her brothers' wives after they married. She would never be able to have a house of her own, or do anything with her life. Everyone would pity her if she was denied the one role that all women were born to take on: that of wife and mother. She burst into tears. She would rather die than end up alone. *Hadi has got to marry me,* she told herself, *he has to*!

A few days later, with Khadja's final blessing and Hadi's consent, Abdul Hussein paid Bibi's grandfather Sayyid Nassir a visit to ask for her hand in marriage on behalf of his son.

After Abdul Hussein had left, despite his weathered bones Sayyid Nassir wasted no time in rushing to find Rumia and share the good news. Bibi burst into the room. 'What is it, *Jiddi?*' she asked excitedly. 'What did the Chalabis want?' She knew their business of course; what terrified her was the possibility that they had found her wanting. Her grandfather calmly explained that they had come to ask for her hand for their son Hadi.

'Yes, yes, I accept!' Bibi grinned. 'He's the one with the blond hair and the blue eyes, isn't he? I saw him once in the market and then another time during a procession at the shrine.'

Rumia was mortified that Bibi could talk to her grandfather with such a lack of respect; she reprimanded her, and Bibi rolled her eyes. Choosing to ignore the look on her daughter's face, Rumia continued, 'If you say the boy is of good character …'

'I believe he is, and I will ask around. You don't have any objections in principle?'

Before Rumia had time to reply, Bibi butted in, 'Why should she, unless she wishes me to stay facing her all my life?'

Her grandfather looked at her and said tenderly, 'All that is good will happen. Have faith.'

Bibi was sleepless from excitement and anxiety. It was a few days before her wedding day and she was worried, although she knew that Rumia had done her best to take care of the elaborate preparations needed to ensure that she went to her new home with all that was required.

Creating the trousseau had proved to be quite a production, from buying the material for Bibi's new clothes and furniture, to finding seamstresses and embroiderers, to getting the best mattress upholsterer. Locating the necessary wares had been complicated by the wartime shortages, and Bibi was concerned that Rumia had dealt with the details with her usual degree of detachment. Her interest in earthly things was limited, and her enthusiasm for minutiae only went so far. Consequently, Bibi was fearful that she would be judged by her in-laws and their acquaintances as somehow inadequate, and Rumia's calmness only irritated her further. She bit her fingernails to the quick.

What on earth would Hadi think of her?

4

Sugared Almonds and Jasmine

Bibi and Hadi's Wedding

(1916)

IN ADVANCE OF the day of the *mahir*, the official religious ceremony that was held on a different day to the wedding itself, servants carried huge *zanabil* from Abdul Hussein's house to Rumia's. The enormous baskets were filled to the brim with sugared almonds, pistachios, dates, fruit jellies and *mann al-sima*, a prized local delicacy made from the boiled bark of trees, mixed with nuts and covered in icing sugar. Each basket was so large that it required two men to lift it. In addition, the Chalabi family sent several trays of *shakkar borek*, thinly layered sheets of pastry stuffed with almonds and baked. Rumia also received the small gifts of candy that were to be handed out after the *mahir* to the men.

Ever prone to dramatic outbursts, Bibi was once more beside herself with worry about whether her mother would attend to all the necessary arrangements. She paced the house, pestered Rumia with questions and locked herself in her bedroom for hours at a time, smoking furtively and ignoring her brothers when they tried to talk to her through the closed door. The death of her father had taught her that happiness could be snuffed out in a moment, and she couldn't stop herself from fretting over what could go wrong.

The cuisine, at least, was no cause for concern. Besides God, Rumia's passion was her kitchen, and she threw herself into planning the feast. Despite having several servants, she reigned supreme in her kitchen, where she took an active part in the preparations. The formality usually present between master and servant was absent in her house-

46

hold, dominated as it was by women. Several of the servants had been there for such a long time that they had become part of the family. This did not mean that any of them had lost their respect for the unstated hierarchy.

Nevertheless, Rumia barely slept a wink herself the night before the *mahir*. There was so much to do, and she was exhausted by Bibi's endless demands and tantrums. She was also depleted by the great efforts she had taken to fund this ceremony. Ever since her husband's death, money had been scarce. Her wily brother-in-law had laid his hands on his brother's business and assets. He gave Rumia a stipend, but it was barely enough to feed two, let alone a family of four.

Rumia's lifestyle had changed dramatically over the past few years. Her house became less and less frequented by guests; she had never been as sociable as her husband. Nor had she shared his enthusiasm for collecting antiques, porcelain, opaline and silver, but now she was grateful for the pieces that remained. They had come to her rescue whenever she found herself with a large bill to pay. To fund the wedding ceremony, she had passed a chandelier discreetly to her brother Raouf to sell on her behalf.

She was loath to ask her brother-in-law for anything. After her husband's death he had become the financial guardian of the family, taking over his brother's estate, his shops and his capital, because Rumia was not versed in business and her own boys were still minors. Her father-in-law was officially her children's moral guardian, as religious custom dictated, but he was old, and left all monetary concerns to his son. It was humiliating enough that her brother-in-law had robbed them of their rightful inheritance, but to have to ask him for what was rightfully theirs ... that was simply too much for Rumia. So she lodged her complaint with God, certain that in His infinite wisdom He would see the injustice and punish her brother-in-law accordingly.

In the meantime, she wanted to make Bibi feel as confident in the arrangements as she could, so she had spared nothing to make the banquet as fitting and sumptuous as possible. Rumia was keenly aware that, of her three surviving children, it was Bibi who had felt her father's loss the most, and she had a deep fear of financial insecurity.

All the ingredients for a splendid celebration were in place. Rumia had prepared all the desserts the day before, with the help of her two servants – Saeeda and Laleh, a pious young Iranian maid from Kuzaran – and a few women who were regular visitors to her kitchen. The kitchen was filled with the trays of *burma*, dark vermicelli covering glazed pistachio nuts; walnut- and sugar-stuffed pastries; *claytcha*, date-stuffed round cakes; *mihalabi*, a rice pudding flavoured with orange blossom essence; as well as her signature *halawa*, with shaved carrots, cardamom and saffron.

The morning of the *mahir*, Rumia rose in time for her dawn prayers, but she couldn't concentrate properly, and knelt three times instead of the required two. Every time she recited a verse, her mind would wander to the kitchen and she would lose the train of the sacred words in her mouth. After her prayers she went to the kitchen, where she was expecting to find Saeeda and Laleh already up and working. But the house was silent. Everyone was sleeping in besides her.

She crept up to Saeeda's room, which was located behind the kitchen, and stood outside, debating whether she should wake her up or not. She took a deep breath and gently tapped on her door. There was no sign of life. She whispered Saeeda's name softly. Nothing. Rumia opened the door and gave a gentle cough. Saeeda croaked, rubbing her eyes when Rumia told her the time and urged her to get up as soon as she could.

'But we worked so late yesterday,' Saeeda complained. 'Just once it would be nice to get a proper lie-in.'

A couple of hours later Rumia left the kitchen and walked to her daughter's room, where Bibi was still sound asleep. She went over to the window and opened the shutters. Mosaics of light flooded in through the *shanashil*, the wooden lattices that framed the windows. Bibi sat up in alarm. 'What's happened, what's happened?'

Her two *mahir* outfits hung from the side of her closet. There was the cream embroidered kaftan that she would wear for the actual ceremony, and a light-pink silk dress, with gold embroidery and a round neck, for the lunch afterwards. Lying on the dresser opposite her bed were the ornate metal hair combs, with flowers painted in lacquer, which Bibi had insisted upon and which were to go on either side of her parting. Her custom-made high-heeled shoes were set out on the floor.

Rumia murmured her approval as she ran her hands over the outfits, but Bibi complained that they made her look short. 'What if they realise how short I am and change their minds?' she asked. Taking a deep breath, Rumia calmly told her to stop her nonsense: she was going to look lovely, but she should get up now as there was still a lot to do. She paused as she turned to leave the room. 'Oh! You nearly made me forget to get out the pearl earrings your father – may he rest in peace – gave me. They will go with your outfits perfectly.'

Bibi perked up at this piece of news. She had always coveted her mother's jewels. She had often thought that if they were hers she would wear them all the time, in contrast to her mother, who never let them see the light of day. Bibi didn't know how precious the gems were to Rumia; that she was relying on them to save her from dreaded rainy days ahead.

As with all *mahirs*, many of the ceremony's details were filled with symbolism for the imminent marriage. Bibi sat on a chair in the loose-fitting kaftan, which had a large round neck. This garment had to be free of any clasps, tied knots or fastenings, which were considered to be symbolic obstacles that might prevent her from speaking the truth when asked if she wanted to get married. Her hair shone and she wore kohl around her eyes, and *sibdaj*, a paste used as a blusher, on her cheeks. She hated the effect of the strong red and wiped most of it off immediately. However, she didn't mind the *diram*, a walnut-based lipstick applied with a finger.

Her feet were in a bowl of water, which symbolized a plentiful life, in which jasmine flowers floated. On either side of her candles burned, garlanded with tiny flowers. Two women stood behind her. They were known to be happily married, and each held a pair of enormous sugar cubes which they rubbed against each other onto a delicate lace cloth that was held above Bibi's head, so her marriage would be sweet like sugar. Facing her was a mirror so that she could see herself speaking the truth. A Quran lay open on her lap.

There was some commotion from the women as the mullah approached the door to the room. He stood outside and asked Bibi loudly whether she willingly accepted Hadi to be her husband. He slowly repeated his question fourteen times, so that she would be sure to consider her answer carefully. Her mother stood near her, counting with her fingers. When he reached the last time, Bibi quickly replied, 'Yes.'

Two of the women in the room started ululating, but they were quickly hushed. Having obtained Bibi's consent, the mullah continued the rest of the ceremony in the room next door, the *dawakhana*.

As was the custom, Abdul Hussein whispered to the mullah the sum of money his family was giving to the bride for her dowry; having agreed to Bibi's grandfather's request, he also disclosed the *mu'akhar* – the amount of money to be paid to Bibi in the event of a divorce. Hadi was standing next to his father, but as he still lived under his father's roof, it fell to Abdul Hussein to conduct the financial part of the marriage. This having been dealt with, the mullah recited a prayer before finalizing the marriage contract verbally. The groom and Bibi's grandfather shook hands, and the mullah put his hand across both of theirs while reading a final short blessing.

After greeting all the female guests and receiving gifts, mostly jewellery from the groom's family, Bibi went up to change into her pink dress before joining the lunch banquet. She felt shattered as she sat momentarily on her bed to catch her breath. She had been very anxious for days now, and her fatigue was finally catching up with her, as was her hunger.

The day was a success, judging from the little food that was left over. Everyone in the household was exhausted but satisfied that all had gone well. Even the servants took Bibi's marriage personally, wanting to give the groom's side the very best impression in order to ensure that Bibi's worth was truly appreciated. Yet the marathon was far from over. Now there was the dowry to prepare, in advance of the actual wedding day when Bibi would move into her husband's home. Although she was officially married to Hadi, she had still not been presented to him.

Rumia now had the *mahir* dowry money offered by the groom's family. The fifty gold coins given by the Chalabis was a generous amount, but she was aware that she would have to exceed it if she was to provide her daughter with the best possible trousseau. A trousseau was often looked upon by in-laws as a barometer of a girl's background and her family's ability to provide for her. A luxurious trousseau suggested a cared-for girl, who might be treated with more consider-ation than a girl who came to her groom without much. A meagre trousseau was like a licence that allowed a husband to get off lightly:

he could provide his new bride with little, as she had little to start with. Moreover, the contents of the trousseau would be put on public display as part of the wedding festivities.

Rumia knew all too well that there was still much to do. As she helped her daughter choose the items for her trousseau, she brought out a metal-studded chest she had kept stored for years. In it were several pieces of beautiful silk that her husband had brought back from one of his trips to India. Looking through the unfolded cloth, she reminisced about those lost days when her husband had filled the house with life. Rumia decided to give Bibi the chest and all the silks in it. She would never wear them. She had taken to wearing black ever since her husband died, and would continue to do so.

Of course, tradition also dictated that the bride offer gifts to Hadi, his father and brothers. For the men, a shaving set each was chosen, along with embroidered towels and underwear, as well as silk-embroidered shaving aprons. It was an expensive business, especially in these difficult times.

In the run-up to the official wedding day, Bibi overheard her mother talking to Saeeda in the kitchen. Rumia was asking her to pray that the silver candelabra she had sent to be sold would fetch a good price.

'What's this, Mother? What are you talking about?'

Rumia glanced across at her. 'Absolutely nothing for you to worry about, my dear.' She turned her attention to some pots of spices and resolutely refused to meet Bibi's eye.

'Yes, it is; of course it is – what candelabra?' Bibi demanded, her insecurities bubbling up in her chest.

Saeeda stepped forward. 'Bibi, there's something you should know.' Calmly, she explained the truth about their financial situation, glancing across from time to time at Rumia, who let her continue.

While she wasn't afraid to challenge her mother, Bibi listened to Saeeda respectfully. She knew from many years' experience that Saeeda was loyal to the family and acted in all their best interests. Moreover, she was wary of provoking Saeeda's temper, which could be fearsome when roused.

The news jolted Bibi's world. She had taken it for granted that her paternal uncle, who was her legal guardian, would be covering the trousseau expenses. Troubled, she looked at her mother, searching for the words with which to express her thanks, but before she could speak Rumia hugged her and told her not to worry.

After a little while Bibi climbed the stairs pensively to her bedroom, where she reached for a hidden cigarette. She understood now more than ever just how important her marriage into the affluent Chalabi family was, and she was relieved that the process was well underway. As she considered the actual wedding, she found herself wondering what she would say to her husband when she first saw him, and how he would greet her. What if he didn't like her? More importantly, what if she didn't like him?

She worried about the first night, when the marriage had to be consummated. Now that she was betrothed, many married women in her circle had offered her advice about her wedding night, telling her what she might expect and warning her of the pain she might experience. She tingled with the mere thought of what might happen, strangely thrilled yet nervous at the same time.

Her mood improved over the next couple of days. The all-female henna party the night before the wedding was an especially joyful occasion. Rumia and Bibi's relatives and close friends were all there. Bibi Istrabadi, the daughter of Saeeda's old employer the Pivot and one of Bibi's dearest childhood friends, teased her mercilessly about her new in-laws and the prospect of her wedding night, while a woman decorated their hands with henna patterns. The women sang and laughed, jumping from song to song, exhausting all the verses they knew. Even Rumia relaxed, momentarily abandoning her piety. With one hand covering the other, they clicked their fingers together, and the resulting rhythm accompanied their ululations around the room. Bibi could never manage this clicking business. As much as she tried, she could never make a sound.

The morning of the wedding day was rainy and grey. Bibi woke up very early. When she looked out of her latticed window and saw the downpour, she was filled with worry that the *hafafa*, the sugaring lady, would not come because of the weather. She could not take her bath until the *hafafa* had removed all of her body hair; depilation was something else she knew she would have to get used to as a married woman. Flustered, she rushed down to see whether the maid had brought her the *tinkhawa*, a mineral hair conditioner bought from the perfumer at the bazaar, and discovered to her immense relief that the *hafafa* had already arrived.

Between beauty treatments, she wondered briefly what her fiancé was doing at that precise moment. She knew that he would go for a ritual bath at the public *hamam* with a party of his friends. Having spoken to the other women about what was customary, she suspected that the men would all be treated to a succulent kebab lunch in the baths' antechamber after they had finished their massage and sauna. It all seemed highly relaxed, compared to her compulsory beautification rituals. Bibi bit her lip in discomfort as the *hafafa* pulled another strip of sugar from her skin. Wincing, she reflected again on how lucky men were.

Hadi went through his wedding preparations with rather more detachment than his bride-to-be, but then there was less for him to do. Nonetheless, he was looking forward to the arrival of his bride that evening. He was anxious and excited in equal measure, trying to imagine what Bibi might be like. He had painted an image of her from the various descriptions that had been provided by his female relatives, concentrating on the more positive picture that his mother had created. Jamila had told him that Bibi was a lively girl, with a pretty face and soft skin. And certainly he liked the image he had now conjured in his mind.

There were a multitude of things for him to absorb in readiness for this great change in his life. Marriage marked his entry into the adult world and the responsibilities that it entailed. He was much less worried than Bibi about the prospect of disliking his spouse. As for the pressure of the wedding night – this had its unpleasant side, especially when the house would be full of people, all expectant that the deed be done. But he was confident that he would prevail. He smiled to himself, grateful for his experience of a few stolen amorous encounters with a young singer from Baghdad.

In any case, first he would have to wait at home while a procession of twenty or so of his male relatives went to fetch Bibi. As it was raining he decided to send her his favourite horse, Najma, who was tacked up by Ni'mati with a bright, colourful bridle and a *kilim*-like blanket under the saddle. He carefully instructed his younger brothers Abdul

Rasul and Muhammad Ali to make sure that his new bride was comfortable.

Bibi was ready. She sat waiting on the edge of her bed, her legs too short to reach the floor as she bit her nails. She was wearing a white wedding dress embroidered with gold circular patterns, which had been made at a Jewish atelier in Baghdad. Her eyes were lined in kohl, and her skin was very soft after the morning's rituals, perfumed with her favourite jasmine essence. She wore a wedding veil which would be pulled down over her face when the men arrived.

Despite the sound of the rain, the singing of the men carried through the street as they walked towards the house. Then it stopped. A loud knock was heard. As a privileged member of the household who was close to the bride, Saeeda took up the challenge, asking loudly, 'Who is it?'

'We are here to take the bride,' the men replied.

'We don't have any bride to give you,' said Saeeda. 'You've come to the wrong house.'

'We won't leave until you give us our bride!'

'I told you – we have no bride for you here.'

The exchange was part of the custom and, as in a play, each side remained faithful to their part. The men knocked again, and Saeeda repeated her reply. This formality continued for a little while longer until one of the men offered Saeeda a few coins to open the door. Saeeda pushed it ajar, just enough to peer out. Looking at the men in mock disdain, she closed the door and said, 'Our girl is too precious for these pennies!'

The men knocked again, crying out, 'We will give you all the pennies you want if you open the door.' This time Saeeda obliged, and took the rest of the coins.

Bibi had meanwhile come downstairs and was standing in the sitting room with her mother, her brothers, aunts and cousins. Rumia gently pulled Bibi's head towards her and whispered a prayer in her ear. Bibi then kissed her two brothers. She said goodbye to everyone else, before bidding her mother farewell. Rumia was crying and Bibi followed suit. Her father's absence was sorely felt, and they both pined for him.

'No, no, you'll ruin her face with all these tears,' soothed one of Rumia's sisters-in-law. 'Come on, my dear, she'll still be near you.'

Bibi's trousseau had been sent ahead a few days earlier, along with the presents to the groom and his family. Her father's sisters, Fahima and Aminah, had already been to the Chalabis' house to prepare her new quarters. They would accompany her to her new home tonight.

When she caught sight of the animal standing at the door, Bibi exclaimed in horror. She was terrified of horses, and she felt the colour vanish from her cheeks as she hesitated before agreeing to mount it. As she was helped up by Abdul Rasul, her heart was thumping so loudly she couldn't hear the four-man band that led the way with drums and trumpets. Sitting sidesaddle, she struggled to hold an umbrella with one hand while she clung very tightly to the colourful reins with the other. Even though the horse walked at the pace of a snail, Bibi was terrified of slipping.

Her relatives and friends sang alongside her as they moved slowly towards the Chalabi house at the other end of the quarter. The entire neighbourhood came out in the rain to watch the *zaffa*, the wedding procession. Although it was not as spectacular as it would have been during peacetime, it nevertheless brightened up the streets and afforded the onlookers the opportunity to put aside their sorrows for a few precious moments.

When the horse finally came to a halt in the courtyard of her new home and she was helped down, Bibi looked up at the figures that stood waiting for her in the main courtyard of the house. She could only recognize her husband's aunts and Khadja, her new grandmother-in-law.

Her own aunts, who had followed her on foot, appeared on either side to help her enter her new home. It was their duty to hand her to her husband. As she took her first step towards her new family, Bibi remembered her attempts to prepare a suitably smiling face in front of her mirror, in anticipation of this moment. She looked up towards the men and, lifting her veil, flashed her most winning smile at them. A ripple of shock went through the group, who had been expecting to greet a shy, demure woman – especially on this, her wedding day. It was a small mercy that her mother had missed the moment, for Bibi's gesture was simply not the done thing. Her *faux pas* was made even worse owing to the fact that, in her nervousness, she had directed her dazzling, flirtatious smile not at her new husband, but at her father-in-law.

Nevertheless, both bride and groom greeted each other appropriately amidst a chorus of ululations from those standing around them. Dinner was served, which gave time for Bibi to relax a little. She was still very nervous about what lay ahead that night. The only man who had ever held her close had been her father, many years earlier. She couldn't stop her stomach churning from anxiety, remembering all that she had been told about what went on between a man and a woman.

Unusually for Bibi, she was lost for words as she sat absorbing her new surroundings and watching her new family. The air was alive with stolen glances as she and Hadi shyly scrutinized each other. Eventually the meal came to an end, all too soon for Bibi. After bidding her new in-laws goodnight, she was ushered upstairs by her aunts to her newly furnished bedroom. In silence she was helped out of her wedding dress and into a delicately embroidered nightdress. She was momentarily comforted to see the large hand-beaten copper bowl that her mother had picked as part of her trousseau. As she leaned over it to wash her hands, she heard a knock on the door. It was her husband.

Her aunts giggled, kissed her and left. Staring at her hennaed palms, embarrassed to be seen in a negligee, Bibi stood frozen as Hadi approached her. He hesitantly stroked her hair as he held her hand. She looked up at him and smiled. She was relieved to see that he was indeed handsome, with kind eyes set in an open face. His arms enclosed her. She didn't resist.

Ever mindful of her duties as a new mother-in-law, Rumia was determined to supply a sumptuous breakfast for the morning after the wedding night. A huge basket duly arrived early the next day for Hadi and Bibi, with freshly baked bread, *gaymar*, honey, a variety of home-made jams, lemon curd, marmalade, rice pudding and Rumia's delicate *shakar borek* pastries. She added a few gardenias, freshly cut from her small courtyard.

There was a lot for Bibi to take in that first week, and the days flew by as she focused on an approaching social function, the prospect of which terrified her almost as much as the wedding night had. On the seventh day after the wedding, her mother-in-law Jamila would be hosting a tea party during which the guests would take a good look at

Bibi and Hadi as a
married couple.

the new bride's trousseau, which would be laid out on her new bed in her new room.

It was a rite of passage that Bibi had previously enjoyed participating in as a spectator – but now she would be the subject of it herself. Several women who were not invited entered the house with their faces covered except for their eyes. It was an open house, so they could not be turned away, and they rummaged through the hand-embroidered silks and lingerie like crows picking over delicacies.

Bibi sat enthroned in the *dawakhana* in another bridal dress, while all the visiting ladies scrutinized her – her hands, her smile, her hair, her nose, her eyes, her manners and, of course, her outfit. Usually Bibi loved attention, but that day she could not wait for everyone to leave. Now that she was getting to know him a little, she longed for the opportunity to be alone again with her new husband. Aware of the women's examining gazes upon her, she smoothed her dress and sat upright, determined that she would not be found wanting.

5

A Giant Broken

The End of the Ottomans
(1917–1918)

AS BIBI EMBARKED on married life in early 1917, Mesopotamia had been at war for three gruelling years. The British forces were slowly but steadily advancing along the Tigris, creeping ever nearer to Baghdad, yet people still struggled on with their lives in the areas that remained under Ottoman control. In Kazimiya, the war manifested itself predominantly in the continued influx of refugees from southern towns, where many of the battles between the Ottomans and the British raged. In these turbulent days, there were more refugees in Kazimiya than there were visitors to the town's famous shrine.

While Hadi continued to work for the military in Baghdad, Bibi busied herself with setting up her new home. The novelty of marriage overwhelmed her at first, as she adapted to her new family and surroundings. She tried to accustom herself to her new role as a wife; she was the ward of her husband, even if he was out of the house all day, and she had to behave accordingly, giving him and his family the respect that was due them. From sharing a room with Hadi to waiting up for him if he came home late, she had to get used to life in this large house, which was composed of several annexes. Each was effectively a set of living quarters for a different section of the Chalabi family. This represented a great change from her mother's house, which was smaller and was occupied only by her mother's little household.

As elaborate as the most intricate Persian carpet, the Chalabi household was patterned by its many different personalities. There was Bibi's

58

dour grandmother-in-law Khadja, who had not forgotten Bibi's 'eaves-dropping' episode and who consequently scrutinized her every move; her parents-in-law Abdul Hussein and his meek wife Jamila, and their five children; as well as her uncle-in-law Abdul Ghani and his family.

Bibi missed her mother's house. She had become used to a smaller household since her father had died, one that was organized and quiet, and in which a great deal of emphasis was placed on food and its presentation. By contrast, the Chalabi household was a bustling hive of activities and disparate demands.

She also missed her former independence terribly, and realized that she had failed fully to appreciate it while living under her mother's roof. She was now answerable in all things to people who were scarcely more than strangers. Such were the customs by which a young wife was bound. Khadja ranked above all the other women in the household, and never let anyone forget it. These were the customs for everyone, but Bibi loathed them.

Unlike Rumia's sheltered home, the Chalabis' was an open house. Abdul Hussein still ran an active *dawakhana*, receiving guests daily, and Bibi disliked the commotion this caused, especially as custom dictated that she and the other women had to stay out of the sight of men to whom they weren't related. She often found herself secluded in the *andaroun*, the private quarters shared by all the family.

The visitors to the *dawakhana* often loudly aired their grievances about the declining state of Baghdad in the war, or were otherwise boisterous. At first, Bibi was more irritated than impressed by the social prominence of her father-in-law, and responded to his status by glorifying her own father's house and her mother's impeccable management of it, initially to herself. She conveniently forgot all of her criticisms and any discomfort she had felt while living there.

A more pressing concern was the cuisine of the Chalabi household. The kitchen catered for many people, and the cooks didn't have the refinement or talent of Rumia and her staff. Such was Bibi's longing for her mother's cooking that she arranged for Rumia to send food to her secretly. Saeeda would visit with bundles hidden under her long black *abaya* so as not to offend the in-laws.

Although Rumia liked Hadi and was happy for her daughter, she found the Chalabis' lifestyle too chaotic for her liking. She worried about Bibi becoming lost in such a crowded house. Shaking her head, she would complain to Saeeda, 'I don't like to speak ill about people, but that house is so busy, there are so many people coming in and out, and their food, my dear ... well, it might do for an army, but it's certainly not tasty.' On the other side of town, Bibi ate her secret supplies with relish and wistfully nursed her sense of loss.

Bibi was very aware that her entire *raison d'être* now was to produce an heir quickly. It was expected of her, as it was expected of every married woman. Until Bibi was able to fulfil her part of the bargain by falling pregnant, preferably with a son, her insecurities and fears of failure overshadowed all else. If she could not produce a child, she feared that Hadi might divorce her and find someone else to marry. She was less concerned that he might take a second wife, as polygamy, although lawful, was frowned upon in his family.

Despite her worries, the issue of motherhood was not a subject Bibi felt comfortable discussing with anyone else in the household; she had not known any of the women long enough to engage in such an intimate conversation with them. But her obsession with conceiving was like a sore tooth that she couldn't leave alone, a wound she was forever probing. She would immerse herself in dark thoughts, praying to every conceivable saint and visiting the Kazimiya shrine regularly, pleading with the Imam to grant her wish. She imagined bringing shame on herself and her mother, becoming socially ostracized. *Barren, barren, barren.*

She was disturbed when one evening Jamila told her the tale of Hadi's dead aunt Burhan, the Nawab's first wife. Shaking her head at the awfulness of it all, Jamila explained that Burhan had died young from an infection; apparently her illness had been caused by an infertility remedy prescribed by a backstreet quack.

On a visit to her mother's house Bibi gave vent to her worries. Saeeda had to calm her down, holding her when she started to hyperventilate. 'You must be patient, my child,' Saeeda said softly. 'You mustn't do this to yourself. Have faith in God; you will become pregnant.'

Bibi groaned in despair and would not be comforted, so Saeeda suggested that they visit Imam Musa at the shrine again. Saeeda firmly believed that he was the granter of all wishes. Bibi was less sure; the Imam hadn't listened to her prayers so far, even though she was his relative. As one of the Ahl-ul Bayt, she was a descendant of the Prophet and the twelve Shi'a Imams she prayed to.

Saeeda tried another approach. She suggested that Bibi go to Samarra, to the *malwiya*, the spiral tower at the Great Mosque, where she should climb all the way to the top. Then she should make her wish and throw her *abaya* to the ground below. 'Like Fahima did, and she was pregnant soon after,' Saeeda said confidently.

Bibi burst into tears. It was impossible for her to go to Samarra; if she did the Chalabi family would surely guess her predicament, and she would never recover from the shame. Even so, she spent many sleepless nights debating the pros and cons of such a trip, before finally deciding against it.

In spite of her anguish at not conceiving a child and her reservations about her new home, a warm relationship began to develop between Bibi and her husband's family, who came to view her very fondly.

She developed a particular bond with her father-in-law. Rather than shy away from him like the young bride she was, Bibi engaged Abdul Hussein in conversation, finding in him an attentive ear to her stories and questions. She was hungry to learn about the world outside, and he indulged her curiosity. With her humour and quick wit, she soon gained a place in his heart and he embraced her as one of his daughters. Just as he had bought his daughters shares in the Kazimiya Tramway company that he now chaired, he bought Bibi some too, so fond of her had he become. At the time, shares were an unusual asset compared to land, which both men and women could inherit or own.

Hungry for paternal affection, Bibi became very attached to Abdul Hussein, and enjoyed his company tremendously. She may even have placed more importance on her relationship with him than on that with Hadi, who was often away at Military Headquarters. She also realized that her closeness to him earned her more respect in the household. A politician at heart, she slowly but surely made her mark in her new home.

She was perhaps fortunate that Jamila, her mother-in-law, was such a subdued creature, who remained at the mercy of Khadja. And the layout of the large house permitted Bibi to build a good relationship with Jamila and her children, especially her daughter Shamsa, while staying out of the dowager's way. Jamila was very different to Rumia. She didn't have the same oppressive piety; she enjoyed a good laugh, and when left alone by her fractious sisters-in-law, Munira, Amira and Shaouna, was very jolly. The wrath of God did not loom over her in the same way that it did for Rumia.

Hadi's three aunts, who had come with Khadja to inspect her as a prospective bride, did not live in the big house, but hovered about its fringes. Bibi succeeded in winning them over too, appeasing them with flattery and politeness.

Yet for all that Bibi gradually settled into her new home, a secret corner of her heart remained in her mother's house, the home of her childhood memories. One day Abdul Hussein overheard her singing a popular new song about the war:

> Chalchal alayah al rumana, wil numi fiza'li
> Hatha il hilu ma rida, waduni lahli ...
> Ya yumah la tintithrin, batlili al natra
> Ma juz ana min hwai, wu maku kul charah ...

> The pomegranate tree engulfed me, and the sweet lemon
>   came to my rescue;
> I don't want this sweetness, take me home ...
> Oh, mother don't wait for me, there is no point in waiting;
> I will never abandon my home, there is nothing to be done.

The song was about the Ottomans, referred to as bitter pomegranates, and the British, who were the sweet lemons. The words expressed the longing of a soldier or prisoner for his home, with the implication being that he wanted neither the sour Ottomans nor the sweet antidote to their bitterness, the British. Hearing Bibi sing, Abdul Hussein could tell that the song might just as well be about her own longing for her mother. The wistfulness in her voice said it all.

When she finished, he came into the room clapping his hands, praising her for her fine voice. Bibi laughed, and asked him teasingly whether he thought she could make a career of it.

He started laughing too: 'No, no, Bibi – you're too good to be wasted in the cafés!'

Bibi smiled. Neither of them spoke of the sentiment that echoed through the verse.

Outside the Chalabi house, change was bearing down on the city gates. Ever since 1914, British forces had been a hostile presence in Mesopotamia, capturing the southern cities of Basra, Nassiriya and Samawa. Although they had suffered a crushing defeat at Kut in 1916, losing 23,000 soldiers to disease and famine, a new military commander had rallied the troops and resumed the advance towards Baghdad.

In March 1917, under the instructions of their Commander-in-Chief Khalil Pasha, the Ottoman troops spent several days digging trenches outside Baghdad in preparation for the British offensive. On 10 March a sandstorm descended on the city. In the face of the relentless storm and the advancing British army, Khalil Pasha realized that he had insufficient men under his command, that too many had been diverted to fight the Russians in Persia. He ordered a retreat.

However, first he issued orders to destroy all the major factories and arms depots in the city. The pontoon bridge was burned and factories were blown up, as was the telecommunication station the Germans had built. Supplies of ammunition were dumped in the Tigris, and the retreating Turks blew up one of the four main gates to the city, the Talisman gate. Most of the Ottoman administrative personnel fled, taking with them official documents and registers. Baghdad was left exposed, the corpses of dead Ottoman soldiers brought in from nearby fronts littered the streets, and a mob of impoverished city dwellers and tribesmen from the outskirts took over the city, looting what was left. Some, such as the Bedouins, were motivated to loot simply as one of their own customs of war, while others were expressing their resentment of the Ottomans and the harshness they had witnessed during the war.

Across the River Tigris in Kazimiya, people heard explosions but could not see what was happening because of the sandstorm that blacked out the sun. At the Chalabi house, family members gathered on the rooftop. In the darkness they could not tell what time of day it was, and sand seeped into their noses and ears even when they retreated indoors. Bibi and her sister-in-law Shamsa clutched each other's hands as they heard a thunderous sequence of the explosions set off by the retreating army.

In the streets outside, people dashed around, desperately seeking shelter, crashing into each other or into walls because of the sand that stung their eyes. Their cries added to the clamour. The thickness of the air, dense with sand, increased the panic.

Amidst this chaos, townspeople watched in disbelief as Khalil Pasha and his officers arrived at Kazimiya station and brazenly boarded a train for Samarra. Luckily, the town was spared looting by the nearby desert tribes, the ruffians and bandits who appeared during chaotic times, as the town's notables came together and took Kazimiya's security into their own hands.

The next morning, General Sir Frederick Maude's Anglo-Indian troops entered Baghdad and marched along the avenue built by Khalil Pasha's troops to facilitate their own military manoeuvres in the city. The avenue cut through the very heart of Baghdad, from the northern gate at Bab al-Mu'adam near the Citadel, all the way towards the central Maidan Square and the river beyond it.

British troops entering Baghdad,
March 1917.

Maude's troops met no resistance. They found a vanquished city, its buildings looted and destroyed. Many Baghdadis were uneasy about the possibility of a Turkish reprisal, so there were few cheers and little applause in the streets. The British flag was hoisted at the clock tower in the former Ottoman government complex in Qishla, the old city, and public order was swiftly restored. It was soon apparent, however, that there was no clear plan for political action. Maude read out an elaborate and flowery speech to the people of Baghdad six days after the city's capture, penned earlier by a British adventurer and diplomat in the region, Sir Mark Sykes. In it, he emphasized that the purpose of the British presence was not occupation, but rather emancipation:

> *Our armies come into your cities and lands not as conquerors or enemies, but as liberators ... Between your people and the dominions of my King there has been a close bond of interest. For 200 years have the merchants of Baghdad and Great Britain traded together in mutual profit and friendship ... Many noble Arabs have perished in the cause of Arab freedom, at the hands of those alien rulers, the Turks, who oppressed them. It is the determination of the Government of Great Britain and the great Powers allied to Great Britain that these noble Arabs shall not have suffered in vain ... I am commanded to invite you, through your nobles and elders and representatives, to participate in the management of your civil affairs in collaboration with the political representatives of Great Britain who accompany the British Army, so that you may be united with your kinsmen in North, East, South, and West in realizing the aspirations of your race.*

Maude's mention of emancipation had some echo in Baghdad, especially among those officers who had participated in the Arab revolt further south against the Ottomans in 1916. Discontent with the Ottomans had grown ever since the Young Turks' revolt of 1908, and was now shared by many. However, their notions of 'liberation' also implied independence and self-rule – features that appeared to be lacking in British plans, which seemed to be more concerned with economic factors such as their ability to maintain control of the oil fields in Iraq and Iran and to secure a route to India, where they continued to have an imperial presence. It seemed that the British were in no immediate hurry to make up their minds as far as the future of Mesopotamia was concerned. Uncertainty filled the air.

The war had taken its toll on Baghdad's population. The Ottomans had confiscated many essential goods, such as agricultural produce, livestock and pack animals, as well as less vital but nonetheless desirable goods such as women's clothing and printing presses. In return, merchants had simply been given signed slips of paper, offering an unspecified amount in reimbursement after the war.

Within the Ottoman army, officers had sometimes made profits on the goods they confiscated. The war had created the perfect conditions for bureaucratic corruption and the abuse of power; it was in effect official looting – of market goods, animals, crops, everything and anything the army and its associates could lay their hands on.

In Baghdad, food shortages had been made worse by the onslaught of a cholera epidemic, as well as river floods which had forced many out of their homes in 1916. Those in Kazimiya, such as Abdul Hussein and his family, had been luckier, as the lie of the land was higher and more protected from the rising waters of the Tigris. The town received many refugees following the floods in Baghdad.

Although there was a sense of relief that the war was over, this alone could not subdue the shock caused by the collapse of the Ottoman administration and the subsequent British occupation. The war had brought with it many aspects of modernity that had rattled the population, most notably the mass mobilization of civilians and military. It had also brought profound loss, and now an important layer of their identity was being stripped from the inhabitants of the Baghdad Province. They had, after all, been Ottomans for centuries.

For Hadi, the biggest shock came days after the fall of Baghdad, when he saw British soldiers on the upper deck of a tramcar approaching Kazimiya, sitting casually on the same benches that had been occupied only days earlier by Ottoman soldiers. Having worked for the past couple of years in close proximity to the Ottoman Military Headquarters and its staff, Hadi could not but feel some attachment and loyalty to his erstwhile colleagues.

Not long afterwards, while walking along the path to the Citadel, he was unsettled by the sight of thousands of Ottoman soldiers who were being detained in a large camp by the British. It seemed only yesterday that the opposite had been the case; he remembered the public celebrations following the siege of Kut, which had lasted for 140 days from December 1915 to April 1916. Then, 13,000 British and Indian prisoners of war had been paraded on foot through Baghdad on their way to detention centres in Anatolia.

Hadi wondered what this British army had in store for Baghdad. He already felt a stranger in his city. Everything that he had taken for granted as a Muslim living in an Ottoman Muslim province was collapsing around him.

As an older man with a longer memory of his Ottoman heritage, Hadi's father had an even stronger reaction to the recent developments. For Abdul Hussein this went much deeper than mere politics – the Ottoman Empire had defined who he and his family were for several centuries. The Sultan had always existed in his memory and imagination, whatever the shortcomings of his rule in Mesopotamia. The situation had been unbearable for the population during the war, but even so he had never desired this outcome. Like many others, Hadi and Abdul Hussein had to grapple with the implications of the Ottoman defeat.

After securing Baghdad, Maude moved north towards Mosul, fighting the remaining Ottoman army there. Several other battles were being fought by the Ottomans against the Allies, who were moving in on many fronts in the Mediterranean, Mesopotamia and Palestine. A blockade of the Dardanelles incapacitated the Ottomans, and Istanbul became awash with refugees fleeing the fighting. Unable to withstand further pressure, the Ottoman government surrendered unconditionally at Mudros, a harbour on Lemnos, a Greek island in the Aegean, on 31 October 1918.

Four hundred years of Ottoman rule in Mesopotamia had ended. What would replace it?

NOVEMBER 1999, BEIRUT

It's a Saturday night and I'm out for dinner with a couple of friends. Another friend calls on his cell phone, inviting us to join him at a place nearby where there's music. Slowly we make our way through the lively night to join him. I haven't bothered to ask what kind of music it is or who will be performing.

On a small stage in the middle of the room, a short, unassuming and serious-looking man takes his place on a chair with a guitar on his lap and starts singing. I discover he is a well-known Iraqi artist, Ilham al-Madfa'i. To my surprise he sings a popular old Iraqi folk song fused with flamenco beats.

> Mali Shughul bil Soug, maret ashufak
> 'Atshan hafn issnin, warwi 'ala shufak ...
> Wu as'al 'anil mahbub, minhu ili shafah ...
> Wi shlon anam il layl, winta 'ala bali
> Hatta il simach bil may, yibtchi 'ala hali

> I have no business in the market, I just came to see you.
> I've been thirsty for years, only the sight of you will quench it ...
> I want to ask about the beloved, who saw him? ...
> How can I sleep the night, when you are on my mind?
> Even the fish in the water cry for me.

Something deep inside me is moved as I hear the familiar words about a lover pining for his sweetheart, sung by Madfa'i with such longing and weariness. Images from the song bring to life a childhood memory of Bibi singing it to me as a taste of her country. I am overcome by floods of tears, which puzzles my friends, and myself as well. My 'foreignness' has come to the fore – the other half of my identity which is usually well concealed beneath my comfortable outward Lebanese appearance.

I can't explain, let alone understand, this deep homesickness that I feel. What am I homesick for?

BOOK TWO

Replanting Eden

٣

SEPTEMBER 2005

My aunt Raifa invites me to lunch at her house in Putney, south of the river in London. She knows I am going to ask questions about her life in Iraq. My other aunt, Thamina, is also invited, as well as two cousins of mine, Raifa's daughter Zina, and my uncle Rushdi's daughter, Nadia. Both my aunts are in their eighties. Although, like me, Zina and Nadia are Bibi's grandchildren, they belong to my mother's generation, since my grandparents had my father relatively late.

I'm the first to arrive. As my aunt sets the table in the kitchen while we wait for the others, I sit in her reception room. I see many valuable pieces, antique *objets d'art*, silver and precious glass – a museum curator's fantasy. Many of these inherited pieces have made a long journey, often from London or Paris (where they were purchased) to Baghdad, and back again to London via Beirut. Each has a story, like the silver tray table engraved with my grandfather's initials which was bought decades ago from Mappin & Webb. My aunt likes to tell their tales. I think they reassure her.

A sadness comes over me as I think of how the emotional charge of these objects has changed over time. Here in my aunt's flat, I feel almost as if they have been reduced to something grotesque. It is as if they have been dragged across history and then forced to fit into more cramped circumstances than they were once accustomed to.

Later we sit in the kitchen, three generations of women discussing Bibi over dishes such as *timan za'faran*, saffron-flavoured rice, and *sabzi*, green herb stew. It seems amazing that Bibi's influence is so pervasive that nearly twenty years after her death she has managed to bring us together.

Culturally and psychologically, my aunts are very different to my cousins and myself. They seem stuck in a time warp, preoccupied with private and public decorum, their values inherited from the world of their parents. This is less the case with my two cousins, but there is also a gap between them and me. I think we are outwardly less governed by the tribal mores and allegiances that dictated my aunts' identities. This is perhaps most vividly manifested in the old-fashioned way they dress, in silk gowns and pearl necklaces with diamond clasps.

Thamina is describing her idyllic childhood in Baghdad. She recalls life then as a constant round of activity. She keeps repeating how 'beautiful, beautiful it was'. She has barely started her reverie when Raifa interrupts her: 'Let me tell them about my mother, God rest her soul.' She launches into a soliloquy about how Bibi loved life, how she charged towards it with her arms wide open so as to embrace as much as she could. She then says that Bibi loved to gamble, but warns me not to mention this, and that she loved to indulge herself with material possessions: jewellery, furs, clothes. Bibi loved herself, Raifa declares, quickly adding that she was also magnanimous with those less fortunate than herself; she was very generous, forever giving alms to the poor. When she got older she started to feel guilty about her indulgences, and worried that God would punish her in the afterlife. She asked forgiveness from him, and died a pious woman.

Raifa stops, and Thamina, not to be outdone by her sister, reaffirms that her mother was never hands-on when it came to rearing her children. She was famous for delegating. Yet Bibi had an outrageous story that she loved to tell her tailors: that one of her shoulders was lower than the other from all the children she had had to carry.

'How preposterous – I don't think I ever saw her carry anyone!' Thamina exclaims.

I ask why Bibi would say that, and Raifa replies that it was because she was fat in later life, and wanted an excuse for it, so she blamed the number of her children. Quietly, I think that Bibi was probably right, having given birth on nine occasions.

Nadia embellishes Thamina's story by saying that she thought Bibi looked like 'a walking onion'. We quickly discover that it is one thing for her daughters to say certain things about their mother, but quite another for the grandchildren to do so; Raifa snaps back at Nadia that she clearly never liked her grandmother.

'Yes, I did,' answers Nadia defensively.

'Well, I loved Bibi,' I interject.

'She loved you too,' says Thamina.

'Ooze away, why don't you,' growls Nadia.

I carry on: 'She had lovely soft skin.'

'I remember she always used rose water,' Zina says.

We continue to share our memories, but after a while Nadia has to leave. However, she still wants to learn a little bit more from her aunts before she goes, so she asks them about Baghdad's markets. They list several, including Soug al-Saray, where they sold books.

'Really? What did the book market look like?' asks Nadia with childlike enthusiasm.

'They were all covered. Like the ones in Istanbul or Damascus. Have you not seen them?' replies Thamina. She explains that Soug al-Saray was a big market, but it was burned to the ground during the war in 2003. They said at the time that civilization had been destroyed yet again in Baghdad, with all those burned books.

Zina remembers how there used to be men with typewriters sitting on the pavements outside Soug al-Saray. They would type letters and requests for people who were illiterate.

'Yes, they were called *ardahaltchi*; it's a Turkish word. "*Arda Haltchi*",' Raifa confirms.

'Ah, and what's the English word?' wonders Nadia.

'Can't you see there's no English word? I don't think they even existed here. Use your imagination!' I snap at her, irritated by her repeated cultural mistranslations. Of all of us, she remains the most nostalgic for her pre-exile childhood, yet she is also the most removed culturally from anything Iraqi, or non-Western for that matter.

'The market was near the courts, so the *ardahaltchis* wrote complaints for people,' says Raifa.

My aunts continue to reminisce about the markets. 'The textile market had the best range of cloth. Your grandmother loved it. There were lots of different silks, devoré and taffeta, brought from everywhere, from Italy, France, India, everywhere.' Raifa concludes sadly: 'There was everything in Baghdad.'

6

Café Chantant

The British in Baghdad

(1918)

WITH EACH WEEK that passed, it became increasingly obvious to Abdul Hussein and Hadi that whatever was going to replace the Sultan's Empire would be markedly different. For some time Abdul Hussein and many other men of his generation had felt out of place in their own land; now they were experiencing profound pangs of nostalgia for what they had lost as they watched the British set about establishing their new administration. Everything the British brought with them was thoroughly alien: their soldiers, their police, their mannerisms and their language. The only flag the people had ever known, with its familiar crescents and stars, had been replaced by the Union Jack. Ottoman Baghdad appeared to have retreated into the shadows, leaving few traces of its existence behind other than old buildings and street names. Yet its soul lingered on in the people and the language.

The pillars of the British administration were erected swiftly. The political vacuum could not be filled at once, but the pressing issue of security had to be addressed without delay. Each day administrators poured into the area from the four corners of the British Empire, these new figures of authority including (amongst others) Egyptian policemen and Indian civil servants. Before long the Baghdadis began to resent these newcomers for what they seemed to represent: colonized peoples – the foot soldiers and lackeys of the British. Baghdadis feared that they themselves would be next in line to be subsumed into the British Empire.

Despite this suspicion and distrust, Abdul Hussein was saddened by the dishevelled state in which Baghdad was found by the British when they took occupancy of the city. While he hadn't exactly welcomed their arrival, he would have much preferred Baghdad to be looking its best when any stranger set foot in it, whatever that stranger's business happened to be.

Certainly from afar the city remained a compelling sight, with its thick palm groves and its minarets glittering with bright mosaics, answering the glow of the golden domes of Kazimiya across the Tigris. To have marched proudly into Baghdad only to be confronted by the sight of looting and filth everywhere must have been a huge let-down for the British troops.

The former *pachalik* or territory of Mesopotamia, which included the provinces of Mosul, Baghdad and Basra, was now under British control. Familiar with India, and the extent to which its various faiths had to be accommodated in order to ensure the smooth running of the Raj, the British sought to impose a similar policy in their newly acquired territory. They made concerted efforts to appeal to the different communities in the provinces, to the Arabs, Kurds, Turkmen, Sunnis, Shi'a, Jews, Christians and Madeans, among others. Order was reinstated throughout the region, with each community allowed to follow its own rites under the umbrella of the new administration.

This was not lost on Abdul Hussein, as a prominent member of the Shi'a community. The annual Ashura processions to commemorate the death of Imam Hussein, the grandson of the Prophet, had at best been ignored by the Turks in the past. Ashura was of great importance to the Shi'a, as in Hussein's final speech before his death he had uttered the words that would become the central tenets of their faith, emphasizing the duty to fight tyranny and oppression, and the importance of seeking truth and justice. In 1918, the last day of Ashura in Muharram in Baghdad was finally given the public prominence that such an important event deserved, and Hadi and his brother Abdul Rasul rode out among hundreds of young men to celebrate it as the crowds cheered them on the street. They felt very proud to be a part of the proceedings. Besides the young men on horseback, the processions were formed of men from the different guilds and neighbourhoods who chanted and beat themselves with metal chains and swords as they marched with others waving flags and striking drums. At the centre of the activities was a passion play performed by a troupe of amateur actors. The ceremonies were concluded with an elaborate expression of thanks and gratitude in the local press by Baghdadi Shi'a notables.

In contrast, the predominantly Sunni Baghdadi politicians were cautious of British overtures towards them. They were very conscious of the fact that with the defeat of the Ottoman state they had lost their official status, from whence they had derived their power. Moreover, while some Baghdadis were delighted with the possibility of progress and modernity, others were fearful of change, and didn't trust the Christian *ingliz*, attributing the recent flowering of the city's nightlife to British corruption and decadence. Baghdad's street cafés were the best places in which to gauge the strength of these feelings, amidst the smoke and the slurping of over-sugared tea, and the rhythmic sound of dominos being slapped on tables.

One afternoon Ni'mati returned to the Chalabi house troubled by a story he had heard from a merchant friend he had visited in Kazimiya's main bazaar, Soug Istrabadi. The merchant's neighbour had lost her temper one morning when she couldn't find even a scrap of bread in the house with which to feed her children. She had grabbed her *abaya* and marched down to the cafés by the square. There, her husband, a carpenter by trade, was sitting idle (as had been his habit for months), complaining to his friends about his lack of work and money because of the war, as he whiled away the hours playing *tawli*, backgammon, and smoking his *nargilleh*.

Undeterred by the decree that the cafés were a male-only domain, the carpenter's wife stormed over to him and harangued him about his attitude in the loudest voice she could muster. She had sold the last of her few gold bangles so that she could feed the children, she shouted, and her fingers were worn out from all the sewing jobs she took on to keep the house going while he sat about in the square, drinking tea.

Her husband, shocked and humiliated, turned a strange shade of aubergine. The men sitting nearby pretended to be invisible; they'd never witnessed such a spectacle, as domestic woes were always dealt with in private, never in public. The man's wife took advantage of his

Men in a popular café, sitting on traditional
high wooden benches.

confusion to carry on. She told him in no uncertain terms that jobs
were in plentiful supply now, since so much building work was being
commissioned by the British – and he had to get one, *now*. On that
note she stormed out, leaving him to pick up the pieces of his shattered pride.

Ni'mati was sure that the woman's uncommon boldness had been
inspired by the sight of all the British soldiers around the place. He
shook his head in disapproval, and concluded that – as if it were not
bad enough that they had commandeered the country – the British
presence was now threatening the masculine basis of authority in the
town.

Still, as the furious wife was aware, the British presence in Baghdad
was creating an economic and cultural boom, as it already had in
Basra. Suddenly, there was so much more to purchase and even more
to build, sending many enterprising Baghdadis into a frenzy of activity as they adapted to their new circumstances and familiarized themselves with the rupee paper notes that had replaced the Ottoman coins
– the *akce*, *para* and *kuruş*.

Unlike the Ottoman administration, which had openly looted from
the people in its troubled days, the British were keenly aware of the
pacifying power of money, and ensured that the abuses of their pre-

decessors were not replicated. The recently empty shops were now bustling with activity, as merchants competed to supply the British army with food, cloth and building materials. Unlike their Turkish predecessors, the British had cash and they spent it.

Even Bibi wasn't immune to the effects of the economic revival the British inspired, as she was very conscious of the fact that Hadi and his brothers had taken to exploring the resuscitated bazaars, hungry for novelty and ideas. Bibi, aged only eighteen, felt vulnerable when she watched her husband go off on these expeditions.

Her fears of infertility had proved to be ill-founded, for at long last, after many supplications at the shrine, she was expecting her first child. Upon discovering she was pregnant she had rushed to the shrine to thank Imam Musa, to whom she had prayed again just as Saeeda had suggested. Standing by the shrine's north-west Murad gate with other believers whose wishes had been fulfilled, Bibi had distributed coins amongst the sea of urchins who gathered there, while Saeeda inspected the livestock to be donated, chiding one man for his emaciated sheep. The slaughtered sheep would be translated into food for the poor.

Now that she was pregnant, though, Bibi worried that Hadi might no longer find her attractive as she had put on weight. She noticed that her husband had started to come home later in the evenings, singing tunes that she had never heard before. Sitting alone in her bedroom, she worried about his absences, biting her nails as she let her fertile mind carry her off to dark places. Imagining the worst, she became determined to find out where he was, and drew upon all the resources that were available to her, foremost among them the steadfast Saeeda.

Following some tactful enquiries, Saeeda was able to inform Bibi that Hadi had been spending his evenings at the Shatt Café, a *café chantant* in Baghdad where two sisters – Rosa and Lilu Numah – sang and danced to popular music played by a live band. Before the outbreak of the Great War, such cafés had started cropping up around the Maidan Square, near the bazaars and departmental buildings. Now reopened, these establishments were thriving, and once more offered performances by women singers who came from near and far.

Other cafés offered more traditional performances, in which travelling *hakawatis* – professional storytellers – narrated the tales of Scheherazade and her king, of warriors and beautiful maidens. Many of the stories were adapted from old Arabian epic poems. Entranced, men of all ages let the *hakawatis* transport them to a bygone age.

However pleasing these performances were, they were but a faint memory of the ninth and tenth centuries, when Baghdad had flourished under the rule of the Abbasids. Then, wine had flowed, poets conversed in courtyards and the sound of lutes filled the air. Highly valued female entertainers had sung, danced and played music for the delectation of their audiences; words of love were inscribed in henna on their bare limbs, calligraphy that came to life when they moved, bending the words and letters with their curves, like waves in the sea. Something akin to geishas, these women came from many lands, such as southern India, Sind, Georgia, Bosnia, Armenia, Aleppo, Ethiopia and Egypt. Known as the *Qian*, they were trained by specialist merchants and kept in grand city houses, where they courted the attentions of wealthy patrons. The *Qian* could be more ruinous than gambling, and many men's fortunes were lost on them.

The young sisters to whom Hadi came to listen now were of a different world: free women, yet still performing exclusively for male audiences, since there were no public places in which the sexes could mingle. For all Bibi's fears, Hadi would never linger to watch the sisters for long – just long enough to be carried away by the music and the atmosphere, and to lose himself in the gaze of the singers.

As much as Bibi fretted over Hadi's new pastimes, she too was intrigued by the changes taking place all around her. New goods flooded the bazaars, and businesses were flourishing. However, her mother, Rumia, soon began to view the British arrival in Baghdad as a curse. Rumia had always looked up to her eldest brother, Abdul Raouf, especially after their father's death, but in June 1917 he was banished by the British to Sumerpur in India for his pro-Turkish activities. Abdul Raouf was a cultured man whose social milieu had been strongly Persian. Some years earlier, he and his peers had formed a circle called the *Akhuwat-i-Iran* in support of the constitutional revolution in Iran of 1909, and had published a newspaper of the same name. The group was suspected by the British of promoting Turko-German propaganda, distributing pamphlets among Kazimiya's inhabitants which called for all Muslims to heed the Sultan's call for *jihad*. As a result, Abdul Raouf was now a prisoner of war.

Perhaps it was partly a question of age, but mother and daughter couldn't agree on the presence of the British in their city. To a degree, the divisions between them reflected the general feeling among Baghdadis. In the wake of her brother's banishment Rumia feared that the British would completely overturn the world as she knew it, in all its religiosity and ritual. Bibi, on the other hand, ever apprehensive of chaos, was relieved when order was imposed under the new regime. But she too was upset by her uncle's imprisonment, concluding that the British were just as draconian as the Ottomans had been when they had banished Hadi's uncle al-Uzri a couple of years earlier. She was also concerned by the fact that her father-in-law was attending many more political meetings. Ever prone to paranoiac imaginings, she was afraid that the British might expel him from his homeland too – and perhaps Hadi as well, since he would sometimes accompany Abdul Hussein to the gatherings. Whatever stability the British had restored to society, Bibi feared that the retribution of their justice might fall upon the heads of her family, swiftly and arbitrarily, at any time.

Nevertheless, one particularly disturbing piece of news that Bibi received from Saeeda caused her to wonder why British justice was not more swift and assertive. She waited until Abdul Hussein came home later that day to discuss this news with him, wondering if perhaps he could do something about it.

Unlike Hadi, who had a relaxed way of dealing with people, Abdul Hussein was a stickler for form. The servants would freeze or make themselves scarce whenever he passed them in the corridor. However, he always had time for Bibi.

She heard him come up the stairs and walk down the vaulted corridor. Rising slowly from her seat, she waddled as quickly as she could across the room to greet him before he disappeared into his quarters. 'Uncle?' she enquired cautiously. Abdul Hussein glanced across at her.

Bibi asked him if he had heard about the case of the *mukhtar*, the headman *Hadji* Jawad in the village of Dabbakhana. Abdul Hussein nodded, and replied that he had. A note of outrage crept into Bibi's voice as she continued, 'He murdered his two sisters and his infant nephew and then buried them somewhere ... and no one has done a thing about it! Can't you do anything?'

Abdul Hussein shook his head sadly and agreed that the news was shocking. He told her that the British superintendent at the police station was investigating the incident, and had already interrogated several men about *Hadji* Jawad's character. Then he sighed, stroked the ends of his moustache, and said cautiously, 'You know there was a lot of talk about them, they say they weren't good girls ...' He stopped short of uttering the terrible word 'prostitute', which would have been unimaginable to pronounce in front of a female family member. All the same, Bibi understood the euphemism. She knew what had been said, but it was still murder, and a *child* had been killed. She wanted to know what the British were doing if not establishing law and order. Impulsively she clenched her fist. 'It's not acceptable, I'm sure he did it!'

Abdul Hussein looked at her thoughtfully. 'My dear, you know how it is with honour here,' he said slowly. He agreed that the *mukhtar* might have been responsible, but in the absence of proof there was regrettably little that could be done. Bibi bit back her tears; it seemed to her that nobody cared because the murder victims had only been women. Seeing her distress, Abdul Hussein promised her that he would have another word with the Police Commissioner. He patted her on the shoulder as he would a little girl and smiled indulgently. 'Come now, tell me, how was your day?'

Bibi's face brightened: 'I'll tell you if you tell me what you are plotting with your friend Sayyid Muhammad.' Sayyid Muhammad al-Sadr was an old family friend of the Chalabis', a religious cleric from an eminent family who, like Abdul Hussein, possessed astute political sense and practicality (unlike many of the Shi'a *ulama*) and who directed their group's political activities. Abdul Hussein had great respect for him, and always listened to his advice in political matters. He also had the gift of the gab, and was very vocal about his dissatisfaction with the political status quo.

Abdul Hussein laughed at her request; he was unused to his women-folk taking such an interest in politics. It rather pleased him that his daughter-in-law found the subject so fascinating, so he indulged her with some information about his plans.

A year and a half into her marriage, Bibi's prayers were answered when she gave birth to a healthy boy, Rushdi, in late 1918. Her joy was unlimited as she felt she had achieved her first duty, and most importantly she had borne a son.

Hadi was clearly delighted with the birth of his baby boy; however, Bibi felt that she finally possessed something that was hers alone. She marvelled at the little infant, cradling him in disbelief. Unsure of how to care for him at first, she was temporarily relieved by the many women in the household, who gave her advice and assistance. Soon after Rushdi's birth, Saeeda left Rumia's service to be with Bibi and her newborn, and quickly became indispensable to Bibi.

Oozing with pride, Bibi was pleased to receive visits from the women friends and relatives who came to congratulate her. As always on such occasions, *naringi* – Seville orange sherbet – was served, as well as the succulent sweets that Rumia had prepared to mark the birth of her first grandson. But whenever her guests asked to see Rushdi, Bibi was secretly filled with dread. She thought him so beautiful that it seemed only natural to her that people would be jealous and would try to put the evil eye on him. Fuelled by superstition, she whispered prayers to ward off curses.

Rushdi was already carving an exalted place in Bibi's heart, one that would forever be his alone.

7

Rebellion

Fighting for Freedom

(1919–1920)

AFTER THE END of the Great War, Abdul Hussein became increasingly involved in politics. He had always been an influential man, but his talents were now in greater demand than ever outside the confines of Kazimiya; he was needed in the official halls of power and by the tribal leaders of the land, as well as by his own Baghdadi peers.

Thus in early January 1919 he was present when Colonel Frank Balfour, a political officer in the British Baghdadi administration, came to Kazimiya to meet the town's notables. Balfour's mission was to discuss a forthcoming plebiscite intended to determine what form the people wanted their future governance to take. He nonchalantly informed his audience that 'most people' in the region wanted the British to stay, but that he was nevertheless there that day to listen to the opinions of the townsfolk.

If the Colonel had hoped for acquiescence, he was met by a volley of harsh words. Abdul Hussein, especially irate, told him, 'If you are so sure of the opinion of the people of this country, why are you hindering our communications?' There were murmurs of approval from the other notables at the meeting, several of whom complained bitterly about the censorship the British had imposed. Over recent months, the postal services had been disrupted and telegrams had been forbidden in an attempt to reduce communications across the area, thereby limiting the possibilities for anti-British incitement and political organization. It seemed to Abdul Hussein that in some respects the British were far more disruptive than the Bedouins had been six years ago, when the nomads had taken to cutting down the telegraph lines.

The meeting broke up to the immense dissatisfaction of all parties, as there had been no clear outcome. It certainly hadn't yielded the results that Balfour had hoped for. The majority in the room, including Abdul Hussein, wanted an Arab Muslim government led by Amir Abdullah – the brother of King Faisal of Syria and the son of the Sharif of Mecca – and to be bound by a legislative council.

Walking back home afterwards with his friend Sayyid Muhammad, Abdul Hussein confided that he was sure the British had something nefarious up their sleeves. He was certain that they wouldn't go through the motions of discussing the plebiscite if they hadn't already decided its outcome; they were merely manipulating the people into trusting them.

At dinner that evening, Hadi and Abdul Rasul asked their father to recount the evening's events. He described how irritable Balfour had appeared. When Sayyid Muhammad had reminded him that the British themselves had proposed the idea of an Arab kingdom, after the Sharif's leading role in the Arab Revolt of 1916 against the Ottomans, Balfour had refused to respond.

Abdul Rasul was now eighteen years old, a more bookish young man than Hadi but as passionate about politics as his brother was about commerce. He now insisted to his father that the people had to pursue their demands for self-determination to the very end, even if it meant going out onto the streets and protesting. If their demands were clearly and eloquently expressed, he felt, they could be published and so used as a means to educate the people. He became increasingly excited as he told his father they *had* to insist on the right to self-determination, and perhaps appeal to the Americans, who would surely be sympathetic to their plight. President Wilson's Fourteen Point speech of January 1918 had specifically raised the point of self-determination – the right of nations to rule themselves – and it was a highly topical subject in international circles. Hadi, in agreement with his brother, chipped in that he was sure they could mobilize the people for this critical cause, even in the face of the British Army.

Abdul Hussein told them that he and half a dozen other men had signed a petition in support of self-determination at the meeting. It had been prepared in advance, and it had taken them some time to get the language right. Yet for all their pains the resulting text was rather short, and was almost dwarfed by the list of elaborate signatures at its foot. It declared:

> *In accordance with the liberty conferred on us by the Great Powers, the principal of which are the two esteemed Governments of Great Britain and France, we, being the local Arab nation, choose a new Muslim Arab Government to be ruled by a Muslim King, one of the sons of our Lord the Sharif, bound by a local* majlis, *and the question of protection will be considered after the [Versailles] Peace Conference.*

In writing their petition, the Kazimiya group had been partly inspired by Wilson's speech, and had met the US Consul in Baghdad in order to express their views regarding the future of their land.

The next morning the petition was taken to the shrine, where more than 130 more signatures were added. A rumour was spread by the Shi'a mullahs that a vote for continued British occupation would result in the equivalent of excommunication for any Shi'a who cast it. However, the Mayor of Kazimiya was a good friend of the British authorities, and drew up a petition of his own in support of direct British rule. He managed to obtain an equal number of signatures, including those of a number of merchants and British Indian nationals. Angered by the Mayor's actions, Abdul Hussein could only curse him for his wiles.

In the event the results of the plebiscite in late 1919 were fixed by A.T. Wilson, the British acting Civil Commissioner, yielding the outcome he needed in order to justify his intentions: according to his figures, 96 per cent of the votes were cast in support of continued British occupation. However, in reality Baghdad and the holy cities were mostly opposed to being ruled by Britain, while elsewhere opinions varied depending on the degree to which the British presence was considered to benefit the locality in question. Whatever order and security the British might have established, the people were growing restless about their political future. Many felt that Britain's policies in India should not be re-enacted in their region, and these dissenters vied for greater influence and power.

Nevertheless, a greater game was being played out over the heads of the people, and a more significant reality was in the process of being forged: the creation of a new country – soon to be called Iraq – from the former territories of Mesopotamia. There were parliamentary debates in Westminster about the future of the new nation, with telegrams flying to and fro between Cairo, Baghdad, London, New Delhi and Simla, but no decision could be reached. Messengers also travelled back and forth between the Shi'a holy cities and among the southern tribes in and around Basra, stretching down to the Arabian Peninsula. Frustration grew in Baghdad when the Arab delegation to the Versailles Peace Conference of 1919 – led by King Faisal – was thwarted by the British.

In truth, Britain owed Faisal quite a bit. In 1916 he had led the Arab revolt against the Turks during the Great War, with his good friend T. E. Lawrence at his side. Better known to the British public as 'Lawrence of Arabia', Lawrence was an Oxford archaeologist who had become a liaison officer to the Sharif of Mecca during the war, working for British intelligence. An unconventional man, he was much respected by his followers as he adopted local customs and dress; like Faisal, he believed in self-rule for the Arab nations. Lawrence had quickly developed a bond with Faisal, and had become his confidant. With his support, Faisal's actions had expedited the loss of Ottoman control over Turkish Arabia, in which the provinces of Mesopotamia, Syria and Palestine had been ceded to the Allies – or, more specifically, to the British.

Faisal was used to taking risks in the course of pursuing a bigger dream and greater power. In February 1919 he was part of the delegation that went to Versailles to argue the case for an Arab kingdom, which he envisaged rising out of the Ottoman ashes. In this he was supported by Lawrence, and was only pursuing the promises that the British had made to his family in return for its role in the Arab revolt three years earlier.

At Versailles Faisal had stood before the Allies, dressed elegantly in his long black robe and male version of the *abaya* cloak, his golden sabre on his belt, his white headdress bound by a thick gold band, hunger burning in his hazel eyes. (Such was the vision that had enthralled Bibi when she heard her brother-in-law Abdul Rasul read out a description of Faisal from one of the many newspapers he scoured in his eagerness to learn more about the Arab delegation.)

The key players at Versailles were of course those statesmen representing the victorious nations: David Lloyd George, the British Prime Minister, who was keen to reduce French influence in the Near East; his French counterpart Georges Clemenceau, hell-bent on ensuring that a defeated Germany would stay on its knees for the foreseeable future; and U.S. President Wilson, carrying with him his young nation's idealism.

Invited by Clemenceau to take the floor, Faisal sat in the centre of the room and read his speech in his native Arabic. He began, 'I am happy to be here in this meeting, which gathers the most illustrious men of the great nations, and I believe that this Higher Court will do justice by the Arab nation in its pursuit of its natural rights.' Standing near to him, Lawrence translated his words in a soft voice, too low for Clemenceau's ears.

Towards the end of his speech Faisal became emotional, appealing for self-determination and the independence of the Arab people. During the exchange of questions, he went so far as to state plainly that his nation had been a great civilization at a time when other nations represented at the meeting were still barbaric and primitive.

Faisal impressed his audience with his bearing and his eloquence, but he failed in his quest. Although the situation in the Middle East was undoubtedly problematic, it was not regarded with the urgency accorded to France's burning need for vengeance against the Germans in Europe, or Britain's dissection of Germany's African interests. And so a final verdict on the fate of the region was delayed until it could no longer be ignored.

In the interim, the Middle East was politely carved up between Britain and France over cups of tea and grand banquets. Politicians drew up new borders on their maps and declared new nation-states. Early grand schemes of a large kingdom extending west to the Mediterranean, and east to the Persian border, were gradually whittled down. Iraq was the result, a set of remote provinces in the former Ottoman Empire which were given a new identity. It was a fulfilment of sorts for the region's people, but hardly an adequate response to the many hopes invested in its creation.

* * *

In the climate of mounting exasperation, a new underground political party emerged: the *Haras al Istiqlal*, or Independence Guard, which wanted an Arab government led by one of the Sharif of Mecca's sons. The party's meetings were attended by both Sunni and Shi'a leading families – the Chadirchi, Bazargan, Fattah, Daftari, Al-Sadr, Arif Agha, Abul Timman, Suwaidi, Kelidar, Kubba, as well as the Chalabi – indicating a Sunni–Shi'a rapprochement that the British had not been banking on; much to the dismay of British intelligence officers, who received endless reports of the meetings, often anti-British in tone, from the well-greased palms of their local informers.

Ever alert to political developments, Abdul Hussein became closely involved with the group, although he was not a member. Along with others from Kazimiya, he was invigorated by its nationalist appeal, and soon found himself attending and hosting meeting after meeting with the *effendis* of Baghdad in an attempt to find a solution to the impasse with the British. To his wife Jamila's consternation, Abdul Hussein's *dawakhana* was busier than ever, and often noisy with raised voices from morning to dusk. From afar, Rumia pursed her lips in disapproval when she learned of the disruption to the household. But Bibi found herself becoming increasingly excited by the presence of political activity at such close quarters; her rapport with Abdul Hussein contributed to her attitude.

The news stories that Abdul Rasul read out loud in the *kursidar* reported that France and Britain, along with Italy and Japan, had summoned a conference at San Remo in Italy to resolve the state of political limbo in the Middle East. The conference concluded in late April 1920, passing a resolution for a set of mandates in the Near East under Allied control. More importantly, it incorporated the controversial Sykes–Picot agreement of 1916, which divided up the region between the French and the British.

The finer points of the mandate system were not fully absorbed in Baghdad, where it was viewed as a veneer for occupation. As public anxiety about the future of the region grew, the Independence Guard's meetings in Baghdad picked up momentum and their patience with the British began to wear increasingly thin.

On 1 June 1920 Abdul Hussein joined Sayyid Muhammad al-Sadr and other Independence Guard members at the Sheikh Sandal mosque in Baghdad, where an effusive poem was read out to welcome the visitors from Kazimiya, poetry being a customary device in rallying the people. The welcome that he and his Shi'a companions were given was especially telling as they were being received in a Sunni mosque. After the collective prayer, Abdul Hussein adjourned with a group of men to a nearby house to discuss recent events and to strategize. It was decided that a rally would be organized at the shrine in Kazimiya to demonstrate the depth of the public's anger to the British. The date was set for 3 June.

On the appointed day, more than 5,000 men gathered in the shrine's courtyard. Abdul Hussein stood in his family alcove, discussing the logistics of the rally with the Kelidar. They found it hard to hear each other speak above the noise, as the town criers beat their drums across town, through the marketplaces and along all the little alleyways, to make sure every man in Kazimiya came to the meeting, together with the *effendis* of Baghdad, who had travelled to Kazimiya by tram. The main square heaved with thousands of men, and the applause and shouts that followed the speeches from the pulpit echoed across Kazimiya. The message was sent loud and clear to the British: the people wanted Arab rule, not British occupation.

By July more than 20,000 people had been called out by the Independence Guard to demonstrate in Baghdad's mosques, and a full-scale rebellion broke out against the British in the south. The tribes were in revolt – the Kurds in the north and the Arabs in the south – but not as a single coordinated unit. The Independence Guard had acted cohesively, whereas the separate Shi'a tribes were encouraged by the Shi'a religious establishment in the city of Najaf to fight the British as infidel occupiers. There was talk of a Muslim state, but it was an unformulated vision. Nevertheless, there was a unifying goal: a total rejection of being incorporated into Britain's imperial domain.

In late September, the British began to retaliate in earnest, deploying their modern weaponry against the humble armoury of the tribes. The revolt began to flag in some regions, unsupported as it was by factions such as those tribes which had already reached separate agreements with the British over their lands.

However, by October the security situation had worsened throughout the middle Euphrates. Even the holy towns were in tumult. The British now resorted to overwhelming force to suppress the rebellion. Aeroplanes bombed Kufa, a town near Najaf that held great significance for the Shi'a: it was in the Sahla Mosque at Kufa that Imam Ali, the first Imam and the Prophet's cousin, had been killed during his prayers in AD 661. In bombing Kufa, the British were attacking one of the most sacred Shi'a sites.

At the end of October there was outrage in Westminster at the loss of Iraqi and British lives and the millions of pounds that had been spent fighting the revolt. Over 6,000 Iraqis and 500 British soldiers had been killed. Yet the Iraqis had made their point; they would get their native government. Although the revolt had been short-lived, it left a deep mark on the Iraqi psyche as a major event in the birth of their nation. It also reopened the rift between the Shi'a and Sunni communities.

Whereas the Sunnis, familiar with the realm of politics, found a *modus operandi* that allowed them to interact with the British, the Shi'a were prevented from doing so by their powerful religious leadership. The Shi'a 'divines' (as the British called these religious leaders) closed all doors on dialogue. Back in March 1920, before the outbreak of the rebellion, the Grand Ayatollah Shirazi had already issued a *fatwa* against dealing with the occupying forces. For most Shi'a believers, breaking with a *fatwa* – akin to a papal edict – was tantamount to heresy.

Thus the pillars of the new state of Iraq were built lopsided, with the country's Shi'a majority excluded from power. The hold of the religious establishment in Najaf, the *marjayyia*, was so strong that there was no alternative source of political authority to represent them. Equally, on their side the British were not predisposed to deal with the Shi'a, as they found them too extremist compared to the Sunni. They believed that the Shi'a were intransigent and would not be persuaded; that theirs was a world of principle and theory, ill-suited to the realities of life.

Many Shi'a grew bitter. They felt that it was because of their rebellion that the British had brought an end to their military occupation, but that now they were being punished for it by their fellow countrymen. Just as they had been under Ottoman rule, so now too they were for the most part excluded from power. The Sunni Arabs had taken that prize.

Abdul Hussein was deeply disappointed by the turn of events. Many of his colleagues in Kazimiya and Baghdad fled British reprisals; some who had arrest warrants out in their names joined King Faisal in Damascus, including his friend Sayyid Muhammad al-Sadr. Abdul Hussein knew the British had labelled him an extremist too, and he had to retreat from public life for a while, as he also risked arrest. He wished above all that his community would be pragmatic at this critical stage and engage with the actual state of things, instead of upholding its arcane theological positions. He mourned the loss of the Sultan: at least things had been clear back then.

He tried to put on a brave face in front of his family, especially Abdul Rasul, who was terribly disappointed by the crushed revolt. The depth of his feelings was evident in his shrunken demeanour and uncharacteristic listlessness. He was certain that had all the rebelling forces united, they would have forced the British to accept their demands. Hadi agreed with him.

Abdul Hussein kept on reiterating that everything was not lost – at least something had to be done now, for even the British couldn't continue as before. And he hoped against hope that Faisal would deliver for all their sakes; for everything he had heard about the Sharif's third son pleased Abdul Hussein.

8

A New King for a New Country

From Mesopotamia to Iraq

(1920–1921)

THE FUTURE OF Baghdad, Basra and their environs was a dilemma that not only taxed Abdul Hussein and the members of his household, but also the great heads of state in Europe. The cost of occupying the region had become too high for the British to maintain, and a solution had to be found urgently. Winston Churchill, then Secretary of State for the Colonies, wanted to resolve the débâcle as quickly as possible. Happily for him, the opportunity arose at an international conference in Cairo in April 1921. The conference was being held following the decision of the French to drive King Faisal out of Damascus, which had been promised to them as part of the Sykes–Picot agreement drawn up during the war. Faisal was now a king without a kingdom, and Iraq a country without a king. The combination of the two elements appeared, on paper, to have potential.

The conference at Cairo was a significant occasion, and many interested parties awaited its outcome, among them Faisal's close ally T. E. Lawrence. 'Everyone Middle East is here!' Lawrence announced cheerfully when greeting his colleague, the renowned British diplomat Gertrude Bell, on her arrival in Cairo for the proceedings.

Like Lawrence, Miss Bell had been deeply involved in the region for many years. An unusual and highly educated woman who had travelled through the Middle East for over twenty years in search of love, adventure and relics, she had been the only female political officer in the Foreign Office, where her local knowledge had proved invaluable in Basra during the war, resulting in her becoming Oriental Secretary to the High Commissioner in Baghdad some weeks after its fall. She had subsequently been instrumental in drawing up the borders of the new country that was to replace Mesopotamia, and was already known to the people of the region as a prominent political figure.

Bell and Lawrence were united in their hope for Faisal to become King of Mesopotamia, or Iraq as the region was now officially called. But they were apprehensive about the likelihood of this happening, as pressure to withdraw British troops from Iraq mounted in London.

In the event, Faisal did not attend the conference that was to decide his fate. However, as far as Churchill was concerned, installing him as King of Iraq was clearly the best solution in terms of Britain's beleaguered finances, as well as its ability to perpetuate control over the region. Moreover, it would enable Britain to keep a tight rein on Faisal's family, which occupied positions of power throughout the Middle East. Faisal's would be the central role through which other members of his family could be kept in check.

Installing Faisal represented a wise choice in other respects. The Allies' decision to depose the Muslim Caliph, in the person of the Ottoman Sultan, was still regarded as dubious. The Caliph was, after all, revered by Muslims worldwide, including the millions in India who were under British rule. So it was a sensible move to suggest replacing the Caliph with someone even closer to the founder of the faith – for Faisal was of the Hashemite clan and as such a direct descendant of the Prophet. His father, Sharif Hussein, had even been appointed Emir of Mecca, the holiest site for Muslims (a decision the Ottomans must have regretted later when he sided with Lawrence and the Christian British against the Caliphate).

Thus was Faisal's role settled. Pending national elections in Iraq, which the British authorities hoped to influence without appearing to do so, the first big battle had already been won.

When the telegram arrived announcing Faisal's arrival in Basra, Abdul Hussein was naturally among the delegation from Baghdad that arranged to travel down and welcome him. He set off by train on 19 June 1921. This was his first experience of the railway to Basra, popularly called the *shimandifair*, a bastardization of the French *chemin de fer* (this term would in turn soon be replaced by *reyll*, a bastardization of the English). The track had been built by the British Army during the war, and finally completed using Indian labour in January 1920. Before then the more common method of transport had been by steamboat.

The journey to Basra was twenty-eight hours long and very hot, but nothing could spoil Abdul Hussein's good humour. At first he could scarcely sit still, so full of anticipation was he. At every turn in the conversation with his travelling companions, he laughed heartily. However, the air of excitement was underpinned by a sense of apprehension: this was a new and unknown political road on which they were setting out.

As the train rattled along, following the river through palm groves, Abdul Hussein's thoughts turned to the implications of Faisal's arrival in Basra. He worried that protests might disrupt his reception. He didn't know what to expect of the southern tribes, who had developed good relations with the British, and he was also suspicious of the British themselves. They had toyed with several other local candidates, so he

Rashid Street, one of Baghdad's main
thoroughfares in the late 1920s.

was not sure if they really wanted Faisal. He also worried that Faisal wouldn't be able to handle Iraq's mixed population, which required deft management.

The train arrived at Basra station late at night. Abdul Hussein arrived feeling exhausted and restless, and was grateful to step out into the cool evening air. Unlike the last time he had visited the city, when a horse-drawn carriage had taken him from the port, this time he was met at the station by a driver in a Morris Cowley convertible. It was too dark to see the patterns of the delicately latticed windows of the townhouses by the river, but the moon's reflection on the water moved him to recite the opening verse of the Quran in thanks for such beauty.

The British navy ship HMS *Northbrook* had carried Faisal and his party from Jeddah. Several of them were Iraqis who had fled to Faisal after the 1920 revolt. Abdul Hussein's good friend Sayyid Muhammad al-Sadr was among them. Also at Faisal's side was his British aide of some years' standing, Kinahan Cornwallis, a distinguished diplomat and Arabist who would in time be the object of Gertrude Bell's unrequited love.

The party arrived at Basra port in the late afternoon of 23 June 1921, after a journey of eleven days. A small crowd of onlookers was waiting at the pier for them, but there were no loud cheers: they had come out of curiosity. Only weeks earlier, perhaps inspired by Kuwait's example next door, the people of Basra had petitioned for complete autonomy and a republic. The tribes, meanwhile, resented the notion of a king altogether.

No sooner had Faisal greeted his official reception committee than he was taken aside by the British political officer St John Philby, and advised that he would have to earn the respect of the people, who were going to vote in free elections. In short, it was by no means a *fait accompli* that he would be king.

In contrast with this hard-headed exchange, Abdul Hussein's encounter with Faisal was altogether more amicable. Expecting to be overwhelmed and perhaps a little tongue-tied, Abdul Hussein was surprised to feel at ease. A delicate man, Faisal was very cordial, more approachable and likeable than Abdul Hussein had imagined he would be. Indeed, before he had a chance to greet the new regent properly,

Faisal was thanking *him* for his courtesy in having travelled so far for the purpose of welcoming their party. Abdul Hussein bowed his head deeply and told Faisal of the high esteem in which he held his fore-fathers, the Prophet's family. He pledged his service and loyalty, and extended an invitation to him to visit the Chalabi house when he came to Baghdad.

Faisal accepted Abdul Hussein's invitation graciously, but explained that first he had to travel up through the country to the shrine cities of Najaf and Karbala in order to pay respect to his ancestors – he was referring to the Shi'a shrines of Imam Ali and his son Imam Hussein. It was hoped that this would gain him favour in the hostile cities that remained subject to the Ayatollah's *fatwa*, prohibiting them from inter-acting with the British or their representatives. The Shi'a were also still smarting from the previous year's revolt and from the strong hand that the British wielded against them. Faisal, a stranger in this land, was keenly aware of the need to win them over.

In light of Faisal's clear concerns about the size of the task that lay before him, Abdul Hussein saw his chance to make a favourable impression and instil in the new King a sense of support that, he hoped, would encourage him in his work. Much would depend on the hospitality he could summon up back in Kazimiya.

'Lunch Thursday STOP Munira prepare house STOP Rumia Fesan-joon END.'

Such was the message on the telegram that Abdul Hussein sent to Hadi from Basra after Faisal's subdued reception there. He wanted to host a party for Faisal at his widowed sister Munira's house in Kazimiya, which had a much grander reception room than his own home. He also wanted Munira to supervise the kitchen, as she had magical hands when it came to cooking (he still dreamed of her marvel-lous pickles). Munira had proven an astute manageress of the estates she had inherited from the late Nawab, and Abdul Hussein was sure that with her and Hadi's help, and the finest produce from his own lands, they could create a feast fit for a king.

Not only did Abdul Hussein want to serve up traditional *quzis* – grilled whole sheep served over a bed of the finest seasoned rice – but also the more refined dishes for which Kazimiya was known: *fesanjoon* (pomegranate-and-walnut chicken stew), lamb-in-apricot stew, *timan bagila* (dill-flavoured rice with broad beans), various types of *kubbas*, vegetable stews and stuffed meatballs and, of course, the traditional *dolmas* (stuffed vine leaves, onions and peppers). The list of desserts was similarly elaborate – *zarda*, *halawa*, *baklawa* and *mahalabi*. He was determined to give the King the best banquet of his life.

Aware of the faith his father placed in his organizational abilities, Hadi asked the family to meet upstairs in the main room of the *andaroun* to plan the event. This was much more than a mere lunch party; it represented a bold political statement in support of the new King, given the continuing and widespread hostility with which he was regarded by the Shi'a religious establishment. Hadi saw his family's role as essential, setting an example to other Shi'a by honouring Faisal and helping him to reach out to them during his visit. Hadi had to keep his cool, as the women – more concerned with the logistics of preparing the meal than with national politics – immediately set about deciding on the menu; even Munira was unusually animated, pleased by her brother's recognition of her culinary talents.

As the arrangements were made, Bibi sat nervously between Hadi and his sister Shamsa, now her firm friend. She was very excited by the prospect of the party, despite the fact that none of the women would be permitted to attend. But soon the intense discussion of food began to make her feel redundant, unsure of how she could help in the proceedings. She couldn't cook or make herself useful in the kitchen as the other women did. She had a knack for remembering recipes, but little interest and even less talent in executing them. However, she knew she could rely on her most valuable culinary asset – her mother, whose *fesanjoon* had been singled out by Abdul Hussein in his telegram. *Fesanjoon* was a Persian poultry dish that carried spiritual nourishment in its very name, meaning 'what the soul craves'.

With the menu decided upon, Hadi divided the tasks among the members of the household. He suggested firmly to Munira that Rumia should help her with some of the dishes, as she couldn't do it all on her own. 'And some dishes should not be left to the regular cooks,' he diplomatically explained.

Munira's right eyebrow rose momentarily, just long enough for Bibi to notice. But this wasn't the time for petty rivalries, so, thin-lipped, she turned to Bibi: 'This won't be an imposition on your mother, will it?'

'No, not at all,' replied Bibi. 'She'll be more than happy to do it.'

Unlike most of the other women in the household, Bibi appreciated just how important the King's visit was, and she wished to contribute to its success in whatever way she could. Finally, after what seemed an age, the meeting ended, and she quickly instructed Saeeda to look after Rushdi while Ni'mati accompanied her to her mother's house.

They walked as fast as they could over the uneven cobbles, and when they arrived Bibi knocked hard on the metal door. Her mother's Iranian maid, Laleh, greeted her, but her eyes widened in horror when she saw that Bibi was not alone. She quickly covered her head and clothes with her hands in a feeble show of modesty in front of Ni'mati. Ni'mati sighed and, turning his back to her, said, 'Get me some tea; I'll have a cigarette outside.'

Bibi rushed past Laleh into the house, calling out for her mother and shouting something about a duck. Bemused, Laleh followed her into the hallway. She explained that Rumia was praying, and said she would make Bibi some tea while she waited for her, then retreated to the kitchen.

Bibi headed up impatiently to her mother's room on the first floor. Rumia's head had barely touched the *turba*, the clay tablet on the prayer rug, to signal the end of the prayer than Bibi announced without ceremony, 'We've no duck and the new King is coming to lunch. If you don't cook your *fesanjoon* for him I'm finished!'

Rumia smiled to herself. She was only too used to Bibi's dramatic turn of phrase. She reassured her daughter that she would cook the *fesanjoon*, and that chickens from the market would do just as well as duck. Bibi could not believe her ears; perhaps her mother thought this was a small lunch for her brothers rather than a banquet for the new King. Perhaps she didn't realize how important this feast was for the Chalabi family.

9

Fesanjoon, a Royal Luncheon

Faisal Visits Kazimiya

(1921)

THE FOLLOWING WEDNESDAY, Faisal arrived at West Baghdad train station after a long and challenging journey from Basra. His visits to the Shi'a shrines at Najaf and Karbala had not been an unqualified success, as in both places he had been snubbed by the religious establishment. However, his reception in Baghdad was warm and well attended.

Mindful not to intrude on the occasion, and that his own moment would soon come, Abdul Hussein watched from a respectful distance while Gertrude Bell assisted the British High Commissioner Sir Percy Cox as he introduced Faisal to the notables of Baghdad. Miss Bell's pleasure in the proceedings was obvious; since her arrival in the city she had wasted no time in developing ties with the local dignitaries, in the name of King George V. Her maverick charm and command of the Arabic language had afforded her *carte blanche* to meet anyone of any significance, and gave an altogether unusual face to the occupying authorities. But Abdul Hussein was not quite sure what to make of her.

On the other hand, he did know that the next day would be very important for Faisal, as he was going to visit the third of the holy Shi'a shrines – in Kazimiya, Abdul Hussein's home town. The reception he got there would be of the utmost importance, especially after his disastrous visits to Najaf and Karbala, and Abdul Hussein had done everything in his power to make sure it would go smoothly.

Faisal's convoy duly arrived in Kazimiya at around ten o'clock the following morning. A large crowd welcomed him, chanting and throwing flowers, as he proceeded to the shrine. The long road that led to its main door, the Bab al-Qibla, was decorated with the new Iraqi flag as well as with flowers and palm leaves.

Hadi, taking his cue from his father, was well aware that Faisal's visit to Kazimiya had to be seen to be a success from the moment he arrived in town. He was concerned that there might be a repeat of the muted receptions the King had received in other shrine cities, so he turned to the town's young men to ask for their help in providing a jubilant welcome.

As a result of Hadi's efforts, Faisal was greeted by exuberant young men waving banners with welcoming words for him. Hadi also made sure to have three sheep ready at the door of the shrine, in accordance with the old Arabian tradition of sacrificing a sheep in honour of a valued guest. As custom required, Faisal stepped over the blood of the sheep and entered the courtyard. He stood for a moment with his palms raised and read the *Fatiha*, the first verse of the Quran, before proceeding with his visit.

Meanwhile, the Chalabi household was becoming agitated because the King was running late. Rumia and Munira had produced a wonderful feast, and Munira realized that in Rumia she had found a kindred spirit who shared her appreciation of and talent for preparing excellent food. In the days prior to the visit the kitchen had acquired the status of an army headquarters, bustling with people and activity. Several women from Kazimiya had come to help, in addition to the exhausted household staff, none of whom had ever before had to prepare for a function as important as this one.

To escape the stress and chaos in the rooms below, and too distracted to play with her young son, Bibi decided the best thing for her would be to risk the glare of the sun and go up onto the roof, where she sheltered under a reed umbrella while she waited for the party to arrive. She wished she could be amongst the party welcoming the King in person, but that was impossible because she was a woman. She comforted herself with the knowledge that Abdul Rasul would share with her all the details of the King's visit later. He would be by his father's side throughout the event, unlike Hadi, who would be busy organizing the logistics.

Bibi took a cigarette from her pocket and lit up, peering down over the edge. At least she could get a sneak look at her future king that way. She liked what she had heard and read about him so far, and was much impressed by his career, his travels and his style. She gathered that he switched between traditional attire and Western suits, and that he looked equally graceful in each.

The party finally arrived. From her secret vantage point, Bibi was able to get a quick look at Faisal's face. She was not disappointed: it was indeed a face of a king. She made her way downstairs so she could watch the feast from behind the wooden screen of the covered balcony upstairs.

The lunch was hosted in Munira's large banquet room, which was crowded with people angling to get a close look at their new king. Before the party sat down to the sumptuous meal, several poets stood up to welcome Faisal in the courtyard. The first was Hadi's uncle al-Uzri, who was now back from his incarceration in Kayseri. He took a deep breath before he began, the sweat dripping off his forehead. Then, in a melodious, deep voice, he recited:

> I stand and welcome glory and breeding.
> There was for the Arabs a hijacked throne.
> I bow to greet its revolutionaries
> Who with the sword realized victory – *al araba*.
> They invited you from Mecca to the valley of the valiant saints
> To take the throne of your exalted ancestors.
> You are the most deserving heir of their inheritance;
> For you the hearts overflow with emotion
> In its folds, the sighs are unable to speak ...
> For they are here to give you allegiance
> Willingly without terror or fear ...
> Their eyes rove. It is no surprise
> That they settle but on your pivot.
> This trust – take it in your palm
> And relieve your people of their misery ...
> And live a king in this state,
> A lion reigning from the skies.
> Above all defiance.

After the delivery of the final poem, Faisal spoke with gratitude and heart-felt emotion, thanking everyone for the warm reception he had received, and looking forward to the bright future that lay ahead for Iraq.

The food was a great success. Indeed, so many compliments were showered on Rumia's *fesanjoon* that Bibi's pride reached unprece-dented levels. It didn't matter that she had contributed nothing to actu-ally making the dish; she felt she had, and hugged herself as she sat secretly watching the men devour it.

During the week it had taken Faisal to travel up to Baghdad, not only Abdul Hussein's household had been preoccupied with hosting the King-to-be. The formidable Miss Gertrude Bell had also been rushing around, arranging Faisal's lodgings and his reception.

In the days before Faisal's arrival at Basra port, the most urgent task had been to find a flag for the new Iraq, and it had fallen to Miss Bell to conjure one up. Partly inspired by her vivid historical imagination and poetic temperament, and partly by Faisal's own background, she had settled on the colours red, black, green and white. The red was for the Hashemite tribe from which Faisal came; the black was for the Abbasids who built Baghdad in the eighth century and reigned over the famed golden age; the green was for the Fatimids, a Shi'a dynasty which had ruled Egypt and parts of the Levant in the ninth century; while the white was for the Umayyads, who distinguished themselves as the first Arab Muslim dynasty in Syria after the Prophet's death, and who were much disliked by the Shi'a for killing Hussein, the Prophet's grandson.

Miss Bell was not particularly fond of the Shi'a, and her mistrust of them may have inadvertently helped to shape the physical map of the country, for when she had been negotiating the borders of Iraq she had insisted on including the Kurdish north, to balance the sectarian equa-tion in favour of the Sunnis. Believing that the Shi'a were extremists, and an obstacle to her dream of a kingdom led by Faisal, she preferred simply to circumvent them.

Nevertheless, from Miss Bell's perspective the new flag symbolized Iraq in all its religious aspects, past and present, and she hurriedly asked the wife of one of Faisal's leading supporters to whip up a proto-type on her sewing machine.

* * *

Two months later, before the Baghdad heat had geared up enough strength to inflict its heaviness on everyone, Faisal perched on the chair that was meant to be his new throne. It was Tuesday, 23 August 1921, a date that was charged with symbolism, particularly for Iraq's Shi'a population. According to the lunar calendar that year, it marked the festival of the Ghadir, when 1,289 years earlier the Prophet Muhammad, Faisal's ancestor, had designated his cousin and son-in-law Ali (from whose lineage Faisal also claimed kin) as leader of the community in a farewell speech to his followers. This seminal event had marked the start of the schism between Sunni and Shi'a, with the latter becoming the partisans of Ali.

So high that his feet barely touched the ground, Faisal's throne was situated on a platform in the middle of the large courtyard of the Sarai, the former Ottoman seat of power in Baghdad. Outside, the city was not yet fully awake, but the river next to the palace was already lively with fishermen and with children playing on the banks. He was flanked on his right-hand side by the British High Commissioner Sir Percy Cox, and on his left by the commanding British officer, Lieutenant Aylmer Haldane. Behind him sat his three closest advisers. They had been his companions for several years now, a band of brothers who had gone on the road with him as he pursued their shared dream of a united Arab kingdom. Only one of them had been born in Iraq: Sayyid Hussein Afnan. The others were Rustum Haidar from Lebanon and Amin Kasbani from Damascus. They were a colourful mix – a Bahá'í (a follower of a young religion founded in Iran in the nineteenth century, an offshoot of Shi'a Islam), a Shi'a and a Sunni respectively. They had accompanied Faisal on the long and crushing journey from Syria, where he had reigned as King for four months, until the French had driven him out at the Battle of Maysaloun in June 1920, with little British objection.

That bright Tuesday morning, Faisal was dressed in the Arab Army uniform, a pastiche of Ottoman, Bedouin and European military styles, with a spiked helmet with flaps at the back. British soldiers of the Royal Berkshire Regiment and freshly recruited members of the new Iraqi army looked down at him from the balconies around the courtyard. Along with over 1,500 other local notables, Abdul Hussein had risen to his feet while Faisal inspected the guard of honour on the way to his unfamiliar throne. Now everyone fixed their eyes on their new and foreign King.

The courtyard was filled with the silence of anticipation as Faisal stood to speak. He looked directly at his audience and announced: 'I offer my thanks to the generous nation of Iraq for the swearing of its free allegiance to me and for proving thereby its affection for me and confidence in me. I pray to the Almighty to give me success in elevating the state of this dear country and this noble nation so that its ancient glory may be restored and that it may maintain a high place among rising and progressive nations.'

He gulped almost imperceptibly before continuing: 'My duty today also calls me to express my gratitude to the British nation for having come to the assistance of the Arabs during the critical time of war and for the generous expenditure of its wealth and sacrifice of its sons in the cause of the Arab liberation and independence ... As I have repeatedly stated before you, the progress of this country is dependent on the assistance of a nation which can aid us with men and money, and as the British nation is the nearest to us and the most zealous of our interests we must seek help and cooperation.'

The hardest part over, Faisal's voice recovered and he concluded with dignity, 'I will spare no effort to profit by the qualities of every man of the nation, irrespective of religion or class. All to me shall be equal. There shall be no distinction between townsmen and Bedouins; for me the sole distinction shall be that of knowledge and capacity. The whole nation is my party and I have no other. The interest of the country as a whole is my interest and I have no other.'

Beaming up at him from below the dais sat Gertrude Bell, who had written to her father earlier that week, declaring that she would 'never engage in creating kings again ... it's too great a strain'.

Thus it was that Faisal, as the putative King of the Arabs, became the first constitutional King of Iraq, which was to become the first Arab state to join the newly formed League of Nations in 1932. Percy Cox read out the proclamation before the cheering crowd and a twenty-one gun salute.

10

Banished

Out of Kazimiya
(1922–1924)

BIBI'S PENCHANT FOR imagining cataclysmic scenarios was like a disease that never fully subsided, but which sometimes receded a little into the background. She may have overreacted before becoming pregnant with Rushdi, fearing that she would never conceive, but that could perhaps be excused on account of her being a new young bride. When another three years passed and she had still not conceived a sibling for Rushdi, she wondered whether someone had put a spell on her. One child was simply not acceptable.

What made matters worse was that two years after Rushdi was born, Bibi's mother-in-law Jamila gave birth to a boy, Saleh. This state of affairs taunted Bibi as she struggled to fall pregnant herself. As before, in her anxiety she unloaded her woes onto her mother, who stoically bore her daughter's ill temper.

Rumia's familiar responses frustrated Bibi, and she would mimic them cruelly, rolling her eyes: '*Be patient, my dear – God is generous; He will grant you a child.*' Rumia urged her to do special *nizir*, wish prayers, as well as to continue frequenting the shrine. She also encouraged her to go to the Sheikh Abdul Qadir Mosque, another port of call for fertility prayers.

Despite her irritation, Bibi listened to her mother and resisted the more esoteric approaches that many desperate women resorted to. These included visits to arcane sheikhs who dabbled in alchemy and magic, and who wrote cryptic formulas, *hijabs*, which they wrapped in tiny packets to be carried or buried somewhere in the house and which were meant to grant the wish of the seeker. Some sheikhs threw molten lead into water, which formed bubbles on the lead that were then pricked to break curses.

Another common practice that Bibi was tempted by, but decided not to perform, was to make a knot in the hem of her nightdress, fill it with crystal salts and sleep with the knot in her gown overnight. The next morning the salt would be collected and thrown into the river, accompanied by an incantation to break any spell.

Finally, Bibi got her wish and gave birth in October 1922 to a healthy baby boy, whom she called Hassan after her late father. He had Hadi's colouring, with his fair skin and sparkling blue eyes. With Hassan's arrival, Bibi's conception difficulties were put to rest, as she quickly became pregnant again. For a while her joy was boundless.

Some weeks after Hassan's birth, Abdul Hussein was offered the post of Minister of Education. This was the only such post to be allocated to a Shi'a, as many in government considered the Department of Education to be less important than other ministries.

However, King Faisal felt strongly about appointing a Shi'a minister to the position, as he hoped such a man would take responsibility for educating the neglected Shi'a community. Symbolically this was important: the King was trying to reach out to the Shi'a in spite of the boycott decreed by the Ayatollah Shirazi in Najaf, which continued to prevent many of them from interacting with the government or its associates.

Faisal was aware that the real reins of power had already been monopolized by Baghdad's Sunni elite, foremost among them the Iraqi officers who had fought alongside him in Syria. The British also remained in the foreground as advisers to the King and his Cabinet, and were conspicuous in their military uniforms as they helped to train the Iraqi Army. However, although his coronation speech had confirmed that the real power still lay with the British, the new King had become a symbol of the Arab spirit for the people; he was strong and steady, and he understood what was at stake and what needed to be achieved.

Abdul Hussein's political insight meant that he had been one of the first Shi'a notables to support Faisal openly. He reasoned that the best

A 1920s studio print of Abdul Hussein
taken in Damascus.

hope for the new country lay with the King. He knew he was taking a
risk by accepting his post, given the *fatwa*, and he anticipated some
criticism from certain hostile circles in Kazimiya – but he was not
prepared for what happened as a consequence of his decision.

On hearing that Abdul Hussein was the new Minister of Education,
Sheikh Mahdi Khalisi, an influential cleric in Kazimiya, lost no time
in accusing him of heresy, and banished him from the shrine for breaking the *fatwa*. Posters were glued on the exterior walls of the shrine
explicitly forbidding Abdul Hussein from entering it, making his fall
from grace public knowledge.

The news of Sheikh Mahdi's banishment of Abdul Hussein shocked
the entire household. They couldn't rejoice in his new appointment
with such a judgement hanging over him. Kazimiya was a small place,

and the shrine was the heart of the town. To be forbidden entry was quite intolerable, particularly for Abdul Hussein, who usually went there on a near-daily basis. It seemed ludicrous that he should not be allowed to enter a place where he had been welcome since he was a child, and where his forefathers were buried. There was much anger in the house towards Sheikh Mahdi.

The exception was Abdul Hussein's brother Abdul Ghani, who reacted to the news dramatically, blaming his older sibling for bringing shame on the family by his decision to join Faisal's government. A harsh and loud exchange of words followed between the two brothers. Abdul Hussein was outraged by what he saw as his brother's narrow-minded, inflexible and parochial attitude. In response, Abdul Ghani challenged Abdul Hussein to tell him what was more important in his life: a British puppet such as the King, or his faith and the blessing of the clergy. In a rage, he accused him of dragging the family name through the mud.

Deeply offended, Abdul Hussein reminded him that they were living in a new country and had a new capital, Baghdad, and that it would be even worse for the Shi'a than it had been before if they didn't embrace the changes. It was bad enough that the Shi'a only had the one ministry, but as that was the case it made sense to make the most of the opportunity, rather than resisting and revolting while the caravan passed them by. 'Furthermore,' he concluded, 'when Faisal, a Hashemite Sayyid, a descendant of the Prophet, accepts Iraq's throne, who am I to refuse a ministry?' In reply Abdul Ghani merely folded his arms and turned away in contempt.

Abdul Hussein persisted, trying to persuade him that if they were to have any say in the country it would not be by boycotting the new government and smoking pipes with the mullahs in the shrines. If only for the sake of their own children, they needed to take what was on offer. Abdul Ghani refused to see his point of view.

Abdul Hussein began to feel suffocated in Kazimiya while Sheikh Mahdi's ban hung over him. Furthermore, the old house was becoming too crowded for comfort now that it was home to his son Hadi's expanding brood as well as his and his brother's families. There was only one thing to do, he decided, and that was to move out of his late father's house, the only he had ever known.

'We are moving to the Deer Palace immediately,' he declared to his wife. 'Start preparing – and I don't want to hear another word,' he interrupted, as Jamila opened her mouth in protest.

Servants loaded mule cart after mule cart with trunks packed high with the family's belongings, before making their way through the narrow Kazimiya alleyways out into the countryside. The operation took days, and had a dizzying effect on everyone in the household. Besides Abdul Hussein and Jamila, their sons Abdul Rasul, Muhammad Ali, Ibrahim and Saleh, and their daughter Shamsa, Hadi and Bibi were also included in the move, along with their infant sons, Rushdi and Hassan.

The children ranged from young adults to babies, with the younger of Jamila's offspring being cared for by their nanny, Dayyah Saadah. Like Saeeda, Dayyah Saadah was African and an ex-slave. She had been Jamila's nanny, and had followed her to the Chalabi household when she had married Abdul Hussein. It was inconceivable that she should be left behind in Kazimiya.

A number of other servants moved to the Deer Palace with the family, among them Saeeda and Ni'mati with his new wife Fahima, another former African slave who had been picked for Ni'mati by Abdul Hussein's mother, Khadja. Having arranged her grandson Hadi's marriage, Khadja had moved on to new quarry for her matchmaking. To Bibi's immense relief, Khadja herself remained behind with Abdul Ghani and his household.

Although Abdul Hussein was upset to be leaving the home in which he had been raised, he was more distressed by the thought that he had become an outsider in Kazimiya. Following Munira's decision to move to her elegant townhouse, where Faisal had been entertained on his visit to Kazimiya, Abdul Hussein had bought the Deer Palace from her. Bordering the river on one side, the rural property extended for several kilometres on the other side into orchards and farmland. The elaborate front garden was separated from the main house by the road along which the horse-drawn tram passed on its way from Baghdad to Kazimiya. There was even an optional stop for the tram that was now called the Deer Palace Stop in honour of the magnificent statue that graced the pool in front of the house.

Bibi welcomed the move with enthusiasm, despite its abruptness. She was happy to leave the old house and its crotchety inhabitants behind. However, she was aware that the ban on her father-in-law was no light matter. She could see how affronted he was by it, despite his putting on a brave face to the world. By moving to the Deer Palace, he was in many respects going into exile in a house built by another self-imposed exile, the Nawab.

Nevertheless, Abdul Hussein quickly became busy with matters at the Ministry and the Cabinet. He helped to orchestrate a concerted effort to fight the Shirazi *fatwa* by reaching out to the Shi'a through education – increasing the number of schools across the country, especially in rural areas, where there were few, as there the age-old system of the *kutab*, an informal traditional style of schooling run by sheikhs who taught the Quran and literacy, persisted. The government-run schools that did exist were primarily in the main cities, such as Baghdad, Mosul and Basra, as were the few private schools and missionary schools for Christians and Jews. Although there was no university, there were several colleges, specializing in subjects such as law and education. Now, with Abdul Hussein's support, a new scholarship programme was established to send Iraqi students to leading Western universities for postgraduate studies.

Abdul Hussein attended many meetings with the King, and was soon receiving large numbers of visitors at the Deer Palace, who either congratulated him on his new post or sought some favour from him because of it. Before long it seemed that half of Kazimiya had passed through his new *dawakhana*. He also quickly developed a rapport with the British advisers at the Ministry, whose professionalism and practicality he valued.

However, within a few months of his appointment Abdul Hussein began to doubt whether he had made the right decision, in spite of the prestige his new post accorded him. Faisal's appointment of Sati' al-Husri as the new Director of Education, which meant that he would effectively run the Ministry, was to cast a shadow over Abdul Hussein and other Ministers of Education for many years to come.

Husri was a highly educated former Ottoman bureaucrat and an ardent Arab nationalist. He had followed Faisal to Syria, where he was Minister of Education during the King's brief reign there. He had expected to be appointed Minister in Iraq too, but because Faisal needed to include a Shi'a in his Cabinet, Husri had to sacrifice his role. He held a low opinion of the Shi'a, who he thought were ignorant and incapable. He also considered them unpatriotic, going so far as to accuse them of being Persians because of their religious affiliations.

Abdul Hussein fought Husri to the best of his abilities by pushing for more schools, more scholarships and more access to existing schools for Shi'a students. He believed that state education represented the best avenue available to the Shi'a; Ottoman policy had traditionally excluded them from positions within the government bureaucracy and the military, which had always been the two principal means of accessing education in the past.

As he came home one afternoon in his *tenta*, his official ministry convertible car, Abdul Hussein's spirits were at a low ebb. He had just had an audience with King Faisal during which he had appealed on behalf of several Shi'a students who had been refused entry to the law college in Baghdad. His own son Abdul Rasul was successfully completing his course there, but even he had required help to get into the college because he was a Shi'a.

As far as Abdul Hussein could see, Husri had blocked these young men's admittance for no reason other than that they were Shi'a, and he had sought the King's intervention in order to resolve the matter. He found it distasteful to have to appeal to the King, as he felt that such a matter ought to have been resolved at the Ministry; however, Husri's power and the support he drew upon were such that this had not proved possible. Even though Abdul Hussein prevailed in this instance, and had persuaded the King to overrule Husri's decision, he was demoralized by Husri's relentless attacks. Recently Husri had accused him in the press of being sectarian, and of not being a true Iraqi, but a Persian. 'Me, a Persian – a Persian!' Abdul Hussein repeated the insult out loud. 'And he can barely string two words of Arabic together – how dare he?'

As he got out of the car, he decided he would have his afternoon *nargilleh* under the big orange tree facing the river. Lush vegetation surrounded the house, with climbers weaving up the front façade, past the arched French windows which created vibrant reflections indoors when the sun hit their stained-glass tops. Abdul Hussein walked along the *madarban*, the long corridor that led from the front door through the two reception rooms of the *dawakhana*, which formed the front part of the house on each floor. The more formal reception rooms were upstairs.

It was siesta time, and the house was quiet except for the faint mewling of his infant granddaughter, Thamina. She was Bibi's third child and Abdul Hussein thought she was delightful. He made his way to his quarters, where he washed before going back outside to relax.

Dispirited as he was, Abdul Hussein always found peace in the gardens, which he loved. As usual, Zein al-Abidin, the head *baghwan* or gardener, was there to greet him while he made his afternoon tour of the grounds. Ever fastidious about appearances, Abdul Hussein had recently built up a wardrobe of tailor-made clothes in diverse Western styles, from the *bonjour* and the *frac* to the smoking jacket. On his head he now wore the *sidara*, an elongated narrow hat which had become the national headdress of Iraq's notables. When it was eventually adopted by King Faisal himself it became known as a *faisaliya* in deference to him.

Amidst the numerous orange trees, the aroma of honeysuckle competed with potent Persian red roses and Damascene pink roses – there were rows and rows of them, alongside violet asters and red-and-white busy lizzys, while behind them yellow, orange and pink snapdragons stood to attention. Next to the second pond at the back of the house Arabian jasmine coiled around a sky-blue gazebo, while dark pink geraniums decorated its sides. The gardener Zein al-Abidin's moments of greatest joy came when important visitors sat in the garden he had created and showered praise on the wonderful setting. As the garden was accessible by boat, Iraqi and British government officials would drop in from Baghdad on some afternoons, as did the British High Commissioner.

Another visitor was Miss Gertrude Bell, who wrote to her father in June 1924 that she had 'walked about under the flowering orange trees' in the garden of Abdul Hussein, whom she had described two years earlier in another letter as 'rather a friend of mine'.

Miss Bell had many friends in high places. Having poured her energies into helping King Faisal develop his court and establish public protocol, she had kept him company in Baghdad during the early days before his family had moved from the Hejaz to join him, and was even alleged to have been his lover. Bibi was shocked and intrigued in equal measure to hear visitors talk with raised eyebrows and pursed lips about the King's rumoured tastes and Miss Bell's apparent lack of inhibitions.

Bibi contrived to meet Miss Bell on several of her visits to the Deer Palace, and in spite of all she had heard, was very impressed by the force of her personality and the respect she commanded among all the men. As a fellow smoker, she also admired the confidence with which she enjoyed her cigarettes in public. She could not quite see what physical attraction this Western woman might hold for the King: she had a rather long, equine face, a pointed nose and a very piercing stare. Her clothes were well cut, yet not to Bibi's taste. However, she was quite tall – always a plus in Bibi's opinion – and had an abundance of fair hair, which she pinned up elaborately on the top of her head and which intrigued Bibi, who was thinking of cutting her own dark hair short, in line with the latest fashions that were filtering through from Paris via the Egyptian magazines she read.

Whenever Miss Bell visited the Deer Palace, Zein al-Abidin would always present her with a bouquet of fresh flowers. Inhaling their fragrance before setting them appreciatively to one side, Miss Bell would then discuss business with Abdul Hussein in her fluent Arabic, which she spoke like a *bulbul*, or nightingale. He was often tempted to raise the issue of Husri at the Ministry with her, and to discuss the political problems he foresaw, but he knew it would be futile. Besides being extreme in her likes and dislikes, Miss Bell was now struggling with her own situation in Baghdad, despite being regarded as the most powerful woman in Iraq. As the new state structure and bureaucracy fell into place, Miss Bell had become increasingly removed from the day-to-day affairs of running the country. Abdul Hussein was not surprised when she returned to her original love, archaeology. It seemed that the future was uncertain for everyone, and he wondered where his own path would lead.

Two years after Abdul Hussein's exile from Kazimiya, Sheikh Mahdi was exiled for his relentless opposition to the King. Only then was the ban lifted. Abdul Hussein heaved a sigh of relief when he heard the news, and was anxious to go at once to the shrine and forget the humiliation he endured.

11

Accidents of Nature

The Baghdad Boil

(1925–1926)

DAYS IN THE Deer Palace took on a distinct pattern that reflected the often parallel lives of its many inhabitants. Abdul Hussein's political and social standing continued to improve, and the house served him well with its large spaces in which to receive visitors and host official functions.

Equally, his wife Jamila had to fulfil her share of social duties as a minister's wife. She began to make social calls on Baghdad's elite, during which she was often accompanied by her daughter Shamsa and by Bibi. More important than these visits were the weekly female get-togethers called *qabuls* that she hosted in the men's *dawakhana* on the first floor. *Qabuls* were the customary practice in all elegant house-holds, and now that Abdul Hussein was in government it was appro- priate that his wife should hold them in their new home.

The Deer Palace *qabuls* took place on Wednesday afternoons in the long and narrow upstairs room, which was covered with the Persian carpets Abdul Hussein had inherited from his father, Ali. Along the walls were sofas and chairs upholstered in floral Genoese velvet of red, white and green, on which the ladies could sit comfortably. Hexagon- al coffee tables were interspersed across the room, their marble tops in frames of coloured glass that sparkled with the reflections of the large chandeliers overhead.

Freed from the despotic influence of her mother-in-law and her husband's sisters, Jamila dressed up for the *qabuls* in rich silks and a typical Baghdadi headdress, wearing her hair in two braids which hung down over her shoulders. The area above her forehead was covered by an *usbah*, a band underneath a *boyana*, another piece of cloth that covered her crown. Unlike Rumia, Jamila was not strictly devout, and didn't cover her hair completely as pious women did; Bibi wondered if this was because she had grown up in Baghdad rather than in conservative Kazimiya.

Dressed in her finery, Jamila always welcomed her guests dutifully and with good humour. The ladies would come into the room and greet those already seated with elaborate hand salutations before sitting down themselves. Bibi always delighted in company, and was especially animated on these occasions. The conversation ranged from the latest British contraption to appear on the market, to the scandals of Baghdad's musical divas, to the tragic death of someone or other.

Abdul Hussein's servant Habib Chaigahwa, who got his name from his job of serving *chai* (tea) and *gahwa* (coffee), would be on hand to serve the ladies. Habib Chaigahwa prided himself on his new concoction of *chocolat chaud*, which he served in winter in dainty green porcelain cups with gold rims. In the summer, when the *qabul* began a few hours later, he created frozen sherbet with orange-blossom water made from the flowering trees in the orchard, which was drunk at speed by the heat-exhausted women.

Abdul Hussein's two grown-up sons, Hadi and Abdul Rasul, flourished in the new country. Whenever he saw them set off early in the morning on horseback – one to college and the other to the new lands he was cultivating – Abdul Hussein felt deep gratitude for his good fortune in having such fine boys.

King Faisal's reign marked a vibrant new era in Iraq's development, despite the political challenges with which he struggled as he tried to balance Iraq's interests with the continued British presence. The country's independence could now be glimpsed on the horizon, albeit at some years' distance. Building activity continued to flourish, the market buzzed with new products and the port at Basra was full of agricultural produce ready to be shipped out to the world.

Hadi was, as ever, full of enthusiasm for new projects, and the times were particularly receptive to his kind of energy. In addition to improving the management of his family's lands and organizing the sale of their produce, he took advantage of a new law that awarded notables and tribal sheikhs the right to acquire empty *miri*, state-owned land, with a view to cultivating it for a period of time before ultimately being granted the right to own it. Besides the advantages of being his father's son and having the resources of his family's lands to draw upon, Hadi had accumulated some money from a small business venture when, a few years earlier, he had organized caravans of mules and donkeys to transport goods back and forth from Iran.

The new legislation was the perfect formula for Hadi, who managed to acquire many large plots of land. His hard work and good luck enabled him to cultivate fertile lands near Baghdad, in Diala province and near Kut, where the laws of the desert persisted. Many of the tribes in Diala lived off raiding travelling caravans or cultivated land. Water was a much coveted commodity in these regions, and whenever a well was dug, or an irrigation canal set up, raiding Bedouins would soon move in to take control of it.

On many occasions Hadi was involved in these incidents, and as a result he always carried a rifle with him. He would set out before dawn on horseback with his weapon on his back and his money tucked inside his vest to pay the farmers, accompanied by Ni'mati and a few other armed men, and ride for several hours into the desert. The roads were unsafe, and many robberies took place along them. As they approached his lands, such as the large plot he rented in Jurf al-Sakhir near Musayib, the men would invariably hear the firing of rifles in the distance, an attempt by the Bedouins to intimidate them. Hadi would respond by shooting back.

Sometimes bullets flew close to Hadi and his party, occasionally grazing someone's cheek or ear. Sometimes the horses took fright and unseated their riders. The important thing was to establish their presence in the land, and this Hadi succeeded in doing. He was popular with the farmers, and particularly with the tribal chief of the Jurf al-Sakhir region, who was grateful for his help in combating the marauding tribes.

Often Hadi and his party returned home late at night, muddied and bloody from a close scrape with the Bedouins. Bibi would look at him in horror, and chastise him for risking his life. She made a scene every time he set out, convinced that he would be killed.

Hadi showed an uncanny ability for managing his new properties, organizing people and production in an efficient manner that soon attracted notice at the royal court. Advised by his contacts in the Baghdad markets, he was always interested in the new types of agricultural machinery that were becoming available. He was among the first to invest in water pumps for irrigation, which proved far more efficient than any prior method. These new products made their way to Baghdad via local agents who represented the foreign companies that imported them from Europe – and from Britain in particular.

The same old trading system continued as before, albeit upgraded to more modern methods, with money lenders replaced by banks issuing letters of credit. Like any other market, that in Baghdad functioned on a basis of supply and demand. And where Hadi perceived a demand, he was willing to help create the supply.

Whereas Hadi devoted himself wholeheartedly to commerce, his younger brother Abdul Rasul was of a different temperament entirely. He had already broken with the family mould by leaving the *kutab*, the basic classes taught by a sheikh in one of the shrine's alcoves in Kazimiya, to attend one of the modern government schools. First he went to the primary *Amiriye*, then the secondary *Mulukiye* schools in Baghdad. He was now in the process of completing his final year at the former Ottoman Rashidiye Law School, which had been established some years earlier in order to train local bureaucrats. He was one of the few Shi'a students there.

Abdul Rasul distinguished himself amongst the family and his father's *dawakhana* regulars by taking part in political discussions, in which he argued persuasively and knowledgeably. He always wanted more for the country, and was often frustrated by the pace of change.

He observed a daily ritual of reading all the available newspapers, including the English-language *Baghdad Times*. Sitting on the terrace at the back of the house after college, he would lose himself in the press despite the background noise from the younger children and the chatter of the women. Whenever she passed through the terrace, Bibi would always interrupt him to get the latest news.

One day he came home enthusiastically with a new publication, *Layla*, the first women's cultural weekly to be published in Iraq. It included poems, literary essays and articles on science and child-rearing. He found his mother sitting on the terrace, and started reading the opening page to her. Jamila stopped him for a second while her granddaughter Thamina clambered up onto her lap. Irritated, Abdul Rasul glared at them both. 'Honestly,' he declared, 'if the progress of women were left in the hands of the ladies in this house, nothing would be achieved in a hundred years!'

Jamila chuckled softly: 'But, my dear, you've hardly chosen the right candidate in me for this mighty task. What do you expect me to do?'

One afternoon Rustum Haidar, King Faisal's Chief of Staff, dropped in at the Deer Palace for an informal visit. He was a close friend of Abdul Hussein's, yet whenever they talked their conversation inevitably steered towards politics. On this occasion Haidar told his friend that the King was under extreme pressure owing to the complicated Kurdish situation, as the Turks and the British fought over who was to control oil-rich Mosul province. He confided to Abdul Hussein that relations with the British were tense, and he feared they might even attempt to remove Faisal, as the French had done in Syria five years earlier.

Their conversation was interrupted when Abdul Rasul entered and greeted Haidar, whom he held in high esteem. The talk quickly moved on to the subject of Lebanon, Haidar's homeland. Abdul Rasul asked him a few questions about the renowned American University of Beirut, which had been founded by Protestant missionaries in the mid-nineteenth century and which was regarded as the elite educational establishment in the region. He also wanted to know, more pressingly, about Haidar's own university education in Europe.

Watching Abdul Rasul carefully, Abdul Hussein noted his son's unusual line of enquiry, and heard the underlying tension in his voice. He understood that this was his oblique way of communicating his desire to travel to him. When Haidar left, Abdul Hussein asked Abdul Rasul to take a walk with him around the rose garden by the river.

Stooping to inhale the scent of a deep red rose, Abdul Hussein glanced up at his son and asked, 'So tell me, what do you intend to do now that your college course is ending?' Before Abdul Rasul had a chance to reply, he continued, 'Would you like to go abroad, maybe travel around a bit?'

A studio print of Abdul Rasul
in his scholar's gown.

Abdul Rasul grinned with relief. He had already discussed his desire
to go abroad with Hadi, who had encouraged him, but he had been
worried about broaching the subject with his father, unsure of how
Abdul Hussein would react.

'Thank you,' he said. 'I have to confess that I have thought about
it. I think we'll only ever get to lead this country if we have some understanding of what lies beyond its borders.' He didn't need to explain that
in using the word 'we' he didn't mean himself or his immediate family,
but their community and countrymen.

Within months of receiving his father's blessing, Abdul Rasul had
left home to study at the American University in Beirut. Shortly afterwards he learned that he had been accepted by Downing College,
Cambridge to read Economics, and in 1925 he was among the first

Iraqi nationals to attend the British university. Abdul Hussein felt that, in going to Cambridge, Abdul Rasul had vindicated his own efforts in the face of his opponent Husri's unyielding animosity.

At the Deer Palace, Bibi was preoccupied with her expanding family. She had given birth to her second daughter, Raifa, in late 1924, and hoped to have yet more children. She also watched with satisfaction as Hadi made a name for himself in public life and began to show others the full extent of his entrepreneurial flair. She exercised her own talents by busying herself with the *qabuls* and other functions she attended both at the Deer Palace and elsewhere. Her maternal duties never restricted her from pursuing her other interests, as she was fortunate to live in a house with many servants.

In 1925 a sandstorm hit Baghdad. One of the effects of such storms is that sand grains seep through people's hair, nostrils and clothes, causing chaos, as had happened on the night before Baghdad fell to the British in March 1917. On this occasion, Bibi's three-year-old son Hassan appeared to have got some grains in his left eye, and developed a mild infection.

A doctor was promptly summoned to treat him. Misdiagnosing the little boy, he gave him the wrong treatment, which damaged the eye. Too young to understand what was happening to him, Hassan exacerbated the problem by rubbing at it. Another doctor examined him, and gave him a different prescription. Yet as the weeks went by, his eye still would not open, although the initial infection had disappeared. A third doctor was called, who found that Hassan had completely lost the sight of his left eye. Bibi was guilt-stricken, despite all the words of comfort her mother offered her; Rumia insisted that she should be grateful that Hassan still had another eye to see with. Hadi was equally pained by his son's suffering, although he tried to remain optimistic.

A few months later, Hassan was playing with his five-year-old uncle Saleh and his little sister in the back garden when he suddenly started to shriek with pain. He fell to the ground and covered his right eye with his hands. Ni'mati rushed over to him, but couldn't touch him, so violently was he shaking and kicking his legs. Bibi summoned the doctor, who confirmed what everyone had been dreading: Hassan had been bitten by an *ukhut* sand fly in his right eye. In most cases the sand fly bites people on the face or the limbs, leaving an ugly scar or 'Baghdad boil' which some think of as almost an authenticating mark of having lived in Iraq. Until a vaccination was discovered in the fifties, almost everyone in Iraq had one somewhere on his or her body. For Hassan, the sand fly cost him his remaining good eye, and he lost his sight forever.

It was heartbreaking for the family to watch as Hassan struggled to cope with the darkness that suddenly surrounded him. Hadi and Bibi felt powerless as their little boy attempted to connect once more with the spaces and faces that had once been familiar to him. Hadi had always been proud of his ability to resolve problems swiftly, but he was helpless in the face of Hassan's plight. There was nothing he could do to make his son see again. He could only ensure that Hassan was cared for; and to provide him with all that he required physically.

Bibi reacted to her son's tragedy with far less stoicism, taking it very personally. She wept every day for weeks, prayed to God for help, and continued to nurse the hope that one day he would be cured. Finally, she withdrew into herself for months, unable to sleep properly or to nurse her infant daughter Raifa.

Watching Hassan slowly feel his way along the walls of the house or up the railings on the stairs, Bibi couldn't shake off her despair. She was far too overcome by Hassan's handicap to notice the determination with which he fought to be like everyone else, as he tried to keep up with his siblings and Ni'mati's growing brood. She became terrified of glimpsing her reflection in his hollow eyes, their once-glittering blue now muted as though a fog had settled over them. Hassan's blindness taunted her: it was a betrayal to her because he was a part of her. She had failed him, and the shame embittered her.

It was fortunate that Bibi was not alone, as the rest of the household quickly adapted to Hassan's predicament. However, she wished that Saeeda was with her to help her cope. Although very supportive of Bibi, Saeeda had always been prone to angry outbursts of her own – indeed, she had come to Rumia's household after a falling-out with her previous mistress – and a couple of years earlier she had left the Deer Palace in one of her fits of rage, returning to the household of Bibi's old friend Bibi Istrabadi in Kazimiya. Bibi suspected that the real reason for Saeeda's departure was that she wanted to get married, and believed that being in Kazimiya rather than stuck out at the Deer Palace would increase her chances of finding a husband.

In spite of Bibi's troubled attitude to his difficulties, Hassan's childhood was happy and warm, as he lost himself in the crowd of children in the household, who took his affliction in their stride so unquestioningly and lovingly that he almost forgot he was in any way different to them. Moreover, Bibi insisted that no one ever mention his blindness to him, with the result that it was never acknowledged verbally by anyone in the house.

12

In Between

A Home Between Two Cities

(1926–1929)

ONE DAY, SEVERAL months after Hassan became blind, Zahra al-Duwayh arrived at the Deer Palace, holding a small boy by the hand. Bibi received her and invited her to sit down.

Zahra declined Bibi's offer and remained standing, still holding on to the boy's hand. 'I've come to your house to find work, *Khanum*. People have told me what kind people live here, and I very much hope that you and your family can find it in your hearts to be kind to me too.' Bibi noted her soft accent, which was from the south of Iraq. With tears welling up in her eyes, Zahra explained that she was on her own with her little boy, Rahim. Her husband had taken a second wife and had begun to beat her. Not knowing what else to do to protect herself and her child, she had run away.

Bibi's heart went out to her; she liked her sweet face and her gentle demeanour. She made up her mind there and then: 'Yes, I'm sure we can find work for you here. There's plenty of space for you both,' she assured her, and immediately made arrangements for Zahra and Rahim to be housed in one of the staff cottages on the estate.

When Zahra had settled in, Bibi told her about Hassan's plight and the need to give him special attention. Zahra assured her that she would put him in her eyes – *akhali bi 'yuni*, as the saying went.

Zahra and her son would stay at the Deer Palace for many years, becoming an integral part of the household. Rahim played with the other children and went to school; he eventually became an army officer after Hadi sent him to military college. Zahra proved true to her word, as she cared deeply for Hassan and also for the girls, all of whom became extremely attached to her.

Abdul Rasul had been at Cambridge University for a year, and was greatly missed by his family. His father loved sharing news of him with his friends, especially with Miss Bell, who always asked after Abdul Rasul whenever she visited the Deer Palace.

Miss Bell had opened the new Iraqi Museum earlier in the summer of 1926. The museum was a largely British enterprise, but Abdul Hussein had been proud take part in the opening ceremony, welcoming King Faisal to the premises. In principle, the new museum and Directorate of Antiquities fell under the auspices of the Ministry of Education, so Abdul Hussein was theoretically in charge of Miss Bell's domain, although he knew better than to try overtly to control her.

Through their relaxed conversations as they sat in his gardens, Abdul Hussein was better able to imagine his son's alien surroundings. Miss Bell described her homeland in detail, and he used her descriptions to flesh out the university customs and magnificent buildings his son described in his letters.

Overwhelmed at first by the beauty of the colleges and the weight of history, Abdul Rasul had nonetheless applied himself to his studies to the best of his abilities. He thought it particularly auspicious that Downing College's motto was *Quaerere Verum* – 'Seek the Truth'.

The British climate, food, people and aesthetics could not have been more of a contrast to his world in Iraq. The few other foreign students who attended the university were mostly from India, so Abdul Rasul had to carve his own way socially. He enthusiastically adopted plus-fours and tweed jackets – much to his mother Jamila's concern when she received a photo of him posing in this strange new costume – and although he didn't really take to cricket, tennis became a passion for him.

Immersing himself in the cultural as well as academic opportunities, he enjoyed going to the Festival Theatre in town, where he saw Aeschylus' *Oresteia* performed by actors in masks. His delight in being at Cambridge was stronger than any loneliness or difficulty he faced. His letters home were always cheerful, asking for everyone's news,

Abdul Rasul and friends in
Cambridge, circa 1927.

although of course he omitted to recount in them the stories of his
meeting young ladies and the drunken evenings that often followed
formal college dinners. Yet Iraq was never far from his mind, and in
his letters he always included ideas that he wished to be adopted in his
home country.

Bibi particularly missed his daily newspaper ritual on the terrace,
when he would often read out loud to her. She tried to keep up with
the news herself, but much preferred listening to Abdul Rasul, who
injected life into the driest of reports.

On 14 July 1926 she was leafing through the paper to see what had
been written about Gertrude Bell, whose fifty-eighth birthday it would
have been had she not died two days earlier. Talking to the ladies in
the *qabul*, Bibi heard there was even talk of suicide – that Miss Bell

had taken an overdose of barbiturates – but this rumour was quickly quashed by the authorities, and Miss Bell was to be given a respectable burial in the Protestant Cemetery in Baghdad. Nevertheless, Bibi believed the rumour. Projecting her own world view onto Miss Bell, she was convinced that she simply couldn't bear to live alone, single and childless, any longer.

Miss Bell's funeral was a small, private affair. King Faisal could not come, having left a few days earlier for his summer holidays in Europe. Unlike General Maude, who had died of cholera the same year his troops had captured Baghdad, Miss Bell was not given an elaborate monument to mark her grave, despite her crucial role in shaping Iraq. Hers was a simple, discreet tomb.

Abdul Hussein came home from the funeral service feeling pensive and quiet. It was the first Western service he had attended, and he had been impressed by the order and dignity of the ceremony. The restraint of Miss Bell's colleagues was in marked contrast to the hysterical funerals he was used to.

The details of her life were entirely alien to him, yet her devotion to her work and her passion for Iraq were unquestioned. She had been a controversial figure; even when she had redirected her attention from Iraq's political future to its archaeology she had ruffled many feathers by granting permission to teams of European and American archaeologists to excavate Iraqi ruins. These foreign missions were allowed to retain half of what they discovered on the proviso that they gave the rest to the Iraqi Museum, the domain of Miss Bell. The policy had resulted in many important discoveries, as the Iraqis themselves were not deeply concerned with their Mesopotamian past, being preoccupied with their new country's future.

More than anything, Abdul Hussein felt sorry for Miss Bell for having died alone without her family. He wondered how her father, who he had once met briefly, would take this bad news. When he settled into a chair in the family sitting room it was a relief to find Bibi waiting for him, wanting to hear about every detail of the service. He found it endearing that she had taken such a keen interest in Miss Bell, and had been determined to be present at many of his meetings with her.

Although Abdul Hussein had been raised in a conservative town, where women had no role in public life, he was able to appreciate Bibi's social skills and ambition. His mother, Khadja, was a strong-minded woman whose influence extended well beyond the walls of her home in Kazimiya, yet hers was a purely domestic domain. Bibi was different; she enjoyed people and the limelight, and she had an enquiring mind. Abdul Hussein was convinced that Bibi had brought good luck with her when she joined the family, in spite of the turmoil they had experienced in recent years.

It was an especially hot July, and everyone was grateful for the elaborate summer roof arrangements of the Deer Palace. The roof was in many respects just as important as the interior of the house, because it was here that the family slept under mosquito nets in the summer months, while taking their siestas in the cool of the *sirdab*, the basement. The roof, covered only by the sky above, reflected the family structure downstairs, with a brick wall separating Abdul Hussein's quarters from Hadi and Bibi's.

The creeping light of dawn would nudge each member of the household awake at different times, and one by one they would climb down the stairs to their bedrooms on the first floor, to snooze away a few more hours until the day was officially launched. Bibi's ascent and descent to and from the roof had become a rather cumbersome business, because she had to take her jewels up with her every night at Abdul Hussein's insistence. He suspected Ali Akbar, the generator man, of having light fingers, as Bibi had recently lost a ruby ring and a *mina*, a gold enamelled bracelet. Bibi was also in the early stages of pregnancy again, and she was relieved when the summer was nearly over so her ascents to the roof could be suspended till the next year. Najla, her third daughter, was born in the late spring of 1927.

Despite the summer nights on the roof and the continuous activity in the *dawakhana*, the true heart of the house lay in the *andaroun*, the private rooms where the family lived. On the ground floor, just behind the *dawakhana*, the *andaroun* rooms were situated on each side of the house in two parallel rows, leading to a large winter family sitting room, and finally to the entrance to the house's Turkish bath, which extended into the garden. Abdul Hussein, Jamila and his younger children had separate quarters in the *andaroun* from Hadi's growing family.

Abdul Hussein was much loved by his grandchildren, who saw more of him than of their father. He gave each of them a nickname that he teased them with. Rushdi and Thamina were his favourites among Bibi's brood, although he tried to bestow his affection on all the children equally. He playfully demanded that they stand in line once a week, when he would give them each an individual share of the farm produce, which they promptly took to the kitchen. This distribution of goods was his way of making them aware of the family's main source of income.

However, he was quite strict with them when it came to what parts of the house they were allowed to enter. One that was out of bounds was the formal dining room upstairs, which fascinated three-year-old Raifa because of the crystal candelabras she could see shimmering through the doorway. The magnificent dining table could seat twenty-four, and was used for the official receptions Abdul Hussein held for personages such as the King, the members of the Cabinet, official foreign visitors or the British High Commissioner.

In the hours before a formal event the dining room became the sole domain of Mahin Najafi, an enterprising Iranian woman who had an *atelier* in town where she produced embroidered napkins and elaborate floral arrangements. She had a great sense of style and elegance, and was especially talented at arranging flowers and folding the ornate napkins, which she also did in preparation for the *qabul*. Soon so many parties were held at the Deer Palace that she became a fixture in the house, arranging vase after vase of flowers that she picked from the garden.

The garden and the orchard behind it were open to the children, who were free to run amok and to swim in the small pool behind the terrace, under the supervision of Ni'mati or Zahra. There was little need for them to set foot outside the Deer Palace. Besides all the visitors and workers on hand to entertain them, there were many activities on the estate that were more typical of what took place on a farm. Every morning Fatima the bread-maker came to bake the bread in the special oven by the kitchen, and a daily delivery of fresh vegetables from the family farms arrived in baskets on two donkeys led by a tall man called Ibrahim. He soon acquired the added name of Daraza – tall – given to him by the Persian cooks to signify his size. The donkeys also delivered two large copper containers of fresh milk from the Dutch cows Hadi had recently imported for one of his farms. Thamina and Saleh looked out for these, so that they could steal the cream on the top. They would then break into the pantry to steal bread – much to the dismay of old Dayyah Saadah, who was responsible for storing the bread, and who would shout out to Jamila, 'Come, come, they've burgled your house!'

Rushdi, who was a few years older than his siblings, preferred the company of his school friends and the visitors to his grandfather's *dawakhana*. As Abdul Hussein smoked his *nargilleh* and conversed with his guests, Rushdi sat near him, listening and observing quietly, but rarely speaking. He was a slender, delicate child and his conversations were mainly with his mother, which whom he shared his opinions on everything.

Hassan, however, enjoyed playing with his siblings, braving scrapes and bumps as he felt his way around the garden, following the sounds around him in order to determine his next movement. When Hadi bought the children bicycles, Hassan sat behind Saleh while he rode. The kitchen and its annexe, separate from the Deer Palace, formed a vibrant world for Hassan. Like a second *dawakhana*, the annexe received visitors from Kazimiya, such as the friends of the staff who called by. They brought the news and gossip of the town, as well as their own stories, which Hassan enjoyed hearing.

His favourite member of the kitchen staff was *Hadji* Hadi, the Iranian cook. As a token of his loyalty to the Shah, *Hadji* Hadi always wore a Cossack hat when he went out, in contrast to the *faisaliya* that Habib Chaigahwa wore. He bickered constantly with Ni'mati, who didn't like the Shah, and also with Habib, whom he considered haughty.

Although he was too young to understand their debates, Hassan was struck by one particular argument between *Hadji* Hadi and Habib about the latest political flashpoint: military conscription in Iraq. *Hadji* Hadi criticized Habib for objecting to the new decree, declaring that he personally would be proud to serve the Shah if required to do so. Habib told him in no uncertain terms that he should mind his own business, and not meddle in the affairs of other people's countries. He berated *Hadji* Hadi for being unsympathetic to his fellow Shi'a, whom conscription was going to affect the most, as they were the largest group in the Kazimiya community. Moreover, he couldn't see why the Shi'a should help the government in any way, let alone by joining the military.

Habib's voice grew louder and louder as he unconsciously mimicked some of Abdul Hussein's constituents who came to the *dawakhana* to complain about this very issue – that the Shi'a were still not proportionally represented in the state. The two only stopped arguing when Zahra came looking for Hassan, and chided them for bickering like children.

Hassan only began to feel different from the other children when they started to go to school and he didn't. Only then did he realize consciously that he was blind.

One day he was playing ball with his sisters in the garden, and threw the ball to his sister Thamina a bit too hard, hitting her in the face. She shouted at him, 'Careful, blind eyes!' Horrified by what she had said, she covered her mouth with her hands before running away. It was the first time Hassan was aware of anyone referring to his disability.

Around the time his siblings started school, a private tutor was hired for Hassan. In those days there were few ways for blind people to earn a living, and the only school for the blind in Baghdad was for the Jewish community. It was recommended to Hadi that Hassan be taught to memorize and recite the Quran as he himself had done as a child, so five days a week a certain Sheikh Abbas would come to the Deer Palace and recite verses which Hassan would repeat after him. They would sit together in the pantry next to the kitchen, and while Hassan had his lessons many people would come into the room, including Habib Chaigahwa, Abdul Hussein's loud driver Karim and Ali the clerk, who smoked and drank tea whenever they could. They always talked loudly, often about the recent discovery of oil in the north and how this would flood Iraq with money. Hassan was so bored with his lessons that he started playing about with the words, swapping them around until he was reprimanded by his teacher for messing up God's sacred verses, although his antics made the others laugh. Finally, one day the Sheikh announced to the household that Hassan had 'completed' the Quran. To celebrate, a dinner was held at which Hassan stood up and recited verses out loud.

Hassan was left frustrated by the experience, especially as each day he heard new stories from his young uncle, brother and sisters about their time at school. He wanted to be like everyone else, and didn't understand why he should be stuck at home reciting the Quran. He confided in his uncle Saleh, who encouraged him to come with him to school, promising to help him and to introduce him to all the teachers. Hassan asked him many questions – about the classes, the teachers, where the pupils sat, what they did. Having considered Saleh's replies, he concluded that he probably needed more help than Saleh alone could offer him.

One childhood friend of Hassan's was Muhammad Ni'mati, the son of Hadi's old companion Ni'mati. Hassan asked Muhammad whether he would help him with getting to and from school. When Muhammad agreed, Hassan announced his plan to his mother. He didn't ask for her permission; he simply informed her that both Saleh and Muhammad Ni'mati would be helping him, and that he had it all sorted out.

Bibi didn't oppose Hassan, as he was so determined. She wanted him to be happy, but she worried about whether he could really manage at school. Nevertheless, she took him to the shops to buy school clothes as she had done for the others. Hadi and Abdul Hussein decided not to interfere, although both of them worried about Hassan too.

The night before his first day at school, Hassan couldn't sleep. He tried to imagine what it would be like – his teachers, the boys, his lessons, how he would find his way around the building. Filled with excitement and anticipation, he knew he couldn't let fear get the better of him, or he would be stuck at home forever.

The school was in an area called Shosa, near Kazimiya. The morning of his first day, Zahra helped Hassan dress before he went downstairs to brush his teeth and wash his face in the outside sink near the kitchen – the one sink he knew how to reach on his own. Then he joined the other children for a quick breakfast of tea with milk, freshly baked bread and cheese.

Then Muhammad Ni'mati and Hassan walked behind his siblings to the tramway stop across the road from the Deer Palace, where Muhammad flagged the tram down. Over the following days his jumping and shouting for the tram to stop became a signal for Hassan, who learned to ready himself to climb on board. The carriage was always noisy with the clacking noise of the horses' hoofs and the wheels. The trip took half an hour, and Hassan always enjoyed it because of the variety of people on board, whose conversations he listened to intently.

At school, Hassan and Muhammad Ni'mati were placed in different sections, which meant that it was very hard for Hassan at the beginning. When Saleh introduced him to his teacher on his first day, the teacher commented loudly, 'You're bringing me a blind boy to teach? What use do I have for him?' Until he became familiar with the spaces and distances, Hassan sat at his desk even during breaks, listening to the other boys play outside. In the early days he could only go to the bathroom if Saleh took him.

Soon Hassan made the teacher regret his words, as he surpassed most of his classmates in his studies. He was gifted with a powerful intellect and an excellent memory. Overcoming the injustice of his handicap became the driving force of his life. He compensated for the loss of his eyesight with an insatiable curiosity and attention to detail, always asking for information, memorizing spaces by counting his footsteps and paying careful attention to any variations on the surfaces of walls in order to guide himself from room to room.

He also learned to read people well, recognizing them by their voices, which also told him about their movements. He could detect whether a person was rising from a chair or approaching him by the sound of his or her voice. Soon he would be able to tell whether someone was thin or fat, tall or short, by their voice alone.

He never referred to his blindness as such, only to 'my difficulties'. He felt a lot of anger – or as he put it, 'Anger was born onto me' – but it pushed him forward, prompting him to find solutions to his problems as he navigated his way through life with courage and persistence.

13

Stolen Hopes

A Young Life Lost
(1928–1929)

BY THE TIME Abdul Rasul returned to Baghdad with an Economics degree from Cambridge, a great future was being predicted for him. Surely the old communal grievance of being Shi'a would not be relevant to him, as he was clearly as well – if not better – educated and equipped than any of his Sunni counterparts. Several luncheon parties were held to welcome him back, and many visitors came to the *dawakhana* at the Deer Palace to congratulate Abdul Hussein on the safe return of his son.

Following Abdul Rasul came a large container of his belongings, crammed with books as well as the pieces of Georgian-style furniture that were to adorn his quarters at the Deer Palace. He set about building a tennis court at the back of the house so he could continue with his game, and entertained the household with his gramophone and his new records.

Before long he was offered the post of Financial Controller at the Ministry of Finance, then headed by his father's friend Rustum Haidar. He threw himself into his new work with passion. His political career had begun.

An eligible and appealing young man from an influential family, Abdul Rasul didn't lack for female admirers, who made themselves known to him through a series of intermediaries. His time abroad had given him an air of confidence and a charming insouciance, especially with women. He had an ease about him that many Iraqi men lacked, having had little interaction with the opposite sex. News of her son's many admirers soon reached Jamila's ears, as young ladies accompanied their mothers enthusiastically to her Wednesday *qabuls*. These were the perfect venue for matchmaking and bride-hunting.

In early summer Bibi would throw lunch parties. These gatherings were jolly affairs, and included men and women, affording the young people ample opportunity to flirt. Sometimes the guests included teachers, who were mostly recruited from Lebanon and Syria during Abdul Hussein's time as Education Minister, as there were still insufficient numbers of locally trained Iraqi teachers – a situation which was being addressed with the founding of a teachers' college in 1923. Abdul Rasul was always a hit at Bibi's parties, flirting with ease and confidence.

Walking in the garden down by the river one day, Bibi was struck by a wonderful idea. Her attention was drawn to the *jizrahs* or small islets, the mounds of rich earth that popped up in the middle of the river in summer. While the *jizrahs* were highly sought-after spots on which to grow garden vegetables, Bibi realized that they would also be wonderful places on which to entertain. She soon began hosting parties on them.

These events required the entire household's participation in order to transport furniture and cutlery, umbrellas, the various dishes and drinks, and obviously the guests themselves, out to the *jizrah*, where one of the farmers would skilfully grill the renowned local delicacy of *simatch masguf*, carp with a spicy tomato filling, on an open fire.

King Faisal's son, Crown Prince Ghazi, came with several of his friends to one of the parties for Bibi's Baghdadi ladies. The Prince was a friend of Hadi's younger brother Ibrahim, who still had rooms at the Deer Palace but was now in the army. The young gentlemen kept at a respectable distance from the crowd of ladies, but they were nevertheless an exciting distraction for them. One of the bolder girls recruited Bibi's pretty daughter Thamina and drew a heart on her palm. She then signed her name underneath and asked Thamina to show it to the Prince. Thamina duly did so, to the amusement of the men.

Abdul Rasul's return to the household signalled a period of even more activity at the Deer Palace, as a new crowd of young men began to visit him, staying till the small hours as they painted their visions for Iraq with eager words. Inspired by his experiences in Europe, Abdul Rasul encouraged Bibi to attend concerts deemed acceptable for women with his sister Shamsa, and even to go out without her *abaya*, then an unheard-of act among Muslim women. On this last point, Bibi would protest in mock horror and call him mad, although she too secretly wanted to throw off the garment. For all her reprimands, Bibi liked her brother-in-law very much; he had become the personification of modernity for her. She even forgave him when he teased her about the fact that she always seemed to be pregnant these days; at the end of 1928, some months after his return, she had given birth to her sixth child, a son called Jawad.

In late 1929, about a year after his return to Baghdad, Abdul Rasul started to complain of bad headaches, poor vision and bouts of dizziness which left him immobile for hours at a time. This caused much worry in the family, and several resident European doctors came to the Deer Palace to examine him. With the help of the Baghdad hospital's new x-ray machine, the doctors all gave the same diagnosis: he had a brain tumour, which explained the deterioration of his eyesight, hearing and balance. They recommended that he consult specialists in Europe, particularly in Vienna and London, who had greater expertise and the resources with which to treat such a case.

Abdul Rasul's illness had come upon him so suddenly that its severity was not at first fully absorbed by his family. Jamila shrank into herself, and while it broke Abdul Hussein's heart to see his son suffer like this, he summoned all his energy and money to organize a party to travel with him to Europe.

From Baghdad, Abdul Rasul flew to Beirut, accompanied by Hadi and two friends who had some knowledge of English. The next day they boarded another flight to Marseilles, then took a series of trains to Vienna. Abdul Rasul's situation was deteriorating so quickly that his brother had to nurse him constantly, helping to feed and bathe him. Hadi felt enormous pain as he watched his brother wither before his eyes. Abdul Rasul tried to put on a brave face. He had reacted to the original diagnosis calmly, and his stoicism surprised his travelling companions.

In Vienna, a biopsy confirmed that a cancerous tumour was causing Abdul Rasul's symptoms. Radiation therapy would be futile, the doctor concluded. In his opinion, the only treatment would be a very dangerous operation to remove the growth. Otherwise Abdul Rasul would die in a matter of months.

Already unable to see or hear properly, Abdul Rasul had begun to retreat into his illness, speaking little and complaining even less. Desperate, Hadi decided to consult the other doctors who had been recommended in London.

He waited until Abdul Rasul was well enough to travel again. The train journey across Europe felt endless, and Hadi spent insomniac nights in their sleeping compartment, terrified of losing his brother, who lay quietly on the lower bunk bed. The burden of responsibility weighed heavily on his shoulders as he desperately prayed for Abdul Rasul's recovery.

In London Hadi knew he could rely on the support of Iraq's Ambassador, Jaafar Pasha Askari. Some years earlier, when Jaafar Pasha had been King Faisal's first Defence Minister, he had asked Hadi to help recruit young men from Kazimiya for the First Battalion of the Iraq army, which he had successfully done. Now Hadi looked to him to return the favour.

However, their stay in London was harrowing for Hadi and his companions, as Abdul Rasul continued to slip away before their eyes. The opinion of the London doctors was that, although the chances of success were slim, surgery to remove the tumour represented the only possibility of saving him.

The telegrams Hadi sent to his father at the Ministry in Baghdad were brief, and always ended with the words: 'God willing, he will be well.' It was as much a supplication as a figure of speech. What could he say that his father didn't already know? He tried to block his mother out of his mind, because he knew how devastated she was at her son's unexpected illness, so he simply wrote her a few lines, sending Abdul Rasul's greetings and asking her to keep praying for him. Abdul Rasul duly underwent the operation.

The news came suddenly, as bad news always does. One cold day when Hadi and his friends entered the ward to visit Abdul Rasul, the matron informed him with regret that his brother had died earlier that morning. Hadi didn't need a translation to understand what had happened, and a sensation of nausea hit him that was so overpowering he nearly lost his balance. One of his friends held his arm and guided him to the nearest chair, then asked a nurse for a glass of water, which he held for him to drink. Hadi sat frozen in anguish, repeatedly muttering the familiar saying, '*Inna lillah wa innah ilayhi i raji'un*' – We are from God and to him we return.

Eventually Hadi asked to see his brother. Abdul Rasul's eyes were shut, his head was bandaged and his face was pale. Hadi collapsed over him, sobbing and holding his cold body tight. He had hoped against hope that his brother would recover, that he was simply too special to meet this fate. He couldn't imagine a world without Abdul Rasul. Their lives together flashed before him: their childhood in Kazimiya, playing with their horses and racing each other over the open fields – the images were endless.

The next few days passed in a daze as Hadi's travelling companions arranged the formalities, with the help of Jaafar Pasha at the Iraqi Embassy. There was no question of leaving Abdul Rasul behind; he had to be brought home, and that could only be achieved by boat.

The hardest task for Hadi was to inform his parents of the terrible news. He decided it would be best to telegram one of the British advisers at the Ministry, Somerville, who could tell Abdul Hussein in person. When the telegram arrived at noon on the day after Abdul Rasul's death, Somerville went straight to the Minister's office. Abdul Hussein was standing by his desk, ending a telephone conversation, but one look at Somerville's face told him all he needed to know. His car was summoned and he was helped into it by Somerville and his driver Karim, who drove him home quickly.

At the Deer Palace, the rooms echoed with howls of grief when Jamila heard the news. She changed forever that day. Her interest in life vanished, and she continued to exist merely because she could not die. Soon the entire house was weeping, the women beating their chests and wailing loudly, overwhelmed by the catastrophe that had hit them. Bibi thought it wiser to send the younger children to her mother's home in Kazimiya, away from the tears and hysteria. Before long the house was overrun with people who had come to share the tragedy with the family, and tents were set up in the garden to receive the thousands of visitors, many of whom had travelled from distant places, who came to pay their respects. Professional male orators recited the Quran in its entirety as custom dictated, while female readers did the same for the women, who grieved separately. Cooks were hired to produce enormous quantities of food, rice and stew on a daily basis.

The journey back from London – by train via Paris and Marseilles, by boat to Beirut and then by car to Baghdad – took two weeks. It felt like an eternity to Hadi. When he and his friends arrived in Baghdad he was barely recognizable, having lost so much weight since he left three months earlier.

The burial took place in Najaf, inside the shrine of Imam Ali. This was a huge mark of respect and honour for a young man who had carried the dreams of his entire community, and the ceremony was attended by all of Iraq's prominent men, the women staying at home as tradition decreed. Iraq's most popular poet of the time, Mulla Abud al-Karkhi, recited a panegyric in which he lamented Abdul Rasul's lost youth and potential.

The funeral rites lasted forty day and nights. After that time had passed, the cool yellow corridors of the Deer Palace rang out once more with the sounds of children's voices. However, there was one set of rooms that remained locked: the library and bedroom that had belonged to Abdul Rasul. Only his father had the keys to these rooms, and they were only ever opened to be dusted. Everything in them remained otherwise untouched. Until their deaths, Abdul Hussein and Jamila sent food every Friday to poor families in Kazimiya in memoriam to Abdul Rasul's soul.

14

Bursting Energy

Hadi's Growing Empire
(1931–1933)

ALTHOUGH ABDUL HUSSEIN had always imagined that Abdul Rasul would be the son to follow him into politics, after his brother's death Hadi extended his interests into that arena, and by late 1930 he had been elected a Member of Parliament for the town of Diwaniya.

At a party held by King Faisal in early 1931, Rustum Haidar introduced Hadi to an English visitor. The man was a partner in Andrew Weir & Co., a British company that had had a strong presence in Iraq for several years, and he was looking for an Iraqi agent to source wheat and barley, which – along with a special type of rice called *amba* and dates from the south – were Iraq's most important agricultural exports. Hadi seized the opportunity, opening an office on Samau'al Street which would eventually become the home of the Baghdad Stock Exchange.

His office was opposite Rashid Street, one of the most important commercial and financial centres in Baghdad, which was mainly inhabited by Iraqi Jewish merchants, accountants, bankers, clerks and translators, flourishing under the continued patronage of the British and assisting them as local advisers in the new Iraqi state. The buildings on Rashid Street were made of brick, mostly two storeys high. Many had columns at street level, sheltering the entrances of the buildings. Rashid Street was always busy, with cars and small trucks, horse-drawn carriages and old donkey carts moving in both directions.

From his office window Hadi enjoyed an uninterrupted view of the bustling street below lined with cafés, warehouses and shops. Samau'al

Street was always congested, partly owing to the mobile food sellers who ambled along with their heavy wooden carts, selling boiled green beans in summer, *turshi* pickled vegetables or *abiadh wu baidh* – boiled egg and mango pickle sandwiches – all year round. People from all walks of life passed by: the *effendis* in their Western suits, men in Bedouin dress with daggers in their belts and *yashmaks* on their heads, mullahs with their black or white turbans, men of the city in their traditional Baghdadi dress of long shirt and waistcoat over trousers. The women concealed their bodies under their *abayas*, and sometimes wore a *pushiya* made from stiff gauze which covered the lower half of the face. Unlike the black *abayas* worn by Muslim women, those of the Christians and Jewish women were often brightly coloured and embroidered.

Seated on the corner, small groups of peasant women in long dresses sold fresh buffalo *gaymar* out of clay urns. Most striking were the yoghurt girls, walking down the street in the mornings with several

Samau'al Street, Baghdad.

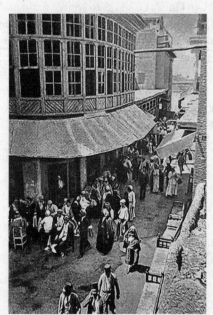

round wooden trays of yoghurt carefully stacked and balanced on top of their heads. They wore long, colourful dresses which they tied with thick bright belts. Many wore anklets. Most had dark kohled eyes and tattooed faces which were visible as they made their way slowly but skilfully, balancing their goods. They always walked together in groups for fear of being harassed or even kidnapped.

The street was a sea of shades and sounds. Although there were specific food markets, farmers would sometimes spread their pomegranates or vegetables on thick rice bags on the ground along the roadside, shouting out to advertise their wares to passers-by. Even louder was the male yoghurt seller, who was usually hunched under the weight of the brass container he carried on his back. Advertising the much-prized Arbil yoghurt from the north, he clacked the glass cups he carried as he walked through the streets and alleyways.

Hadi's office was called 'Offees Abdul Hadi Chalabi'. It was in the Middle Iraq Daftardar building and was managed by Yusuf Zubaida, a talented and loyal Baghdadi Jew. He was a tremendous help to Hadi, particularly when it came to communicating with the nearby offices of Andrew Weir & Co. Hadi's task was to source grain and transport it to Weir's ships in Basra, 10,000 tonnes a shipment. He soon found himself with 90 per cent of the country's grain export on his hands, and he also exported the barley and cotton that grew on the lands between the two rivers, from the north of Baghdad extending all the way to Basra.

It was partly due to his good relations with the people in the regions where the grain was grown that Hadi's elaborate network functioned so well. He had agents throughout the country, from Baquba, Mosul, Hilla, Kut. The deliveries that came from the north by train were stored in the *sif*, grain warehouses, that Hadi built near Kazimiya and later also in Basra, where the grain would be inspected by his employees and repackaged for the journey south to Basra.

Hadi would often stay up late with his staff to greet the train from the north at the station, much to Bibi's irritation and worry. She would berate him, complaining, 'What are you going to do – count every grain? Why can't you sit still just for a minute?' Then she would pout and add, 'I have to sit here all alone; what kind of a life is this?'

Eventually he would explode, 'You really have no idea, have you?' Then he would storm out.

Very soon anyone wanting to sell grain came to Hadi, as he was the most active buyer. To encourage the development of agriculture he started to extend credit to growers in order to enable them to purchase seeds and machines. His children saw him only fleetingly, as he was always busy. Even when he came home late and tired in the evenings there were always guests to entertain, constituents to attend to, people waiting to see him in the *dawakhana*.

One day in 1932, among those waiting outside the house was a *mukhtar*, the head of a nearby district where Hadi owned land. The man was accompanied by a young girl. The *mukhtar* asked Hadi to protect her from her father, a tribal chief, who had sworn to kill her because she had broken his word by marrying a man she loved from outside the tribe. He explained that he could not think of a safer place to hide her than with Hadi Chalabi. Hadi agreed that the girl could stay at the Deer Palace while he and others sought to convince her father to forgive her and give his approval to the marriage.

After a month the man declared that his daughter could come home and be married again in front of him. However, some months after she returned to the district he came to her house and shot her and her husband dead. He showed no remorse when he was taken to jail, because he had not forgiven her for the humiliation she had inflicted upon him.

Bibi was deeply shocked by the incident, and reproached her husband for believing the man. She took the girl's murder very person-ally, but could only vent her anger by criticizing Hadi's handling of the matter. In turn, Hadi was beginning to lose his patience with her. 'How was I to know that the girl's father would harm her, after all these months?' he yelled.

Bibi wasn't convinced: 'You of all people should have known better when it comes to the actions of angry tribal fathers and brothers towards their womenfolk,' she argued.

Although no longer as involved with the tribes as he had been when he had first started to acquire the stretches of empty *miri*, state-owned land, Hadi still had links with grain-growers that obliged him to travel the length and breadth of the country. Yet his attachment to Kazimiya never diminished as his national reputation grew. He always lent an ear to the people's concerns there, even when his commercial activities were met with criticism by some – especially the growing number of communists in the town.

The Deer Palace was in some ways a world of its own, but it was not immune to the harsh realities that many people faced in Iraq. The variety of visitors who came to the house, and the colourful mixture of its staff, who were from all corners of Iraq as well as from Iran, meant that the Deer Palace was in many respects a microcosm of the entire country, with all its good and all its bad, embracing the diversity of Iraq's many regions and attitudes.

Although she was generally content within the confines of the Deer Palace, where she enjoyed a measure of independence, Bibi longed to experience more of the world. She found herself becoming tearful when, in 1933, she went to her first musical concert with her sister-in-law Shamsa. She couldn't help remembering how, nearly four years earlier, Abdul Rasul had encouraged her to go to a public performance. She felt he would be proud to see her now, although he would have laughed to know that she was pregnant again.

When Bibi had heard that the Egyptian diva Umm Kalthoum was coming to Baghdad to give her first concert in Iraq, she was adamant that she wanted to go. She had immediately sent a message to Abdul Hussein at the Ministry of Education informing him of her wish, as she knew he was often given tickets for such events.

Having secured the tickets and her father-in-law's approval, Bibi and Shamsa went to the performance at the Hilal theatre. Decked out in their *abayas*, which Bibi found increasingly irritating, although on this occasion she was grateful that it concealed her rounded belly, the two women sat in the box reserved for Ministers and their families. As she looked around the auditorium, Bibi could see that nearly all of Iraq's leading figures were there, and realized that she had unwittingly placed herself in the social limelight.

She later learned that she had boosted her father-in-law's standing, as the fact that Abdul Hussein was willing to let his daughter-in-law attend such an event was seen as evidence that he had a progressive attitude. Although Jamila entertained guests within the confines of the Deer Palace, she was entirely absent from any public function (especially after Abdul Rasul's death), as were the wives of many politicians. Like most couples, Hadi and Bibi didn't go out together in public, as most socializing was still done inside the home in the separate quarters of the men and women, the *dawakhana* and the *andaroun*.

Encouraged by her experience, Bibi went to more concerts and relished her newfound sense of freedom. But in the summer of 1933 her pursuit of music received a setback when the country was plunged into mourning following King Faisal's death during a holiday in Switzerland, where he had been undergoing treatment for his weak heart.

The King's death at the age of only fifty came as a shock to the entire country. His body was flown back to Baghdad, where the wailing and weeping of women in the streets during his funeral procession was deafening. By the sheer force of his personality Faisal had held together the many disparate factors that contributed to the Iraqi equation. In a memo he wrote to the Cabinet some months before his death, he revealed his disappointment with the slow pace at which Iraq was becoming one nation. He was well aware of the social injustices and the discrepancies between the ruling party and the rest of the population, as well as the frustrated aspirations of the Kurdish people and the sectarian and ethnic issues that continued to besiege Iraq.

Faisal's death represented a huge loss. He had managed to find a workable balance with the British that had allowed for the progress and development of the country as a whole, despite the burdensome Anglo-Iraqi Treaty of 1930, which was intended to pave the way for Iraq's independence but which allowed Britain to maintain military bases in the country and obliged Iraq to provide support in the event of war. Under his reign, modernity had been embraced with enthusiasm, but at a pace that was in keeping with the nature of Iraqi society, rather than imposed upon it from above. Faisal's campaign for Iraq's independence had promised to secure a steady transition towards complete political self-rule.

Faisal's son, Crown Prince Ghazi, succeeded him. Far more nationalistic than his father, he was close to the army, but lacked Faisal's deft political and diplomatic skills. He was also more outwardly hostile to the British, although he was fond of commodities associated with them such as whisky and fast cars. Perhaps it was partly owing to his youth, but his weak handling of government affairs eventually led to a very public eruption of clashes between prominent personalities.

Greatly saddened by Faisal's death, Abdul Hussein served his last term as Minister in 1933, having belonged to nine different Cabinets, but he remained a Member of Parliament. In addition to his own work as an MP, Hadi's export business often kept him away from home, but he remained keenly aware of his obligations in Kazimiya and at the Deer Palace. When Ni'mati's wife Fahima died of cancer that year, he tried to comfort his childhood friend as best he could.

'Why Fahima?' Ni'mati covered his face with his hands. 'Why my everything? Isn't the King enough?'

Hadi laid his hand on Ni'mati's shoulder. 'My home is your home; you and your family are always a part of this household.'

15

Prison

Uninvited Guests at a Feast
(1935–1936)

IT WAS A Saturday in April 1935, and a religious holiday, the feast of
Ghadir. Abdul Hussein's sisters Munira, Amira and Shaouna had come
to the Deer Palace for their usual monthly breakfast visit, and that
morning the whole family gathered together as the women shared the
gossip they had brought with them from Kazimiya. Their voices and
laughter mingled with the sound of trickling water from the little fountain that decorated the back terrace.

The family always unwittingly followed the same seating pattern
for these visits. Jamila sat in her favourite place on a carpet-covered
mattress with her ornate Russian silver samovar at her side. A gift from
her husband, the samovar was one of her most prized possessions, and
she loved serving tea from it. Near the samovar was a large basket
filled with freshly baked bread from the kitchen, home-made
marmalade, white cheese and fresh milk from Hadi's cows. Abdul
Hussein, in his dressing gown, reclined in his safari chair at a right
angle to Jamila. His three sisters were seated opposite him on the
mattress that the servants set up whenever they arrived. Across from
Jamila were the benches where Bibi sat with her sister-in-law Shamsa
and Hadi, who had taken the day off work to celebrate with his family.
That morning Bibi bounced her baby son Talal on her lap. She was
pregnant for the eighth time.

Most of her older children were playing on the terrace. Raifa was
tugging at her older sister Thamina's dark plaits, trying to persuade her

to run around the garden with her. She wanted to climb the trees and shake fruit to the ground, much to the gardeners' annoyance. Their younger sister Najla was shaking her curly head of hair, refusing to eat her breakfast, irritating Bibi with her stubbornness, while Hassan concentrated intently on his great-aunts' conversation.

Abdul Hussein's sisters seemed like entertaining caricatures to Hassan and his siblings; nearly everything about their great-aunts was parochial and archaic. Hassan in particular took great delight in memorizing Munira's conversation in order to mimic it later. She spoke in long, slow sentences peppered with old-fashioned words, and the girls would squeal with laughter when Hassan put on a falsetto voice, imitating her.

For all that she had an amusing way of speaking, Munira appeared to be the most worldly of the great-aunts; the children knew that she ran her lands herself, and they were secretly envious of the way she spoilt her son from her second marriage. Hassan suspected that she might have her eye on Thamina for the boy, but that Bibi was determined she would never get her way. Bibi was not impressed by Munira's mollycoddled son, and felt that to have a good-for-nothing husband was worse than not having one at all. Besides, Thamina was still only twelve years old, although she was already turning into a

Thamina standing near the deer statue in the early 1930s.

beauty with her pale skin and heart-shaped face. Abdul Hussein had once remarked that she reminded him a little of his sister Burhan.

Munira and Amira were sharing a particularly tasty titbit of gossip when a raucous outburst suddenly shattered the calm. They could hear men outside the front door shouting, 'Where's Abdul Hadi Chalabi? Let him come out now!'

Hadi quickly rose to his feet to see what the matter was, followed by his father and Bibi. After a moment, Hassan and Thamina went after them. As they reached the front door they heard their mother yelling at some policemen, 'Where are you taking him? Have you gone mad? Bring him back!'

Bibi was beside herself. Ignoring her, the officers escorted Hadi to his car, where one of them took his keys from him and pushed him into the back seat. Hadi glanced back at his pregnant wife as they drove away, the gravel spitting under the tyres.

Ushering Bibi back into the house, Abdul Hussein tried to calm her down, then rushed off to get dressed. Thamina started to cry, asking her mother again and again to tell her what was happening, and where the men were taking her father.

Outside, Karim the driver, Ali Akbar the generator man and Zein al-Abidin the gardener were standing by the pool, shaking their heads and discussing what they had seen. Hassan made his way over to them to ask if they knew what had happened. He was joined by his elder brother Rushdi, who had come running down from his bedroom.

The men all started to speak at once, telling the boys that earlier in the morning there had been a bloody incident by the old cemetery at Mughaysil in north Kazimiya. Thirteen people had been killed and eighty injured.

'There were four dead policemen among them,' Ali Akbar observed darkly.

Noticing Hassan's confused expression, Karim explained, 'A funeral was going to take place in the new cemetery by the river. But when the funeral party arrived, they found the ground was flooded. As they couldn't perform the burial there, they decided to bury the body in the old cemetery. But by then they were very angry and upset. And when they arrived at the old cemetery …'

'… which is still religious *waqf* land,' interjected Zein al-Abidin.

'... they found that the foundations had already been built for the new central post office, despite all the protests made against the plan,' Karim continued. He said that the men's tempers had become heated, and several of them had started throwing rocks at the building works, before attempting to burn them down.

Ali Akbar said a police car had arrived at the scene, and the officers had tried to calm the situation down, promising to investigate the matter. However, no sooner had they left than two armoured police cars had arrived and more than a dozen armed police surrounded the area. News of the incident had trickled down to the main square, where people had been milling about, enjoying their day off, and many of them had decided to head over to the old cemetery to find out what was happening.

'And then order broke down completely,' Ali Akbar concluded.

Rushdi couldn't understand what this business had to do with his father. Both boys went back into the house and found Abdul Hussein, who was now dressed and talking quickly to the women on the terrace.

Listening to him, Rushdi gathered that on the instructions of the Minister of the Interior, Rashid 'Ali Gailani, the police had arrested his father on suspicion of inciting the incident at the cemetery, and that now he was being accused of treason, the punishment for which was death.

Rushdi felt sick. He knew that his father had appealed to the *Qa'im-maqam* of Kazimiya several times on behalf of the citizens who wanted to stop the building project, but he couldn't understand how that could have led to an accusation of treason. Even if some of his father's supporters had been among the protesters at the cemetery, how did that imply that he was conspiring against the government?

Rushdi and Hassan were too young to know that there were more fundamental political factors at play. MPs were elected on an individual basis within their districts, but with known political affiliations; once in Parliament they took the side of one of the existing political parties. Hadi belonged to the Jamil al-Madfa'i division, which was in opposition to the ruling Yassin al-Hashemi group. Ideologically speaking, there was little difference between the two factions; their rivalry was all about power. The Hashemi division MPs derived their support from the tribes as well as the army, whereas the Madfa'i camp had more of an urban basis. Madfa'i himself was a respected statesman who was known for his moderation, whereas some of his rivals were not averse to the army's political involvement in their campaigns.

Hadi's early political career had not been entirely independent of that of Abdul Hussein. Duty and family loyalty transcended personal beliefs, and as his father's son, Hadi had faced as many challenges as benefits from the connection. Abdul Hussein's reputation was still not particularly good among the Shi'a clerics, because he did not pay them the homage they thought their due, or cultivate them with flattery and money. He had, moreover, broken ranks with them when he had decided to join Faisal's government over a decade earlier.

Hadi, however, was blessed with a different temperament, and had an ability to communicate with the Shi'a clerics that his father lacked. Possessing humility and a down-to-earth attitude, he had succeeded in building good relations with the *ulama* of Najaf in particular. He put his Shi'a connections to good use when conflicts arose between the government and the clerics, who continued to hold important sway over the local population. The clerics in turn knew that Abdul Hadi was the man to go to when they needed support in publishing their religious works, and they often acknowledged him in their books, praising his philanthropy.

However, Hadi's commercial success and his consequent wealth had become matters of concern for several politicians. Most MPs were Sunni, as the reins of power continued to be held by the Sunni elite, with the blessing of the British. Although the political landscape was slowly shifting as more Shi'a men became educated and joined the economic and political system, Hadi was still viewed as an outsider by many in Parliament. Worse, he was an outsider who had influential contacts and considerable resources of his own.

For all that his wealth and influence might have won him enemies, his detention came as a complete surprise to the majority of people who knew him; such arbitrary arrests were highly unusual, and many concluded that Hadi had been seized for sectarian reasons, rather than because of any legal violation he might have committed.

He had barely been driven away from the Deer Palace before the news spread through the area. Scores of men came to the house. Social duty demanded that they all be welcomed in the *dawakhana* with tea and conversation, regardless of the inconvenience at such a difficult time. They had, after all, come to show their support.

It soon became clear that several other men had been arrested as well as Hadi, including his close friend the Mayor of Kazimiya, Sadiq Istrabadi. Over the next couple of days Abdul Hussein learned that Hadi had been driven to the Qaimaqam's office in Kazimiya, and taken from there in an armoured car with the other men to the prison next to Bab al-Mua'dham in Baghdad's old city, near the Foreign Ministry.

The entire group was housed in the same wing. Their conditions were not overly harsh. They had to dress in pyjama-like suits, but they were permitted to sit outside in the courtyard, and in due course they were even allowed to receive home-cooked food.

Initially forbidden from receiving visitors, after a month Hadi was allowed to have contact with his family, and once a week the children went to visit him in jail. He was always very happy to see them, and put on a brave face despite the seriousness of the charges that had been levelled against him. He assured them that he was fine, asked them about their schools and their everyday lives, and told them to take care of their mother, to comfort her and be good to her, especially as she was expecting a baby.

It took the children some time to absorb the transformation in Bibi herself: all her love and concern for her husband had risen to the surface, and the entire household changed pace under her direction. Although she still couldn't cook, and never intended to learn, she orchestrated a campaign to send enough food each day for fifty people to Hadi and the other inmates in his wing.

Bibi mobilized the household towards one purpose: getting Hadi released and the charges against him dropped. 'They're ludicrous!' she exclaimed to Abdul Hussein. 'Why would he ever want to overthrow the government anyway? Whatever he had to do with that cemetery business, it was just politics – not treason.' She focused her secular lobbying efforts on her father-in-law, imploring him to overturn every stone in the government in order to get her husband released.

She also started going to the shrine in Kazimiya daily, imploring the two saints to help free Hadi. She prayed particularly to her favourite,

Musa al-Kazim, holding on to the brass railings that covered his tomb, knotting her prayers into the strips of material that she tied onto the railings as she pleaded for Hadi to come home.

In her determination, Bibi reverted to the ways of her mother's household, and invoked all manner of otherworldly means to help her cause. She was particularly grateful to have the companionship of Saeeda once again. After an unhappy marriage Saeeda had recently returned to the Deer Palace, and Bibi listened carefully as she explained that the twelfth Imam, whose shrine was in Samarra, would fulfil the wishes of all those who prayed to him.

Bibi set about writing letters daily to the Imam, begging him to prove her husband's innocence and secure his release. She folded her letters into neat squares, which she put in small reed containers along with lit candles to light their way. Then, at twilight, she would walk down through the garden to the riverbank and launch her little craft, certain that the river would deliver her missives to the Master of Time, as the twelfth Imam was called, and that he would heed her prayers.

She did this every evening that Hadi was in prison, her prayers floating alongside other reed boats dispatched by the poor and desperate. Often on a calm night, several of these fragile vessels could be seen flickering and floating on the moonlit Tigris, carrying impassioned pleas to the saint.

Bibi interpreted every event, large or small, through the prism of her husband's imprisonment. Allowing superstition to get the better of her, she even blamed what had happened on the shiny red Dodge that Hadi had recently bought; it was the car in which he had been driven away. One evening she said to Saeeda, 'That wicked red car brought us unwanted attention and bad luck. From this day onwards no one in my family will own a red car ever again!' Familiar with Bibi's stubborn nature, Saeeda nodded, confident that her edict would hold true for generations to come.

Even by Saeeda's standards, Bibi's superstitious beliefs knew no bounds in relation to her husband's predicament. Every morning she examined her dreams for hidden messages. Once, when her seven-year-old son Jawad was singing a song from a *Tarzan* film he had seen at the Khayam cinema, she slapped him when he reached the point at which Tarzan fell off the tree; she interpreted the little boy's hollering as symbolizing Hadi's death.

Bibi asked Abdul Hussein to speak to his friends at court, to the King, to anyone who could influence the proceedings. He calmly ignored her more extreme behaviour and set about rationally investigating why his son had been targeted. It soon became clear to him that Hadi was the victim of a political set-up. Several eyewitness testimonies were produced by Gailani's men asserting that Hadi was at the site of the protest, inciting the townspeople to violence. This was the incriminating evidence that needed to be disproved.

When a date was set for the case to go to court, Abdul Hussein appointed a team of lawyers to defend Hadi, led by the distinguished Jamil Baban, an Iraqi Kurd. One of the most difficult hurdles for Abdul Hussein to overcome was that of friendship. His good friend Nuri al-Said, who had been a close associate of King Faisal, was Foreign Minister in the present government, which had accused Hadi of treason. Abdul Hussein tried to steer a diplomatic course, campaigning on behalf of his son without damaging his relationships with such important allies.

Hadi's lawyer, Mr Baban, requested that his case be dealt with through the judicial system, which would make it less likely that the Deputy Prosecutor General could simply rubber-stamp a verdict in accordance with the government's wishes, as witnesses had to be called in, and British judges assisted the Iraqi judges in both the Primary Court and the Court of Appeals. After many days of the trial, which had been enthusiastically followed by the press, several witnesses who had initially testified that Hadi was at the gravesite inciting protesters retracted their testimonies, admitting they were false.

The case against Hadi was dismissed in both courts, and the papers published the final verdict: Hadi was innocent; he was free to go home. The threat of a death sentence no longer hovered over him.

The three months of Hadi's imprisonment had taken their toll in one way or another on all the inhabitants of the Deer Palace.

Jawad habitually took to overeating whenever he was upset. If he was ignored by his parents, told off by one of the servants, or had a row with one of his siblings, he would comfort himself by hiding in his bedroom, munching on whatever he could smuggle from the pantry. The chaos of the kitchens when they had been churning out six dishes a day for an army of prisoners had meant that he and Thamina could sneak in past the cooks to steal the delicious aromatic burned rice crusts. Now both were rather plump.

His father's imprisonment had forced Hassan to open his ears to the political chatter he picked up from school, from the household staff and from the newspapers. He became increasingly aware of the frustrations of various groups, especially the tribes, who complained that they were being denied political power. It was suggested by some that the old Ottoman system, in which authority had been blatantly monopolized by the Sunni elite, was back in place. A new generation of young, educated and impatient Shi'a men was emerging in the urban centres, voicing criticism of the authorities in far more eloquent terms than its predecessors ever had. This developing tug of war was still in its early stages, but relations between the government and the leading *ulama* of Najaf were tense, to say the least. For its part, the government argued that any criticism of its workings undermined the state.

Both Rushdi and Hassan noted with great unease how their school had become more militant and overtly nationalistic in its teachings, as had other schools. During morning assemblies, the boys winced as the school anthem kept changing, with more aggressive language being added to its verses. Their history textbooks were full of alienating references to those who weren't Arab Sunni, and all students were encouraged to undergo military training. They knew these developments to be the work of Sati al-Husri, the all-powerful Director of Education who had been the bane of their grandfather's years in the Ministry of Education. It was an uncomfortable atmosphere that, they felt, pushed each person towards his basic sectarian or religious identity.

The day in July 1935 when Hadi finally came home was like a carnival at the Deer Palace. The gardens, the *dawakhana*, even the road where the tramway passed were packed with well-wishers. As she looked down towards the river at dusk that evening, Bibi whispered her gratitude to the Imam. The Master of Time had obviously received her letters.

* * *

Before his imprisonment, Hadi had been a somewhat distant presence whom his children revered from afar, rather than a father who was present among them, dictating their everyday discipline and routine. Their grandfather was a more familiar figure to them domestically, but it was really Bibi, their mother, who governed their world. While she kept herself at one remove from them, with the exception of her beloved Rushdi, Bibi showed her love and affection for her children through deeds rather than words. New clothes would appear for them when least expected, and there would be surprise outings that she didn't participate in, but which she had nevertheless arranged through the labyrinth of relationships that made up her world – with her mother, her relatives, the household staff or people who worked with her husband and father-in-law.

The children had been taken aback by the way in which Bibi had dedicated herself to their father's release, as previously they had given little thought to their parents' relationship with each other. It clearly didn't resemble their grandparents' marriage: every day, Abdul Hussein seemed to make Jamila cry. Their parents' relationship was different; Hadi and Bibi were not demonstratively affectionate towards each other, and outwardly Bibi adopted a more dismissive attitude towards Hadi than he did towards her. She answered him back, did what she wanted within the house and was generally critical of him.

However, she always made sure that she was wearing her high heels when he came home at night. She had trained her children and the staff to warn her of his arrival, and had told Hadi's driver Karim to honk the car horn, or *horin* as he called it, a few moments before approaching the house in order to give her time to rush and put her heels on if she wasn't already wearing them. Even though they had been married for nearly twenty years, she was still petrified that her husband might see how short she was. She was equally concerned about him knowing that she now had quite a bit of grey hair; this was clearly hereditary – she had started to go grey in her early twenties. Therefore she always scheduled her elaborate henna dying sessions in the late mornings, when her husband had to work.

For all that she cared passionately about Hadi's opinion of her appearance, Bibi never held back when she felt slighted by him. Then she let her displeasure be known to all.

One morning, just as they were settling down to breakfast, the children were disturbed by the sounds of shouting. Thamina and her younger sister Najla rushed to the doorway that faced onto the courtyard, from where they could glimpse their mother yelling from her bedroom on the first floor.

'Party – I'll give you an all-night party!' Bibi was screaming.

'*Khatuna*, I swear it was just a musical party! At the Sheikh Jamil farm. Just men. Honestly!' Hadi's voice echoed up from the courtyard.

Thamina looked at Najla in amazement: they had never heard their father call their mother 'darling' before.

Suddenly Bibi hurled something down at Hadi. The girls watched in disbelief as two shirts and a pair of underpants billowed out like parachutes, before drifting down to where he stood. His face contorted with fury, he ran out of the courtyard and stomped up the stairs. A moment later, it was raining clothes in the courtyard as Hadi hurled Bibi's dresses out of the window, while she continued to toss his suits and ties onto the flagstones below.

Thamina sighed and bit deeply into her bread and honey. Life at the Deer Palace was back to normal once more.

Even after his release, the fight was still far from over for Hadi. He wanted to ensure that his friend Sadiq Istrabadi was released as he had been. Although he had been found guilty, there was no proof that Sadiq Istrabadi had committed treason, and his sentence had already been reduced from death to life imprisonment on appeal. It took several more months of persistent legal action by Hadi before his friend and the other men arrested with them were finally released. The collapse of the case was like a slap in the face for the government, particularly for the Minister of the Interior, Rashid 'Ali Gailani.

Hadi's imprisonment led to both father and son attempting to distance themselves from the Cabinet. From that time on, Abdul Hussein confined his involvement in affairs of state to his role as Senator. Hadi turned his full attention to his business enterprises. He became a shareholder in the first cement company in Iraq, collaborating with established industrialists such as Nuri Fattah Pasha and Muhammad Hadid. Later he moved into cotton and natural oils. In the spring of 1936 he became the first President of the newly created Iraqi stock exchange.

Although Abdul Hussein and Hadi successfully redirected their energies, the political climate remained uncertain. In October 1936, when Hassan was in his final year at the Karkh Middle School, he was sitting outside during a lunch break when he felt something fall lightly on his head. A friend explained to him that military planes were flying overhead, dropping leaflets demanding the resignation of the Hashemi government. Supported by the opposition National Democratic Party, the popular Iraqi Kurdish army general Bakr Sidqi had led a successful military *coup d'état*. Sidqi had made his name several years earlier when he had led attacks against Assyrians near Mosul who had been demanding self-rule, killing 4,000 of them.

The first of its kind in the Arab world, Sidqi's coup established a disturbing precedent in Iraq, by resorting to military action to influence political events. Many suspected King Ghazi's complicity in the coup, owing to his frustration with politicians who were forever trying to curb his wild behaviour.

Under Sidqi's leadership now, Baghdad was bombarded by the army. Sidqi also ordered the murder of one of Faisal's old hands, Jaafar Pasha al-Askari, a popular figure and a dear friend of Abdul Hussein's, for attempting to defuse the situation.

16

Carefree

Growing Up in the Golden Age
(1936–1938)

LIVING HALFWAY BETWEEN Baghdad and Kazimiya, the children of the Deer Palace veered increasingly towards the capital as they grew older. In Baghdad, the marketplaces and alleyways teemed with possibilities, while Kazimiya was linked to domesticity and the past for them, rather than adventure and the future. Their middle and secondary schools were in Baghdad, as was their father's office. They also went there for trips to the shops, the doctor, the cinema and their friends.

On Fridays, the children often visited their grandmother Rumia in Kazimiya, whose world had remained unchanged. She doted on them, cooked them their favourite dishes and told them wonderful stories of her youth, when she had travelled in her *takhtarawan* – her wooden palanquin – across the desert to Iran. She transported them to other worlds with her tales of magicians, monkeys and the jungles of India. When the children were little, they were mesmerized by her words; as they grew older, they longed to see for themselves the sights she described.

The homely routines of Kazimiya manifested themselves in the Deer Palace through the presence of many of its staff, who came from the town; its cuisine, with its Persian influence – the cooks having been carefully tutored in Rumia's elaborate recipes; and the menacing great-aunts, who arrived for breakfast every few weeks with their news and their lists of demands for their brother Abdul Hussein.

Rumia, seated, surrounded by her grandchildren: from left to right,
Jawad, Thamina, Hassan, Najla and Raifa.

Few questioned the level of familiarity in Kazimiya, which even
included the telephone operator, who knew everyone's secrets by listening to their conversations. When Rumia picked up the phone, she never
asked for '706', the Deer Palace number; instead she simply asked the
operator to 'Give me my daughter's house', which he duly did, knowing exactly who Rumia was by the sound of her voice.

The children appreciated how the world created within the grounds
of the Deer Palace was a magnet to many. It was easy for visitors to
get there by tram or boat, and ever since they had been very little, they
had been used to the important guests who would call on their grandfather and their parents.

Like their grandfather, they enjoyed taking trips on the Tigris in the
little round *guffa* craft expertly steered by boatmen, who came to know
the children's schedule, and would often call by on late summer afternoons on the off-chance that they wanted to go out on the water. With
time, the family acquired a motorboat, and the services of the *guffa*

boatmen were no longer required, as the children went out in the boat with Ni'mati.

By far the most exciting time of the year for the young residents of the Deer Palace was the Eid festival that followed the month of Ramadan. Eagerly awaited, Eid was a time of celebration – of visits to friends and relatives, new clothes and other gifts.

In 1936, following Hadi's appointment as President of the new stock exchange, Eid promised to be an even more lavish occasion than usual. Bibi's patience had worn very thin towards the end of Ramadan, as Thamina, Raifa and Najla had nagged her continually about the exact date of Eid in September, which was determined by the appearance of the new moon, marking the end of the religious month. As usual, a few days before Ramadan came to an end the religious authorities indicated the date, much to Bibi's relief.

The preparations started promptly, with Ni'mati taking Rushdi, Hassan and Jawad to the cobbler in Baghdad to get them fitted for new shoes, followed by a visit to Bakir Tahrani, where they each chose a new suit and a new *sidara*. Thamina, Raifa and Najla went with their mother and Aunt Shamsa to pick out material that a seamstress in Kazimiya would make into new dresses. They chose their shoes from Khalil Hamadani, a well-known shop in Baghdad that imported European shoes, considered the best in quality.

The first day of Eid at the Deer Palace was always busy with well-wishers, many of whom were Hadi's employees who had come from his various estates to collect their *'idiyah*, or Eid tip. *'Idiyahs* were usually distributed to servants and workers, as well as children and the poor. Some of Hadi's employees who had received their *'idiyahs* early in the morning gathered in the courtyard in front of the Deer Palace, chatting and joking. The atmosphere was light-hearted and celebratory.

The children were woken by the sound of music coming from outside. Leaving one-year-old Hazem and three-year-old Talal to be tended by their nanny, the older boys and girls hurriedly jumped out of bed and dressed in their new clothes, shoes and hats, then ran downstairs to watch the band that had entered the courtyard.

Composed of professional Jewish drummers, trumpeters and oboists, the band toured the houses of prominent citizens at Eid,

Rushdi.

earning a good day's pay. The children made it downstairs just as the musicians struck up a *chopi*, inspiring some of the visitors to join hands in a folk dance. Thamina, Raifa, Najla and Jawad clapped along, laughing with delight.

Always beautifully turned out (he even insisted on wearing perfectly ironed pyjamas at bedtime), Rushdi leaned nonchalantly against a wall; he would soon be turning eighteen, and was about to enter his final year at school. These proceedings were really a bit beneath him. He turned to Hassan and said airily, 'Do you know, this might be the last Eid I celebrate with you lot here at home for a couple of years, if I go abroad for my studies.' Hassan simply nodded and tapped his foot in time to the rhythm. Unlike Rushdi, he loved all types of music, and he wasn't too proud to take pleasure where he found it.

Once the band had left and the crowd dispersed, Rushdi and Hassan joined their siblings indoors, where they all greeted their grandparents, parents and older aunts and uncles. Standing in line, each child was given an *'idiyah* by their grandfather Abdul Hussein. This gift marked the absolute high point of their day, and made the younger ones feel very rich indeed. Then, with the exception of Talal and Hazem – and Rushdi, who considered himself too old for such childish amusements – the children went with Ni'mati by tram to the large playground in Baghdad.

Impatient to arrive, the girls jumped off first as always, provoking a few harsh words from Ni'mati, who was extremely protective of them. He scolded Raifa and Najla: 'Shame on you for behaving like this – you're Little God's daughters. Cover up now!' Amused by the way Ni'mati still called their father 'Little God' in remembrance of the way Hadi had rescued him as a child, both girls rolled their eyes and ran to the merry-go-round and swings.

For Jawad the highlight of the trip was a drink of Namlet, bottled soda water, which he gulped down with savoury *gargari*, boiled and salted lupin beans sold in a cone of old newspaper by a man on a little cart. The girls preferred pink candy floss, and Thamina ate several swathes in quick succession, making herself sick.

When they returned home, the house was packed with visitors paying their respects in the *dawakhana*. Hadi was out as usual on the first day of Eid, visiting relatives, friends and acquaintances, while their grandfather stayed at home to receive the guests. Some of these, such as the tribal sheikhs, would break into verse as they wished Abdul Hussein well. Listening from behind doors, the children loved spying on these colourful figures in their flamboyant costumes.

Over lunch they interrogated their mother about exactly who had come to the house, what had been said, what had been worn and, and, and ... until Bibi nearly lost her temper with the barrage of questions. She was soon distracted by the splendid Eid lunch, which was even more sumptuous than usual. Even after all the rubbish they had eaten in the morning, the children devoured the festive dishes of *kubbahs*, kebabs and flavoured rice.

The next day was also devoted to visits and guests. Hadi gave the children more *'idiyahs*, knowing that the girls and Jawad had already spent the money their grandfather had given them at the playground. Accompanied by Ni'mati, Rushdi decided to take his younger siblings to the cinema in Baghdad to see *Mutiny on the Bounty*. They loved going to the cinema, as did most of the population. Cinemas had appeared in Baghdad with the arrival of the British in 1917, and had flourished ever since. Usually owned by Christian Iraqi proprietors, some specialized in Arabic films, most often made in Egypt, while others – like this one – showed the latest British and Hollywood hits.

That night, as Hassan rolled over to go to sleep, he sensed that his sister Najla was standing by his bed.

'Hassan,' she whispered, 'I'm scared. What if the people in this year's Ashura parade rebel like the people in the boat in the film? What if they burn down the shrine with us in it? What will happen to Daddy's horses in the procession?'

Hassan laughed. 'The parade's been held for hundreds of years; why would they want to burn the shrine down now?'

The streets of Baghdad were filled with music and verse. In the small cafés in the old neighbourhoods gramophones blared out Egyptian love songs and Iraqi melodies, increasingly performed by female singers. The music of Iraq catered for all tastes: there were popular tunes, Bedouin songs, gypsy songs, songs sung in falsetto by men dressed as women, women's bands for female social occasions, the dirges of official mourners, religious music, the songs of labourers – ranging all the way to the more formal and elegant *maqam* that held a unique position in the high music of the region.

The Deer Palace children had grown up with a gramophone in the house. The selection of records was mostly determined by the grown-ups, and they became familiar with recordings by singing legends of the Arab world such as Umm Kalthoum and Abdul Wahab, as well as classical Iraqi music.

When the first Philco radio box entered the Deer Palace in the early thirties, the impact was instant. The BBC World Service acquired a central role in Rushdi's life as he listened to its daily news broadcast. He became fascinated by the Spanish Civil War, and the way in which it represented the major ideological forces of Europe. From an early age Rushdi had had a negative opinion of Communism, which to his youthful mind represented anarchy and disorder. That is not to say that he was drawn to fascism, but that he believed in the authority of the state.

All the children listened for hours to the radio. They especially liked the Egyptian station, but they couldn't help feeling a little deprived by the fact that all the stations were foreign. However, in 1936 the first Iraqi radio station went on air. With the launch of Baghdad's own station, the leading Iraqi cultural figures of the day came to life on the airwaves for thousands of listeners. Every week people gathered around the radio to listen to Mulla Aboud Karkhi's weekly poem. Much-loved singers such as Sultana Yussif and Zakiya George also sang live on air.

Hassan's world suddenly expanded. He became the keeper of the Deer Palace radio, knowing exactly how to tune it with his sensitive fingertips. Soon he knew the Baghdad station's daily programme by heart, and the radio would be his lifelong friend, accompanying him through the ebbs and flows of his life. At night it became the lullaby that transported him to his dreams, as he listened to the soporific cadences of Umm Kalthuom yearning for a long-lost love. It was the voice through which he could see the world.

Sometimes his grandfather Abdul Hussein would join him and they would listen to the Quran being recited by one of Egypt's famous readers, or to songs of unrequited love. Bibi would sit with Hassan by the radio in her sitting room on winter evenings, while Hadi was away on business. Hassan would entertain her with his mimickry, and they would both sometimes break spontaneously into song. A particular favourite of theirs was the hit song *Amantu Billah* by Laure Daccache, a Lebanese artist who composed her own music, played the lute, sang, and directed her own orchestra:

> Having you in my life is proof to me that there is a God.
> The splendour of your beauty is a vision from God.
> The splendour of your beauty is strange;
> It makes the heart cry from burning desire.
> The healing tears run down and all meet in your wonder.
> The splendour of your beauty is a vision from God.
> Having you in my life is proof to me that there is a God.

Bibi had a good voice, and was never shy of using it. When she had been younger at her mother Rumia's house, she had often sung out of a sense of protest and boredom; now singing became her eccentric way of displaying the fact that she was attuned to life's disappointments

The Haidarkhana Mosque in Baghdad, late 1930s.

and love's betrayals. She was extremely pleased when she realized that her blind son shared her talent for music.

Inspired by the songs he heard on the radio as well as by the musical Egyptian films screened at the Royale cinema in Baghdad, Hassan decided to learn to play the oud, a Middle Eastern form of lute. Bibi gave him her full support, buying him an instrument and encouraging him in his lessons.

His first teacher was Master Yusuf Habib, who lived in an old house in the Haidarkhana district of Baghdad. Upon entering the courtyard, Hassan was always struck by the strong smell of *siraj*, the sesame oil that was used for cooking in keeping with kosher requirements. It had a distinctive sharp aroma that lingered in the air, unlike the ghee he was more familiar with.

Hassan enjoyed his lessons. However, he soon realized that although Master Yusuf had trained generations of musicians, he remained strictly a teacher rather than an artist or composer. After a while Hassan wanted to expand his knowledge, and sought out a new teacher. He eventually found his way to the Kuwaiti brothers, Daoud and Saleh, who were also Iraqi Jews but who were more experimental in their approach. In time they would refresh Iraq's musical traditions with new tunes sung by popular artists such as Salima Murad.

Salima Murad, who was later bestowed with the honorific masculine title of Pasha, was a Baghdadi Jew who became the voice of Iraq. She and her orchestra performed throughout Baghdad, in singing clubs and cafés and in the few theatre halls that existed, as well as the royal court. Soon she had become a leading figure of society and a darling of the political elite. She was also a great favourite among the people, and *Hadji* Hadi the cook saw her perform on the large stage at the Jawahiri hotel, as well as at smaller venues. *Hadji* Hadi knew that Hassan was a fellow music-lover, and took him to many of Salima Pasha's performances.

Hassan's interest in attending these concerts was encouraged by Bibi, who only wished she could go to them herself, but they were purely male affairs. She had heard how Salima Pasha was said to seduce her audience with her voice. The singer was always elegant on stage, with softly curled, shoulder-length dark hair, full lips and tailored knee-length dresses – which sounded thrillingly risqué to Bibi. She performed with a veneer of innocence, yet captured the audience completely with her demeanour, inspiring frenzied applause. One particular tune of Salima Pasha's stuck in Bibi's mind; it was called *Khadri el chai khadri* – 'Brew the Tea, Brew it' – and Bibi would hum it before nearly every cup of tea she drank.

While Hassan was taking lessons with Daoud Kuwaiti, the brothers were invited to form and direct the orchestra of the new Iraq radio station. Hassan was even invited to join the orchestra on a few occasions to play on air; Bibi was thrilled.

In order to take his playing of the oud to a new level, Hassan took classes with Jamil Bashir, one of Iraq's most talented musicians, who had been trained at the national music conservatory in Baghdad after

Salima Marad, aka Salima Pasha,
one of Iraq's most famous singers,
in the early 1930s.

the Great War. Meanwhile the concert halls in Baghdad continued to be filled with eager listeners. The entertainment industry was largely in the hands of Christian and Jewish Iraqis, who were given licences to run these establishments, which was forbidden to Muslims in a tradition going back to Ottoman times.

Hammudi the Arabanchi, or 'tram-man', was a regular driver on the Kazimiya–Baghdad line. He usually drove the evening shifts, and would sometimes stop off for a cup of tea in the servants' quarters in the Deer Palace, especially if he had to wait for the remaining *dawakhana* visitors to catch the last ride home. He struck up a friendship with *Hadji* Hadi the cook, and they discovered that they shared a passion for wrestling.

The *zorkhanas*, underground gymnasiums, were an integral part of Baghdad's popular culture. The wrestling performed in them combined physical exertion with a spiritual dimension, and the matches and training sessions were accompanied by musicians and singers who performed classical *maqams*.

When Hassan, Rushdi and Saleh asked *Hadji* Hadi and Hammudi to introduce them to the world of *zorkhanas* and wrestling champions, the two men were flattered that the boys were interested in their passion, and were very happy to share their knowledge. Sometimes, egged on by the boys, they would have a practice bout in the garden next to the kitchen, Saleh providing Hassan with a detailed commentary on the action. In spite of his fastidious nature and dislike of physical exertion, Rushdi took a keen interest in the *zorkhana* gossip.

One day, Hammudi and *Hadji* Hadi offered to take the three boys to a *zorkhana*. Having gained Hadi's permission, they set out by tram with *Hadji* Hadi to the terminus in Karkh next to the busy pontoon bridge, then took a *gharri*, a horse-drawn carriage, to the Dahana neighbourhood where Hammudi lived. When the alleyways became too narrow for the carriage to pass through, they walked the final stretch to the *zorkhana*, where they found Hammudi waiting for them. He greeted them enthusiastically, proud to show his friends in the club what important friends he had, and ushered them through the narrow entrance and down the stairs.

At first the boys were overwhelmed by the strong smell of incense that pervaded the room. Then Rushdi spotted the man who ran the *zorkhana*: one of Iraq's most famous wrestlers, *Hajj* Abbas Dittch, whose surname meant cockerel, in reference to his greatness, and who was especially revered for his defeat of Herr Kramer, a visiting German champion. Hammudi introduced the boys to some of the wrestlers. Many of them had sweaty palms, and Rushdi regretted not having brought a small vial of cologne with him.

In the centre of the room was a pit, about six metres in diameter, where the wrestlers trained and performed. Hanging on the walls were Quranic verses, images of Imam Ali and his son Imam Hussein, photographs of famous wrestlers and displays of the special costumes that were worn on important occasions, tight-fitting, heavily embroidered *sarwals*, like knee-length kilts. Leaning against one wall were the practice instruments, including large batons carved from walnut.

At a sign from Abbas Dittch, everyone took their seats on low wooden boards covered in worn-out reed rugs. Saleh described to Hassan how the *murshid,* the guide, supervised the entire room from an alcove diagonally across from where they were sitting. He had a smoking incense holder and a large *dunbuk* drum. He pulled a string to ring a bell, then uttered a chant-like greeting that the audience repeated back to him.

The first wrestler to enter the pit bent down to kiss the floor, as if in a house of God. When all the wrestlers had gathered in the pit, they formed a crescent and looked up at the *murshid* to receive their instructions and repeat chants after him. The warm-up exercises began, and silence fell on the room. The exercises were interspersed with calls to prayer, and each was accompanied by a different drumbeat, as well as traditional rhapsodies and elegant *maqams*. The overall effect was hypnotic.

The boys were dizzied by the two-hour event. By the time they left, it was early evening. Hammudi stopped to buy them a snack from a stallholder, who was singing to advertise his wares: '*Shalgham hillu hillu el shalgham*' – 'Turnips are sweet, sweet are the turnips'.

The tram-man said, 'These boiled turnips are simply the best, and the sweet molasses covering them is delicious. You won't find a tastier bite to eat anywhere.'

Saleh and Hassan devoured the sweet purple turnips, which seemed to taste all the better for having been bought on the street. Rushdi, however, was in a state of great inner turmoil. On the one hand, he felt guilty that Hammudi had paid for this food when he knew the tram-man was poor – he told himself that he must give him a tip at the appropriate moment – but more critically, he didn't know how he could eat these turnips when he knew how poorly washed the metal plates on which they were served were. He forced one piece into his mouth, and gave the rest to Saleh and Hassan.

The turnip-seller, noting Rushdi's discomfort and his well-made suit, sarcastically commented to Hammudi, 'Doesn't your friend like turnips? Well, I guess he does look a bit *nazik*, delicate. Yes, that's it: delicate.' The word spoke volumes: the boy had clearly never savoured the rich and robust flavours of life on the city streets, but belonged to a sheltered world.

Rushdi concluded that he preferred to follow the *zorkhana* through Hammudi's gossip and the newspapers. One live performance was quite enough for him.

17

A Dark Cloud

The End of a Generation
(1938–1939)

IN IRAQ, THE first stages of the war that was about to erupt in Europe were played out on the long-wave radio frequencies. Soon, images of uniformed men and the increasingly militarized societies in Germany and Italy were seen in the short newsreels that were played at cinemas.

One morning the children awoke to the sound of shouting downstairs. It sounded as if Hadi was shouting at Rushdi, while Abdul Hussein and Bibi were attempting to reason with their father. Although Bibi and Hadi had their skirmishes, Hadi rarely raised his voice with the children.

Raifa and Jawad crept down the hallway and peered into the *andaroun*, where they could glimpse Rushdi hiding behind a chair, his pale hands raised above his head defensively as Hadi towered over him, shouting at him.

Having already acquired a taste for the good things in life, Rushdi had recently started to enjoy Baghdad's nightlife. Young men often spent their evenings at all-male gatherings where the entertainment was female and often lewd, although in recent years it had come to reflect the audiences' more sophisticated appreciation of music and performance.

Rushdi had gone out the night before, and hadn't returned till six in the morning. Bibi had been beside herself with worry, imagining the worst as usual. When Rushdi walked into the house he had been met by a committee of parents and grandparents in the hallway. His father was in a rage, especially when it transpired that Rushdi and his friends had spent the night listening to a performance by a rising star, Afifa Eskandar.

Coquettish and blessed with a sweet voice, Afifa represented a new generation of singers who were as prominent socially as poets and the literati. She fused different singing styles, could perform in various languages and had the discipline to perform long pieces that required great vocal elasticity. She held a salon that attracted the most influential cultural, social and political figures in the country, and where she captivated them with her beautiful voice. Forever subject to the stinging gossip of Baghdad, she held her own with elegance and dignity.

Hadi was incensed that his eldest son had stayed out all night, especially as it was very near the time of his final examinations at school. His behaviour showed a flagrant disrespect for his elders that Hadi could not tolerate. Held back by Abdul Hussein from striking Rushdi, Hadi turned his anger on Bibi: 'I congratulate you on such a fine

Afifa Eskandar, another renowned
Iraqi singer, in the late 1930s.

upbringing! This is the result of all your pampering: a good-for-nothing, spoilt boy who's a slave to the melodies of an upstart singer. Well done!'

Bibi defended herself and Rushdi stoutly, while Abdul Hussein attempted to broker peace. Hadi's face was red with anger, and he made as if to strike Rushdi again. He reminded them all that at Rushdi's age he was already a married man with responsibilities, not gadding around town like a lazy young fool. Gesturing scornfully at Rushdi, who was peeking over the chair back, he bellowed, 'I won't tolerate such behaviour.' With that, he turned on his heel and disappeared to his office until late that night.

As a young man, Hadi had been no less enchanted than Rushdi by the singers of his day, and as Bibi suspected, he continued on occasion to indulge his interest; but there should be a subtlety to the way these things were done. Hadi had never challenged the authority of his parents. Besides, he had come of age during the war, and had not had much time in which to enjoy himself before a strong sense of duty had taken over his life. Moreover, he had been subject to his grandmother's marriage decree when he was the age that Rushdi was today.

But Hadi's frustration with his eldest son went beyond the generation gap between them. He saw in Rushdi a pampered boy with a fear of the world that lay beyond the comfort zone his mother had created for him. Hadi knew that Rushdi belonged to a different age from his own. His frustration stemmed from the powerlessness he felt when it came to teaching his son to become a man.

In this respect, Hadi felt that he was thwarted by his own father, Abdul Hussein, whose authority reigned supreme in the household, and also by his wife. Bibi had always ensured that Rushdi remained protected from the harsher realities of life. Rushdi's 'specialness' was marked in everything: he had his own room, he ate his own food and received extra care from the staff. His siblings all accepted without question that he was special, and behaved accordingly.

Some months later, Rushdi started to campaign to be sent to England for his university education. Naturally, Bibi gave him her full support. She was even prepared to sacrifice having him near her in order for him to fulfil his desire to study abroad.

Hadi was against this new plan, as he believed his lazy son would not make the most of his time abroad, or actually do any studying. He had certainly not been impressed by the manner in which Rushdi had prepared for his final school exams, with the support of his mother. A man had been employed to read Rushdi's lessons out loud while the youth lay on his day bed and languorously ate grapes from a bowl placed behind him.

However, Hadi was unable to overcome Bibi's determination to make sure that Rushdi got what he wanted. Overpowered by her constant nagging, he relented, and in 1937 Rushdi was equipped for his journey into British academia.

Of course, for Rushdi England meant London – and most particularly those parts of the city where his exquisite silk ties and tailor-made suits came from.

His trip to England was an elaborate one. First he crossed the desert from Baghdad to Damascus by a bus from the fleet of the Nairn Bus Company, founded in the early 1920s by two New Zealand brothers. He was met by family friends in Damascus, who provided a car to take him to Beirut. After several days of regal repose there, he took a first-class passage by ship to Marseilles. The journey to Paris was equally pleasant. Once there, he spent many days exploring the city, and visited the International World Fair, where Picasso's *Guernica* was displayed for the first time.

Rushdi was an eager traveller, and his curiosity about the world appeared to be a family trait, shared by his grandfather and father, as well as his mother. He was in some respects very much a product of his environment, in his obsession with appearance and hygiene, his generous hospitable streak, his political and social conservatism and family pride. But he was a great fan of all that was modern and novel, and was in awe of European – or more specifically French and British – luxuries and the fine life.

Perhaps more than any other member of his family, Rushdi felt an acute sense of frustration over the situation of the Shi'a in Iraq, for which he partly blamed the British. He was preoccupied by the fact that the Shi'a were still second-class citizens when it came to political power, although they were the majority numerically. Equally, he resented Britain's heavy-handed role in Iraq, especially with respect to the economy, which he considered to be abusive to Iraqis, including his father. His almost metaphysical interpretation of history never ceased to surprise others. He had a firm respect for institutions of any kind, and this extended to a long list of establishments, ranging from the *National Geographic* magazine to Harrods and the Stock Exchange.

Newly arrived in London, Rushdi relied on several of his father's friends and acquaintances to help him settle into city life. Some of these were Hadi's business contacts, including several Iraqis, predominantly Jews, who acted as agents for London firms in their dealings with Iraq. His home became the Savoy Hotel, where he indulged his love of luxury. His aim was to attend the London School of Economics, like his uncle Muhammad Ali years earlier. However, first he had to pass the entrance exams.

The novelty of London's clubs and department stores soon wore thin, while the task of studying for the exams became more challenging for Rushdi with each day that passed. The British weather also took its toll on him – he hated the endless grey skies and the rain, and he began to feel bored. He found London cold and lonely, and resented the condescending attitude of the English. He was reminded of his foreignness constantly; he felt like a fish out of water, and longed for sun and warmth.

He started lingering in his hotel suite all day, wondering how he could get out of his predicament without losing face in front of his father. He paced up and down the room, thinking through various scenarios in his head, until one evening he was sure he had found the solution. It could not have been simpler, he thought.

The next day he had lunch as usual at his local, Simpson's in the Strand, next to the hotel, where he tucked into roast chicken, his favourite dish. He then commissioned a photographer to take some pictures of him in his hotel room. They showed him in his usual indoors attire: silk pyjamas, stylish dressing gown and velvet slippers. He was sitting on an easy chair by the hotel window, and was clearly sobbing. Who said a picture wasn't worth a thousand words? When the photographs had been developed he sent the most moving one to his mother, without any accompanying note.

The effect was instantaneous. No sooner had Bibi opened the envelope than she frantically set out to save her son from his terrible plight.

Hadi begrudgingly met her demands yet again, and arranged for Rushdi to return from London. Once he was home he cautiously explained that he would prefer to go to the American University in Beirut. His father agreed to let him go a few weeks later.

In Beirut, Rushdi rented elegant rooms on campus and befriended the elite among his classmates. The city suited his tastes as it had entered a *belle époque* of its own, modelling itself on the French Riviera. Lebanon was a French mandate, and the cultural influences of France were evident in the lifestyle of its capital. Being a Mediterranean port, Beirut was also open to the world, to the large emigrant Lebanese communities in Egypt and beyond, as well as to the Arab elites who passed through it. It was a dynamic, religiously diverse city, although it was the Christians, with their strong identification with France, who most influenced taste and fashion.

Lebanon was home to many schools and several universities that were more advanced than those in the rest of the region. A lot of the schools had been founded in the nineteenth century by Christian missionary groups. It was a popular destination for students, who were drawn to its pleasant climate and attractive lifestyle. There were Iranian, Iraqi, Turkish and even some European students, in addition to the Lebanese.

The picturesque campus of the American University was situated amongst pine trees on a hill that overlooked the Mediterranean. Rushdi thrived on the variety of its students, on the abundance of good restaurants, clubs and beautiful women. But although he was very happy in his new surroundings, he was aware that the forces of Fascism were looming large on the horizon in Europe. He was not to know that the world was on the brink of war; that the fates of Palestine and the Jews would be dramatically reconfigured by events which would segregate communities that had co-existed for hundreds of years.

18

A New Home

The Shadow of Death

(1937–1939)

MANY THINGS HAD happened during Rushdi's absence. One of the most significant of these was a change of address for the family. A new house had been designed for them by an architect with the elaborate name of Sayyid Muhammad al-Hassani Jawa al-Tag, who had studied in Britain.

The move was in some respects inevitable. While the gardens of the Deer Palace were large, the house was getting too small for Bibi and Hadi's growing family. Moreover, the family's social standing meant that they now required a larger space for the many social functions they hosted.

The plot purchased for the new house was vast, and there were no other buildings nearby. The architect, obviously influenced by the Regency period, had created a sumptuous palace of local yellow brick, concrete and marble, surrounded by elaborate gardens, fountains and classical statues.

The prospect of their new home excited everyone, but leaving the Deer Palace was not a simple matter, and there were some complicated logistics involved. First and foremost, the statue of the deer was moved from its rectangular pool and installed in an even more prominent position in front of the new building. It was placed on a large, circular island surrounded by a pond, around which a gravel driveway circled to the front steps of the house, which had been designed as a classical portico with arches.

The Sif Palace, 1930s.

The new house, nicknamed the Sif Palace, was divided into two wings: one for Hadi and his family, and the other for Abdul Hussein and his. The *dawakhana* featured prominently on Abdul Hussein's side of the house, while the *andaroun* sprawled behind it and upstairs.

Bibi was pleased that she didn't have to share her space with her in-laws, while having them live next-door meant that the bulk of social traffic and domestic organization remained with them, allowing her to dip in and out as she pleased. Her lack of interest in running the house remained unchanged, and she was content to delegate most decisions to the others.

Hitler's rise to power in Germany had not gone unnoticed in the Chalabi household, although the family did not share some Iraqis' enthusiasm for him. Bibi believed that they liked Hitler because he was standing up to the British. The Iraqi newspapers were full of praise for him from journalists she considered to be extreme nationalists. The changing climate in Europe worried her, and she engaged in lengthy discussions with her son Hassan and her brothers-in-law.

The Chalabis had not long moved into their lavish new house when thoughts of Europe were pushed from their minds by sad events in their own home. Bibi's long-suffering mother-in-law, Jamila, had suffered from diabetes for a while; her condition had led to gangrene, and she was rushed to hospital. Seeing that there was no hope, Abdul Hussein insisted she be allowed to die in her own home. She had become a living skeleton; even her teeth were a source of terror for her granddaughter Thamina, as they had become too big for her mouth.

When Jamila died, Abdul Hussein was filled with sadness, guilt and regret. After all those years of brutish behaviour, he realized how dear she had been to him, and her death marked him deeply. He confided to Bibi, 'She suffered terribly. She never really got over the death of Abdul Rasul; she loved that boy so much. I'm sure that's when she first became ill.' His voice wavered. 'And she was such a good woman.' For a moment Bibi thought he was going cry in front of her, but he composed himself. However, she knew that he often wept in his private quarters.

In truth, Abdul Hussein was inconsolable. He set out on a trip to Istanbul, where he languished for several months, losing himself in the streets of the city, remembering his youth and nursing his hurt. A passionate lover of all varieties of *dolma*, he tried to consume his sorrows by indulging his large appetite with the sumptuous dishes of the Bosphorus.

When he finally returned home, he quietly resumed his afternoon get-togethers at the *dawakhana* and listening to the radio with Hassan, comparing notes on songs and the Quran reciters. However, his overindulgence in food soon came back to haunt him.

On 9 March 1939, Hassan awoke to piercing cries. He followed the sounds to his grandfather's room, where he found Abdul Hussein screaming: 'I'm burning, I'm burning – help me, I'm burning!' The room filled with raised voices, but there was nothing anyone could do to quell the fire. Abdul Hussein's spleen was infected, which before the days of antibiotics was lethal. His was a very painful death.

His female relatives took turns to keep a vigil by his side, crying and reading out prayers to facilitate his transition to heaven, appealing to God in their wretchedness. Later that morning his male relatives came to collect him for his burial, after his body had been wrapped in a temporary shroud. As was the tradition, the women would not be attending the burial ceremony. With high-pitched wailing, his sisters bade him farewell, crying, 'May God be with you, may He open the gates of heaven to you. There is no God but God, and Muhammad is His Prophet. May God have mercy on you ...' Bibi held tightly to Shamsa's hand. There was nothing left to say.

The men carried the body on their shoulders, and the funeral procession set off towards the Kazimiya shrine, where once Abdul Hussein had been forbidden entry. At the shrine *maghsal*, or special washroom for the dead, his body was washed and prayers were read over him to prepare him for burial. His body was then wrapped once more in a white cotton shroud until only his head remained exposed. Then he was taken on a last visit to the shrine, where he was carried around the tomb of the Imam as more prayers were read.

He was to be buried in the Imam Ali Shrine in Najaf, where most Shi'a desired to be laid to rest, and his body was placed in a hearse for the long journey. The funeral party followed in their own vehicles. At Najaf they unloaded Abdul Hussein's body and took him to visit the shrine there too, circling the silver-encrusted tomb of Imam Ali, before performing another prayer for the dead.

After this, Abdul Hussein was taken to the family tomb inside the shrine, where he was buried next to his son and wife, to the accompaniment of prayers. He was laid to rest with a *turbah*, a palm-sized disk baked from the clay of Karbala where Ali's son Hussein was slain.

Abdul Hussein's obituary dominated the front pages of the main newspapers, and for forty days the house was filled with mourners come to pay their respects. The popular German Ambassador to Baghdad, Herr Grobba, approached at least one Iraqi notable during the funeral proceedings in an attempt to find new supporters for the Reich.

On 1 September 1939 Hitler invaded Poland, and Britain and France declared war on Germany.

The Second World War contributed to the increasingly dual identity of Iraq. Independent though the country was, it was still bound by the Anglo-Iraqi Treaty that had been imposed on it by Britain in 1930, and which stated that Iraq would support Britain in the event of war. But while Iraq was officially an ally of Britain, many Iraqis felt exceedingly hostile towards it and its colonial interests. This schism divided the country.

Hitler had already made his mark in Iraq. Radio Berlin, an Arabic-language radio station, started broadcasting to the nation from Germany in 1938 through the crisp, strong voice of its leading Iraqi anchorman, Yunis Bahri. Suitably blond and blue-eyed, Bahri was an intrepid world traveller with a penchant for the Far East. Originally from Mosul, he was already a well-established broadcaster in Baghdad when he fled the country, it was said on board a German plane that was carrying a press delegation.

Anointed by his good friend Goebbels as a *maréchal*, Bahri bombarded the Iraqi airwaves daily, always with the same introduction: 'This is Berlin. I greet the Arabs. My brothers, people of Iraq, swell out the earth.' In his broadcasts he attacked the British and their imperialism, calling for people to rise up against them everywhere. His appeal was great, and he became a fixture in the new battle of the airwaves.

Perhaps it was a residue of the Ottoman period, but Iraqi attitudes towards Germany were not coloured by the same hostility as those towards Britain. The Germans were not seen as imperial; they had been allies of the Sultan, and had contributed to many modernizing projects in the Empire from the late nineteenth century onwards. With nationalist sentiments running high across the region, many Arabs sought a *rapprochement* with Hitler, as an enemy of their enemies. Many admired his military organization, and sought to emulate it by creating local paramilitary groups such as the *Futuwwah* in Iraq. The irony was that the Arabs were low on the list of inferior races as far as the Germans were concerned, being like the Jews, of Semitic stock.

Sati' al-Husri, Abdul Hussein's opponent during his time at the Ministry of Education and regarded by many as the father of Arab nationalism in Iraq, was greatly influenced by German nationalist theories. His admiration of Hitler's Germany even influenced the Iraq museum, which was then under his direction. Disapproving of Gertrude Bell's concentration on Iraq's rich pre-Islamic heritage, Sati' initiated excavations of Islamic archaeology. When discussions started about building a permanent museum to house the many artifacts that were unearthed, Sati' sought out Walter March, one of two brothers who had built the 1936 Berlin Olympic Stadium under Goebbels. Although the museum was not constructed until after Sati's expulsion from Iraq in 1941, it would be to March's design.

٤

OCTOBER 2006

My father calls from Baghdad to tell me that our friend Imad al-Farun has been shot dead outside his house. I am at a loss for words. I was very fond of Imad, a charming old lawyer who seemed to belong to another era with his neatly combed hair and moustache, Old Spice eau de cologne and pressed suits. We often used to sit drinking tea in the late afternoon in my grandfather's palm garden in Baghdad, where Imad would fondly remember the old days. He was one of the closest links I had to my grandparents' world in Baghdad.

Some months before he was killed, after the December 2005 elections, Imad had helped me locate Gertrude Bell's grave in Baghdad. I had found myself unexpectedly drawn to her story, and had read her letters from Baghdad. I couldn't explain my interest in the life of a prudish Victorian adventuress turned colonial politician who died eighty-odd years ago in Baghdad. I could not imagine having an equivalent fascination with any contemporary figure who was part of the Coalition administration in Iraq.

Locating Gertrude's grave is not an easy mission. Her world, like Bibi's, has been buried underground and in memories, almost as if it never existed. Two Iraqis, Muhammad and Ali, a Kurd and a Marsh Arab respectively, help me in my search. Neither has ever heard of her, despite the fact that her legacy is inextricably linked to Iraq, but they accept my word that she is an important figure. Muhammad, the more urbane of the two, is more taken with her story and therefore more concerned with finding her, for my sake at least.

Our first stop is the British Cemetery. The memorials pay tribute to soldiers of all colours and creeds, but they are all men. The caretaker, Abu Ahmad, is adamant that '*el Meees Bealll*' is definitely not there, but he will not let us see the cemetery records, which are stored in a large box by the entrance. He explains that his son is the official caretaker, and he is on a tour with the Iraqi national fencing team in Morocco, so we will have to wait.

A week later, Imad mentions a forgotten Anglican cemetery in the centre of the city. When I visit it, the graves seem to be strewn around in no particular order. Men from as far apart as Madras and Illinois are laid to rest next to a small group of leftist Palestinian militiamen who sought refuge in Baghdad years earlier.

When I find Gertrude's grave it is a stark plot of earth. I think of the elegant memorial erected at the military cemetery for General Maude, and wonder why Gertrude was buried in this barren spot. The plants on her grave have withered and the ground is a dusty grey. Remembering her love for gardening, for flowers and trees, I feel compelled to honour her memory and restore her grave (which I maintain to this day) by planting flowers and trees and fixing the stone. Thinking about it now, the act of tending to Gertrude's grave marks my first concerted effort to come as close as I can to Bibi's world and to unlock all those memories.

At the Anglican cemetery I meet Ali the caretaker, who tells me that he enjoyed a taste of international exposure when British journalists visited the cemetery in the days after the fall of Baghdad. He bemoans the fact that neither the British nor the Church pays any money for its upkeep, and explains that his job has passed down to him from his father, who worked in the cemetery from the 1940s. Eventually silence descends on us. There is not much to say.

Muhammad and Ali, I suspect, cannot fathom why I have sought out the grave of a long-dead Englishwoman. Most likely they think I am mad. Ali, standing next to me by Miss Bell's grave, asks me what I want with it, then jokingly adds: 'Maybe she can give us the secret of Iraq from beyond the grave.'

Maybe. I wonder how Miss Bell would have reacted to what has become of Iraq, the country she worked so hard to create.

BOOK THREE

A Dangerous
Garden

MAY 1993

It's a chilly spring afternoon in Providence, Rhode Island, as I rush to be on time for my History of Twentieth-Century Architecture class in Brown University's List Art Center, designed by the renowned architect Philip Johnson. The course is taught by a popular German professor and his lectures draw large crowds. I have just finished writing my term paper and am anxious to hand it in.

The subject of my paper has unexpectedly taken me on a journey into my heritage by way of Taliesin, Wisconsin. While leafing through a book on Frank Lloyd Wright, I came across drawings of a dramatic circular building. I read that it is a design for the Baghdad Opera House by Lloyd Wright.

I learn that the opera house was to be one of Lloyd Wright's most ambitious projects. He took his inspiration from stories of Edena, the ancient city that had once lain to the south of Baghdad and which legend claimed was the Garden of Eden. In designing his opera house he set about creating a modern Eden, infusing it with his interpretation of the local architecture and landscape, and drawing upon his readings of the Bible and the *Thousand and One Nights*. The building was to be situated on an island in the Tigris that he had spotted from an aeroplane, and was to be linked by bridges to each riverbank; its axis was to be directed towards Mecca, as all mosques are, alluding to its status as a shrine of culture. I discover that Lloyd Wright spoke highly of the many Iraqis he met, including the King, who gave permission to build on the island. While in the country he lectured the Iraqi Engineers Association on the need to avoid the twin pitfalls of modernity and capitalism, advising them instead to take their cue from the ancient inheritance of Iraq and to use this as the basis for the creation of a modern Mesopotamia.

This seems too good to be true. I realize I have found my topic, but there is something else. Wright's vision presents Iraq as a cultural landmark and places its heritage within a larger landscape, which affords me a sense of the country's cultural importance that is at a remove from both my family's memories and its present-day political conditions.

I find much solace in my research. It is a route into Iraq's cultural history that links to my Iraqi identity. It gives me great hope to know that this place, which projects so much negativity and pain today, once – and not so long ago – confidently embraced the modern world. I imagine that if this could happen in the past, it can certainly happen again in the future.

Frank Lloyd Wright's opera house was never built, because a violent *coup d'état* overthrew the regime that initiated the project.

19

Mountains and Floods

Domestic Changes
(1939–1941)

'THE BUS IS leaving in ten minutes, everybody on board, on board everyone!' shouted the clerk – much to Bibi's irritation, as she had already taken her seat and was waiting impatiently to set out. Nearby, other streamlined stainless-steel Nairn buses – American-made, air-conditioned and as large as railway carriages – glistened in the burning sun. On this day in the late summer of 1939, nearly half of the seats on the bus were occupied by members of the Chalabi clan.

Hadi, the new family patriarch, had decided to organize a summer holiday for everyone in the Lebanese mountains. This was a very welcome trip, as Abdul Hussein's death in March had cloaked the house in a shroud of mourning, and the past few months had taken their toll on them all. In April King Ghazi had been killed in a car crash. Ghazi had been a friend of Hadi's brother Ibrahim, who had been an officer in the army with him, and had visited the Deer Palace on a number of occasions. His infant son, Faisal II, succeeded him under the regency of Prince Abdul Ilah, the new King's maternal uncle. Many people suspected the British of killing Ghazi because of his nationalist politics and overt hostility to them. Uncertainty stalked the streets of Baghdad once more.

The family hoped to recuperate from their grief in Broumana, a picturesque mountain resort above Beirut. The Lebanese capital's heavy humidity in the summer meant that many people retreated to the resort's fresh climate. However, Bibi planned to go on frequent shopping trips to the city, where the fashions of Paris had been enthusiastically adopted and there was an abundance of talented seamstresses. Her love of clothes had only grown stronger with time.

Bibi was also looking forward to the opportunity to show off her clothes. She saw that headwear was being shed by women in Beirut at a faster pace than in Baghdad, and *abayas* didn't exist in this Levantine land. Although Bibi had never lacked propriety – her mother, Rumia, was a strict adherent not only of the *abaya* but of the *hijab*, and religion ran strong in her upbringing – she had long thought of the *abaya* as old-fashioned, an image she didn't want to portray. She was heartened to see that some of her Iraqi sisters had begun to shed their *abayas*, and was determined to join them.

If Hadi minded when she informed him of her decision, he didn't show it. However, Ni'mati, his childhood companion and loyal aide, considered Bibi's behaviour scandalous. The day he first saw her go out without her *abaya*, he covered his head with his hands and lamented the downfall of 'Little God'. He regarded Bibi's decision as treacherous and ungodly, and it changed the way he thought of her forever. He was particularly appalled that she also allowed her daughters to go out without their *abayas*, believing that this would effectively kill off any marriage prospects they might have.

So it was unencumbered by her *abaya* that Bibi settled into her seat for the eighteen-hour bus trip to Damascus. The bus left on time at 3 p.m., and at 9 o'clock the next morning the family disembarked in Damascus, where they were welcomed by friends. They planned to stay for a few days in the city before setting out along the twisting mountain roads to Beirut.

When they finally arrived in Beirut, Bibi instantly fell in love with the sea; this was the first time she had seen the Lebanese coast, and the long corniche promenade near downtown Beirut soon became a favourite place for her. The mountainous Mediterranean landscape, with its aromatic pine trees and wildflowers, was very different to the terrain of Baghdad, and it appealed to her immensely.

Beirut was certainly ahead of Baghdad in the sophistication of its restaurants, the range of its imported goods and the general openness of its people. The women dressed more fashionably, the shops had more variety, the cafés were frequented by both sexes. Bibi loved the

Lebanese sense of hospitality; in Beirut, she felt as if she had come home. Accompanied by her two eldest daughters, Thamina and Raifa, she often went down the mountainside from their villa in Broumana to the city, to the shops and souks, where she took great delight in the myriads of materials, shoes, handbags and jewels.

Thamina held up a swatch of silk as blue as a peacock's feather for her mother to admire: 'Isn't it lovely?' The shopkeeper pulled out more rolls of fabric, including the latest Italian imports and polkadot patterns, which Bibi loved. Looking at Thamina, she smiled and said, 'It might not be too long before we have to start buying fabrics for you ...'

Thamina blushed, aware that her mother was hinting about creating a trousseau for her now that she was sixteen and of marriageable age. Bibi did not see any benefit in encouraging her daughters to pursue their education beyond secondary school, and discouraged their father from sending them to the convent school in Baghdad. There had been several arguments on the subject between Hadi and Bibi, with Bibi always getting the upper hand. She viewed the girls as a heavy respon-

Najla, Rushdi and Thamina in Lebanon, 1939.

sibility which she needed to be rid of as soon as possible. The most obvious method was marriage.

Standing behind her mother and sister, fifteen-year-old Raifa rolled her eyes; she didn't mind shopping, but as she watched them pore over the fabrics she felt left out, frustrated that her sister always commanded more attention. She would have preferred to have been back in the villa playing cards with Talal or taking a walk with her sister Najla. Najla rarely accompanied them to Beirut; she didn't much care for clothes shopping, and when she did join them she would often row furiously with Bibi. Of the three girls she most resembled their grandmother Rumia, to whom she was very close, and even at thirteen she was critical of some of Bibi's more modern ways.

In this respect, Najla was not dissimilar to Saeeda, who had accompanied the family on the trip and who certainly didn't feel comfortable with the modernizing trends of Lebanon. She had spent most of her life in Kazimiya, and her references did not extend beyond the town,

Bibi (fifth from left), surrounded by her daughters,
daughter-in-law, sisters-in-law and other female relatives
in the Sif garden in the early 1940s.

nor did she want them to. She cut an almost fairytale-like figure amongst the pale gold stone of the buildings in Lebanon, like an ancient crone with her black *abaya* trailing behind her. Hers was not a typical look in chic Broumana.

Bibi complemented her shopping sprees with tea parties and luncheons in the summer homes that sprinkled the mountainside. She felt alive again. However, in spite of her many distractions, she could not ignore the talk of a sinister wave that was sweeping over Europe.

Two days after the end of their holiday, the family learned that Europe was officially at war.

Thamina had grown into a traditional beauty, and Bibi had already declined the proposal of a match with Munira's good-for-nothing son. After the family's return from Lebanon, several suitors came forward to ask for her hand in marriage. Islam permitted the marriage of first cousins, and such unions were extremely common, a tradition going back for centuries. Indeed, the majority of tribes and clans believed strongly in the advantages of 'keeping it in the family'. Hadi was therefore under pressure to give his daughter's hand to a relative.

There were two contenders from Hadi's side of the family, but Bibi flatly refused to consider either of them as a potential suitor for her daughter: one was too dim, the other was mean, neither would make Thamina happy. Nothing would shift her position. She and Hadi spent day after day arguing back and forth on the subject, followed by more days during which Hadi refused to speak to her at all. Finally, Bibi's brother-in-law Muhammad Ali moved out of the shared family home in protest, but still she wouldn't budge.

During this battle over Thamina's future, Bibi complained that her blood pressure was soaring. At her request she was attended morning and evening by her strident (and equally short) cousin Saleh Bassam, who was a newly qualified doctor, talking to him for hours on end while Hadi left her to her own devices.

Not only Hadi boycotted her company, but the entire extended family on the Chalabi side. So offended were the suitors' families that many of them turned against Bibi for life; some of them also developed a grudge against Hadi that would in time have political consequences. Bibi looked down on them all; they in turn considered her an awful snob, who clearly wore the trousers in 'that house', as they bitterly referred to Hadi's home.

In the spring of 1940 the Tigris and the Euphrates flooded severely, inflicting great damage in the region. The Euphrates broke its banks at the Sariya dam near Falluja, submerging entire villages and farms, sweeping away everything in its path and bringing the transport systems in Baghdad to a standstill. The orchards near the Chalabis' home were flooded, and the rising waters crept nearer and nearer to the house. Hadi instructed the male members of the staff to remove all precious items from the ground floor, and they quickly set about rolling up his extensive collection of valuable Persian carpets and taking them upstairs to safety.

While all around her were busily moving possessions, barricading the kitchen and making sure that essential supplies were brought upstairs, Bibi stood on the balcony watching the water approach. When Saeeda came to find her, she was smoking a cigarette furiously. Agitated, she turned to her and said: 'We're all going to drown. The whole house is going to be flooded. And there's nowhere to hide. Look.' She nodded down at her high heels. 'How can I even run away, in these stupid shoes?'

For Bibi, the flood had assumed biblical proportions: like the flood of Noah, it was a sign from God. Well versed in the biblical stories of the prophets, she recited the *Surat Nuh*, Noah's Verse from the Quran:

> O my Lord! Forgive me,
> My parents, all who
> Enter my house in Faith
> And [all] believing in men
> And believing in women:
> And to the wrongdoers
> Grant Thou no increase
> But in perdition!

The floodwater entered the house, and for many days everyone had to sleep on the first-floor balconies. Their food was very basic: boiled eggs and potatoes with chutney – not dissimilar to the popular street food sold in carts in the souks. The fountains outside oozed with filthy water; even the statue of the deer was not spared, although it still managed to look dignified with its legs half submerged.

Bibi declared to Hadi, 'This Sif Palace is cursed. Your mother died within months of us moving here, your father soon followed her, and now there's this flood. I'm not living here any more, ever.' She immediately set about packing, and within a couple of days she and the rest of the family had moved to Kazimiya.

The one person who refused to budge from the flooded building was Jamila's aged former nanny, Dayyah Saadah, who stayed on the first floor 'guarding the house' as she put it, although she could barely move. Muhammad Ali, Hadi's brother, came with his young nephews in a small rowing boat to deliver food to Dayyah Saadah and remove any precious items. Worries about looters were never far from anyone's mind.

Even after the waters had retreated and order was eventually restored, Bibi's mind was firmly made up, and as usual nothing could persuade her to change it. After a brief residence in Kazimiya, she and her family moved into a large rented villa across the river from Kazimiya, on Taha Street in the Sunni neighbourhood of A'zamiya, where the shrine of a revered lawmaker, Abu Hanifa, stood. Once more, the deer statue followed them to their new residence, and it was already destined to move again, as Hadi had purchased a large plot of land across the road, where he planned to build a new home in time.

Soon after the family's move to A'zamiya, the mystery of Bibi's refusal to agree to Thamina's marriage was solved. One evening, she casually hinted to her husband that her cousin Saleh Bassam had asked for Thamina's hand before anyone else had approached them about it. Bibi confessed that she had preliminarily agreed to the match, but told Hadi that she would naturally like to know his thoughts on the subject.

For a few minutes, Hadi's face was so red with rage that Bibi cowered behind a chair just as Rushdi had done years before. She was sure that he would strike her, even though he had never before raised his hand to her. 'So that's what it is – it's all about your family and your pride!' Hadi blustered when he could finally speak. 'You want to marry

Thamina to your uncle's son, the same uncle who cheated your mother out of your father's inheritance?'

Bibi dismissed the matter of the lost inheritance as irrelevant – it was so long ago – and said that she knew Saleh Bassam, that he was desperate to marry Thamina and that he would do his best to make her happy. What was more, he would never dare to disobey Bibi, and the match would mean Thamina could continue to live near them.

For all her defiance, Bibi knew she had pushed matters to the absolute limit. When Hadi summoned Thamina to ask for her opinion, she deferred to him, saying, 'Father, you know best,' tears of embarrassment shining in her eyes. However, she knew that her father would never overcome Bibi's will, and was secretly relieved that her fate had been sealed. She welcomed her mother's choice, and was excited to be getting married.

In January 1940 Rustum Haidar, the Minister of Finance, who had been a loyal ally of King Faisal I, was shot in his office by a pro-Nazi policeman who later confessed that he had been in a rage with Rustum for having refused him a promotion.

Rustum Haidar had been by far the most cosmopolitan politician in Iraq. Although Lebanese by birth, like King Faisal he had adopted Iraq as his home country and had become devoted to it. Besides being a pillar of the Shi'a community and an old family friend of the Chalabis, he had been a colleague of Hadi's for some months. Hadi's position as one of the largest suppliers of wheat and barley to Weir & Co., which supplied the British Army, meant that he played a key role in the management of Iraq's agriculture. In 1939 he had been invited to be the only person without a Cabinet post to join the Iraqi Central Supply Committee, a government agency that had been established to ensure the supply of food and goods for the population. The committee was headed by Rustum.

As soon as Hadi heard the news of the shooting, he rushed to the hospital where Rustum lay critically wounded. While the doctors tried to save Rustum, news of the attack spread rapidly through Baghdad, with many Shi'a convinced that the attack could not have been planned and carried out by one man, but that it signified a more sinister plot against the Shi'a as a whole.

The news was also received very badly by the Chalabi household, adding to the despondent mood created by the ominous reports of the war. Rushdi and Hassan were outraged, while Bibi worried that retaliation attacks would follow, and prayed for her family's safety.

When Rustum died four days later, the stability of the country hung by a thread. The Prime Minister, Nuri Pasha al-Said, drew upon all his reserves of political cunning and goodwill to prevent a revolt from breaking out. Having consulted with the Prime Minister, Hadi exerted his influence over the Shi'a mullahs in Kazimiya in an attempt to calm the situation. Although their sway over the town had been undermined in recent years owing to the growing appeal of Communism, he hoped they could help contain the anger felt on the streets.

Filled with anger and frustration, Hadi could not sit still. He sent messages to the shrine cities of Najaf and Karbala and contacted community leaders, then turned his attention to organizing several memorial services for Rustum.

Rustum's funeral was very well attended, and there was a real risk that the crowd would erupt into violence at any moment. The Shi'a community were by now convinced that his assassination had been carried out by Arab nationalists who objected to their monopoly on

Hadi (far right) accompanying Nuri Said (second from left) and his entourage in the courtyard of the Kazimiya shrine in the 1940s.

politics being broken by the Shi'a, who had started to acquire more positions of power in the late thirties and early forties.

Not long after Rustum Haidar's assassination, the growing nationalist element in government meant that Nuri's position as Prime Minister became untenable. Unlike the pro-Axis elements in government, Nuri and his supporters believed strongly that Iraq's interests were best served by working alongside Britain. The country was already on the path of progress and development, and a leader in the region; given what had been achieved in the past couple of decades, they believed that they had as much to benefit from Britain as she had to benefit from them. However, Nuri now found himself in a minority, and he resigned from his post as Prime Minister.

As her own mother had done before her, Bibi dedicated herself to preparing her daughter's trousseau. She took Thamina for fittings at the atelier of Madame Tokatelian, an Armenian seamstress who created a stunning wedding dress in silver and white lamé, as well as some other outfits for her.

Bibi also chose some suits for Thamina from the two modern department stores that had opened in central Baghdad, near Rashid Street. One of these was Orosdi Back, which already had branches in Istanbul, Cairo and Beirut; the other was Hassou Ikhwan. Both offered a shopping experience far removed from the vibrant, chaotic, old-fashioned souks, and Bibi relished the joys of ready-to-wear.

Thamina's wedding took place at the family home. There was a grand dinner party, which was boycotted by the relatives of the rejected family suitors. A Hungarian dancing troupe entertained the guests. At the end of the evening, Hadi turned to his wife and said, 'You've committed a *jarima* – a crime – do you hear? There were so many others she could have married. A crime, I say; do you understand what that makes you?' Bibi refused to answer. That evening, she reflected, Thamina hadn't looked like a victim, radiant in her dress, laughing and smiling with her new husband.

To Bibi's immense satisfaction, her plan succeeded, and Thamina remained in her daily orbit in A'zamiya. Within a few months she was pregnant. She was so embarrassed to show her condition in front of her father, betraying as it did her conjugal relationship with her

A wedding portrait of Thamina and Saleh Bassam, 1941.

husband, that whenever she saw Hadi she would hide her bump with a cushion.

20

Blood and Salons

Mounting Tensions

(1941)

ALTHOUGH IT WAS still being fought many miles away from Iraq, the war had started to divide the country. The Regent, Abdul Ilah, was committed to supporting the Allies, but the atmosphere had almost reached boiling point in Baghdad, where students staged anti-British protest marches, while the local press vented its nationalist frustrations on front pages, and pamphlets of every type littered the streets of the capital. It was unclear where the effects of the German propaganda machine ended and those of home-grown anti-British sentiment took over. But it certainly seemed to many that the German campaign spoke to them more strongly than any British claims to loyalty did, despite Britain's pervasive presence in the country.

The pro-Axis politician Rashid 'Ali Gailani had taken over as Prime Minister from Nuri Pasha al-Said in a national coalition government. Joining forces with a group of Arab nationalist officers who called themselves the 'Golden Square', he objected strongly to Iraq's providing any further help to Britain's war effort, although it was bound to do so under the terms of the 1930 Anglo-Iraqi Treaty. Officially he and his supporters claimed to want neutrality, while secretly they used Iraq's Ambassador to Ankara to appeal to the Germans and Italians for arms.

In April 1941, under the orders of the Golden Square, the Iraqi Army seized Baghdad, while Rashid 'Ali and his followers took over Parliament. The Prince Regent, Abdul Ilah, fled the country to the safety of Transjordan, where he was soon joined by several politicians who were opposed to Rashid 'Ali, including Nuri Pasha al-Said. In Baghdad, Parliament was pressured into voting in favour of the Regent's cousin taking over the throne. It was a coup.

Hadi did not escape Rashid 'Ali's ire. He was accused of funding the tribes in the south to rebel against the new government. It transpired that many of these rumours originated with Hadi's own cousin, Salim, the son of Abdul Ghani, whose argument with Abdul Hussein seventeen years earlier had precipitated the move to the Deer Palace. Salim was a Communist who hated the monarchy. Hadi had tried to maintain good relations with him over the years, and was dismayed to hear that his younger cousin had slandered him in this way.

Salim was a well-read intellectual who contributed frequently to Communist publications. His activism had often him landed in prison, as the Iraqi Communist Party had been banned by the government in 1935. His political purism was so extreme that he often neglected his family in favour of what he regarded as the greater cause. Bibi was highly critical of Salim, and thought he should behave much more responsibly towards his family. Hadi, on the other hand, tried whenever he could to help Salim out, and had often used his connections to get him released from prison. When challenged by Bibi, Hadi justified his forgiving attitude on the basis that his cousin had lost his way in life, and that it was pointless to punish him any further.

However, as far as Salim was concerned, his older cousin personified everything that the class struggle was about. He disapproved of all that Hadi did and all that he stood for. Bibi quietly suspected that his opinions were as much the result of jealousy of her husband's success as of ideological beliefs.

As the fractures deepened between branches of the Chalabi family, the level of tension and terror escalated in Baghdad, with Radio Berlin blaring out of every café in town. Rumours spread of the growing influence of the pro-Hitler, ultra-nationalist Palestinian Mufti Amin Husseini, who was in Baghdad rallying anti-British sentiment. Alarmed by the situation, the British decided to test Rashid 'Ali Gailani's allegiance by asking for permission to land their planes in Iraq under the terms of the 1930 Anglo-Iraqi Treaty. At the same time they withheld their recognition of the new government, which they were certain was backed by military force.

Rashid 'Ali and the Golden Square confirmed these suspicions by demanding several preconditions before the British planes could land. Almost overnight Iraq was transformed into Axis territory, and British troops landed at Basra once more, as they had done in 1914.

In retaliation, to the dismay of the civilian politicians, Iraqi forces surrounded the British airbase at Habbaniya, north of Baghdad. Interpreting this as an act of war, on 18 May 1941 British forces launched an attack on the 9,000-strong Iraqi Army at Rashid Camp, south of Habbaniya. Britain could not risk losing her important supply of Iraqi oil to Germany.

Hitler's interest in Iraq was, however, only fleeting; he viewed it primarily as a means of harming British interests. According to his plans, Iraq would act as a base from which to launch aerial attacks on Russia. To that end, he initially paid lip service to the demands of Rashid 'Ali and his supporters, describing them as his natural allies against the British.

Baghdad felt the heat of the gathering confrontation as the fire-flares of the Royal Air Force lit the sky to the north, while pamphlets continued to be dropped onto the city itself and news began to circulate that a division of the Luftwaffe had been stationed at Mosul.

Bibi was terrified by the sight of British Wellington bombers flying overhead on their way to raid Habbaniya. Until that point the war had been fought in distant lands, and she and the family had seen it only on newsreels at the cinema. Now, the sound of the planes' engines overhead woke everyone early in the mornings. The fighting felt very close, and Bibi was dismayed when she spotted an old man jumping with joy down in the street one day, shouting, 'King George is dead, King George is dead! The war is over!' She was sure he was repeating something he had heard on the radio.

As the situation continued to escalate, Rashid 'Ali's followers lobbied the Germans heavily for support. Several German planes bombed Habbaniya, and Major Axel von Blomberg of the German High Command was sent to liaise with Rashid 'Ali, but was mistakenly shot down by an Iraqi soldier.

The Germans soon abandoned their campaign. Within a few days, Rashid 'Ali and the Golden Square were defeated by the British, and the Iraqi forces surrendered. Rashid 'Ali fled to Iran, before continuing to Berlin. The Prince Regent, Abdul Ilah, and his supporters made plans to return to Baghdad, while the British military surrounded the city; they did not enter it, as they did not want to seem like an occupying force.

In spite of the nationalist leanings of many of her acquaintances, Bibi was vehemently opposed to Hitler. She knew that he had started this war, and for that she would never forgive him. She had also learned about some of his crimes in Europe from her Jewish friends, the Munashis, Sha'shous and Lawis, who had told her of persecution and of the Kristallnacht pogrom of November 1938.

She hated the pro-Axis politicians, and covered her ears whenever she heard Yunis Bahri's voice blaring out of the radio. It seemed to her that these days Iraq was possessed by violence. The language on the streets had changed, becoming harsher and coarser, and the atmosphere rippled with tension. One afternoon towards the end of May 1941, while riding in a carriage on her way back home from Baghdad, she was incensed when she heard men outside a café cheering news of German planes bombing London. She was even more outraged when she overheard Ghaffuri, the driver who had replaced Karim, agreeing with Ali Akbar that Hitler was going to win the war.

Early one evening a man knocked at the door of the Chalabi house. He was thin, with neat greying hair, and was dressed simply in a white shirt and navy trousers. He introduced himself as Nabil, and said he needed to see the master of the house urgently; he had a message from his boss, Mr Lawi, a Jewish friend of Hadi's from Samau'al Street.

Ni'mati noticed that the man was sweating despite the cool of the evening, and wondered why he was so nervous. He was slightly put out by the request, as Hadi was taking a siesta, and Ni'mati never liked to wake him up. Nevertheless, he proceeded to Hadi's room, knocked gently and softly called, 'Effendi, there's a man here to see you. She says it's very important, from Mr Lawi.' Ni'mati always switched genders around. Although he had lived in Iraq for most of his life, he never fully mastered Arabic.

Hadi jumped out of bed at once, and asked Ni'mati to get the visitor some tea. When he joined Nabil downstairs, their conversation was direct and to the point. Nabil's boss, Mr Lawi, was asking Hadi for help. Given the current tensions, he and the heads of several other families felt unsafe in their neighbourhood, and wanted Hadi to find a place where they could lie low until things calmed down. Rumours were rife that the Baghdadi Jews had rejoiced at the attack on the Iraqi force at Habbaniya, and that British-trained Jewish secret agents had been recruited in Palestine to help the British fight the Iraqi Army. The news had not yet spread of Rashid 'Ali's flight, and many of his proclamations were still heard in the streets.

'Please tell Mr Lawi that this is his house and anyone he brings with him is welcome,' Hadi assured Nabil. 'Please hurry and tell him I'm waiting for him to come as soon as possible.' He immediately started to instruct his staff to bring out mattresses and clear out rooms on the first floor in preparation for their guests.

Bibi was woken from her siesta by the commotion. When Hadi announced that Mr Lawi and several families were coming to stay with them because they felt unsafe in their own homes, she readily agreed. She was aware that accommodating these visitors presented a considerable risk to the household, especially as Hadi was known to be on good terms with the monarchy and had not been friendly with Rashid 'Ali and his followers. But she also knew that Hadi could not turn them away.

Gurji Lawi and his family were the first to arrive, followed by Munashi Sha'shou and his family, Salman Zilkha and his wife, and others. Mrs Zilkha was so shy that she found it hard to join in at mealtimes, so a small gas burner was bought for her use. Everything possible was done to cater for the guests' kosher diet, although no one wanted to attract attention to the house by going to the Jewish market in *Souq Hanoun* in old Baghdad.

After a week the families returned to their own homes. Events were soon to force them to seek refuge on a more permanent basis.

The Prince Regent returned to Baghdad by way of Habbaniya on 1 June. Crowds flocked onto the streets to welcome him back, among them members of the Baghdadi Jewish community, who were dressed in their best clothes because the occasion coincided with the feast of the Pentecost, which they celebrated by paying their respects at the tomb of Joshua, in western Baghdad.

As the Jews celebrated Prince Abdul Ilah's return, they were brutally attacked by a mob of angry, defeated and disbanded soldiers and hooligans. One man was killed, and sixteen were hurt. The police took them to the hospital, where the staff were harassed to hand them over to the mob outside. The chief doctor refused.

Further incidents followed in the Jewish neighbourhoods of Baghdad. Soldiers and townsfolk went on the rampage, breaking into Jewish homes, killing and looting, while the police stood by and did nothing. In one incident a bus carrying Jewish passengers was stopped by soldiers, and the travellers were pulled out and murdered. Cases of rape were also reported. Shops were broken into on Rashid and Ghazi Streets, the main commercial streets of Baghdad, and the loot that wasn't taken was strewn on the road.

The rampage continued the next day, with crimes being committed by soldiers, police and bands of youths who belonged to pro-Nazi movements. Furniture was dragged out of homes in the Sinak area, while in other districts the police demanded protection money from Jewish families. In many respects the behaviour of the mob echoed that of the rabble who had swarmed onto the streets following the Ottoman retreat of 1917, taking advantage of the chaos in order to

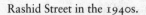

Rashid Street in the 1940s.

pillage and rob, rather than expressing their discontent with the government through more articulate means.

The violence only subsided when the Mayor finally asked the Regent to intercede on the second day. Abdul Ilah mobilized those army divisions that remained loyal to him, and some semblance of order was restored. All in all, between 135 and 180 people had been killed, hundreds injured and many hundreds of Jewish households ransacked. These days' terrible events became known as the *Farhud*, the Great Loot. A government investigation placed the responsibility for failing to prevent it squarely at the feet of the army and the police.

It was clear that many people continued to nurse strong anti-monarchy and anti-British sentiments. Now this hostility had been channelled into acts of violence directed at the most obvious pro-British minority, the Jews, in a shameful act of impotent revenge. Horrified by what had happened to her Jewish friends, Bibi wondered if her own family would be next to experience the rage of the mob.

Against the advice of the Cabinet, the Prince Regent insisted that the officers of the Golden Square should be publicly hanged for their part in the failed coup, thereby widening the divide between pro-Axis nationalist sympathizers and the monarchy.

A new government was formed under the old patrician politician Jamil Madfa'i, and for a few months great care was taken to boost the Jewish community's confidence. Many pro-Nazis and extreme nationalists were arrested, and their non-Iraqi sympathizers were expelled from the country. The Jewish population also received financial support, allowing it to recover somewhat from the horrors of June 1941.

Nevertheless, many Jewish families decided to move elsewhere after the attacks, although older members of the community, who were attached to their homes and their traditional way of life, were not tempted by immigration. Many in the younger generation gravitated towards Communist or Zionist organizations, whose activities had recently seen a marked increase.

The British *de facto* occupation of Iraq continued to aggravate the already hostile population, but censorship and strong security were put in place to ensure that the chaos of June 1941 could not be repeated. Freya Stark, the British traveller who was working for British intelligence during the war, concocted a plan to promote British ideals and principles among the educated classes of the region by bringing the 'fans of democracy' together. Supported by the Ministry of Information in London, she began her project in Cairo by setting up a group she called *Ikhwan al-Hurriya*, the Brotherhood of Freedom. It quickly earned the nickname *Ikhwan Freya*, Freya's Brothers, when she introduced it to Iraq in 1941. In Baghdad she had the support of the British Institute, which organized cultural events and published a newspaper.

Nevertheless, the success of Nazi propaganda and the bitterness of many Iraqis at the army's defeat meant that only twelve people attended the first few meetings of the Brotherhood that Freya held at her house in Alwiyah. She was vocal about not wanting the organization to become a mere British mouthpiece, and believed in following up local initiatives rather than taking instruction from the British. Although many continued to believe that the British authorities were really in charge of the operation, gradually the circle began to expand. Freya travelled across the country to spread the word. Tribal sheikhs were cajoled, lawyers, doctors and engineers were approached, in the name of promoting democracy. She even went to the holy cities to promote her project among the turbaned men of the shrines.

Although by the end of the war the membership of Freya's society numbered in the thousands, and even included some officers from the army – that bastion of anti-Britishness – it is doubtful whether the Brotherhood created any change in public opinion. For the ambitious it was a useful means to gain access to people of influence, and many saw it simply as a social club where like-minded secular individuals could meet and enjoy a drink, ignoring Islam's prohibition of drinking. It was said that one of the Brotherhood's most notable successes was to introduce young Iraqis to whisky.

In Baghdad, Rushdi was among those to join the *Ikhwan al-Hurriya*, having been introduced to Freya by his progressive cousin Abdul Jabbar. Rushdi had recently completed his degree at the American University of Beirut, and was beginning to dabble in politics. He believed that an Allied victory would present the best outcome for Iraq, and was soon hosting *'azimahs* – meetings – at the Chalabi house.

Freya attended some of these meetings, but not all, as the aim was to make them as autonomous as possible. However, Rushdi welcomed a British presence. For him and for many others of his generation, the gatherings represented their first opportunity to interact directly with the British. Rushdi was convinced that if Iraq was to become one of the great nations of the world, much still had to be learned from its seeming rulers.

Bibi was less impressed by Freya Stark than she had been by Gertrude Bell. When she met her briefly, Bibi felt that Freya lacked Miss Bell's charisma and focus. She was also not amused when Rushdi told her that Freya had started holding meetings among women in Baghdad to encourage friendly relations between the British and the Iraqis.

'Bah,' she snapped when Rushdi invited her to attend one of these meetings. 'You want me to drink horrible tea and sit with people who want to tell me how to think? You must be mad. Who are they anyway? She's not the Khatun, this Freya.' Despite Bibi's support of the status quo, she didn't want to spend time with British women, who were notorious for their condescending attitude to the locals.

The trauma of war had superficially removed many of the barriers between the British and their former subjects, but deep divisions remained. Before the war, the British had treated Baghdad much like any other colonial outpost. There was a degree of interaction between the British and Iraqis in government circles, in the professions and in the sphere of business, but otherwise there had been little contact between the communities. With notable exceptions such as Gertrude Bell, the Iraqis and the British had lived in separate worlds.

Bibi may not have been willing to waste her time in the company of Freya Stark and her ilk, but she was delighted to receive an invitation from the Queen Mother, Queen Alia, to a lunch party for the ladies of the political and social elite who had been steadfast in their support of the royal family during the coup.

At the party, Bibi was amused to be served lukewarm Coca-Cola. Clutching her glass, she whispered to her daughters: 'You'd think they'd know it should be served chilled, wouldn't you?'

Najla laughed. 'Why don't you chill it with that?' She nodded at Bibi's large diamond ring, which glittered like ice. Bibi chose to ignore the remark; she loved her jewels and was grateful for any opportunity to show them off, however disapproving Najla might be.

The party was to be the first visit of many monthly visits that Bibi and her daughters made to the royal household, where they also met King Faisal II's maternal grandmother, Queen Nafissa, and her daughters, Princesses Abdiya and Badiya. Thamina struck up a friendship with Princess Badiya that would last a lifetime, and would put her own life in danger.

21

An Education Overseas

Mixed Fortunes

(1941–1945)

HASSAN WAS ADMITTED to Baghdad's College of Law in 1940, just before his eighteenth birthday. His single-minded drive for academic success meant that he had already overcome many obstacles; unable to write his own papers because of his blindness, he had persuaded

Hassan.

schools and educational establishments to find alternative ways of examining him. In order to get into college he sat an oral exam, at which he impressed the Dean and a committee of professors with his sharp intellect.

He quickly settled into his courses, enjoying the intellectual life of the college immensely. He particularly enjoyed taking part in debates, during which he would address each point in an orderly and eloquent fashion, listening intently and memorizing what he heard, then reflecting carefully before offering his opinion.

The war did not change the routine of college life, but outside the classroom it dominated every conversation, with the student body being equally divided between support for the Allies and for Hitler. Hassan was vehemently anti-Hitler. Like his mother, he believed that Hitler's politics were steeped in violence, and that the Führer was a disruptive influence on Iraqi society, heightening the tensions that already existed between the different communities.

The majority of Hassan's professors had trained in Egypt or were Egyptian, and Hassan listened intently as they spoke of Egypt's many decades of interaction with the West. Particularly influenced by the French, they regarded the Enlightenment as the key to progress, and Hassan began to dream of visiting Cairo, and even of experiencing the West for himself, where he felt he would be judged on his own merits.

One day Hassan's *oud* instructor Jamil Bashir arrived at the house with some particularly interesting information. He had found a study-assistant who could read Hassan his lessons, help him prepare for his classes and write his thoughts down on paper. The study-assistant's family came from Mosul, like his own; they were Catholic, and in need of money as the father had recently died. The only difficulty was that the study-assistant was a young woman, and she and her mother were very concerned that her work should not transgress propriety in any way. If she were to accept the post it would be an unconventional situation, and they wanted to ensure that her reputation would remain above reproach. Hassan brushed aside the issue of the study-assistant's gender: if this arrangement worked, he thought, it could offer him the freedom to pursue his studies he had always yearned for.

Jamil brought the prospective study-assistant to meet Hassan. She kept her *abaya* on even when she had entered the house, and she was clearly uncomfortable. To put her more at ease, Hassan interviewed her through Jamil; but still she had a nervous tic, swallowing her tongue before every sentence she spoke, which made a sound a little like a glugging waterpipe to Hassan's ears. When he learned that she shared the same name – Jamila – as his late grandmother, of whom he had been very fond, he asked about her family.

Jamila gave a little gulp, then said, 'We are three girls and my mother. My father died four months ago.' In spite of her nerves, her voice was low and gentle, like a soothing breeze on a hot day; she still had a Mosul accent: 'a' became 'i'. She explained that she had finished secondary school and hoped to take a teaching diploma. She was currently employed as a teaching assistant, but the children's school where she worked was a two-hour bus journey from Baghdad, and the idea of helping Hassan with his studies interested her. She had visited his house once before, when she had been collecting money for the wounded with the Red Cross. Hassan's mother had donated generously.

'Thank God she did, or you would think very badly of us!' Hassan laughed.

Jamila gave a small smile. Hassan could tell that she felt very unsure of herself, and that everything about this situation must be alien to her. He quickly sketched out her duties for her, explaining that she would have to accompany him to his lectures to take notes so that he wouldn't have to borrow them from his friends.

When Jamila politely told him that she couldn't accompany him anywhere in public, Hassan realized that she didn't want to invite slander by accepting his requests too quickly. He continued cautiously, 'Your salary will be paid monthly, and our driver can pick you up and drop you at home, so you won't have to worry about transport.' Jamila told him she would have to think the matter over and discuss it with her family. Before she left, Jamil assured her that she had nothing to fear from Hassan or his family, and that in his opinion she should begin working straight away.

It did not take long for Jamila to make up her mind. After weighing the disadvantages of tongues wagging against the advantages of a well-paid and – more importantly for her – a humanitarian job, she sent word that she would accept the position. Hassan went to his father to discuss her salary, and Hadi suggested that he give her a month's money in advance to encourage her.

Hassan's passion for the intellectual life was addictive, and he had a wonderful wry sense of humour. After a few weeks as his study-assistant, Jamila confided to Jamil that, more importantly than anything else, in working for him she knew she was truly needed, and the realization gave her drive and energy. She was happy to become Hassan's eyes.

Hadi was pleased that his son had found someone to help him, and Bibi was also delighted with the arrangement. Rushdi too was support-ive. He described Jamila to his brother, telling him that she was much darker-skinned than Hassan himself, with long bony hands, short, curly dark hair and a full mouth. He did not add that she also had an unfortunate mole on her nose, drawing attention to the least attract-ive part of her face, and that she was really rather plain.

On paper, Jamila's duties seemed simple, but they became far more complicated as Hassan's dependency upon her grew and she became more involved in his life. Although everyone was pleased with her, her position was always that of a paid secretary, and not a social equal or a family member. Unused to seeing women working outside a domes-tic context, Habib Chaigahwa christened her Jamila Effendi, *effendi* being the respectful title for a man. When she arrived at the house, he would announce: 'Jamila Effendi is here'.

Jamila started by merely reading to Hassan during afternoons at the house, but it soon became clear that he needed more assistance. She gradually overcame her reservations and started to accompany him to college, taking notes for him under the curious gaze of the other students. It was an uncomfortable experience at first, but her stern face and dedication to the task in hand soon quietened the gossip-mongers. A year into their arrangement, Hassan encouraged Jamila to enrol at the college and study for a law degree herself. He thought it would be a wasted opportunity if she didn't, given how much time she spent on the subject anyway.

Hassan's dream was to obtain a doctorate in law, and the obvious place to go to study was Cairo. His Egyptian professors encouraged him and recommended him to their colleagues at the King Fouad University, a Western-style institution built in 1908.

Hassan realized that the only way he could hope to study for a doctorate was if Jamila were to come with him, but he wasn't sure how to convince her. He was very sensitive to her conservative nature, and knew of her strong attachment to her family. He decided not to put the question to her until just before his final exams. Then she would have the entire summer to think about it and make up her mind.

In the autumn of 1942, Hassan set out for Cairo. He did not travel alone: Jamila again put her devotion to him and his education above her personal concerns. The reason was simple: she had fallen in love with him. She knew that by travelling with him she was compromising her chances of getting married in a society that placed a great value on marriage and was scornful of old maids. Yet accompanying Hassan also represented an opportunity for her to see the world and continue her own education.

To begin with, Jamila and Hassan found accommodation in separate boarding houses in Cairo. However, it soon became apparent that for practical reasons this arrangement could not be sustained for long, so Jamila broke yet another taboo by agreeing to find a flat to rent for both of them. After some searching they settled on a furnished apartment on Murad Street in Giza, near the river and not far from the university, which was housed in a grand neo-classical building, without any Islamic or ancient Egyptian references in its architecture.

Hassan found that he adored Egypt, though Iraq was never far from his mind; indeed, Cairo constantly reminded him of Baghdad. This went beyond the obvious fact that a river cut through both cities, and had more to do with the countries' standing in the region. Both Iraq and Egypt were competing for dominance, and their similarities were obvious to Hassan.

Like Iraq, where King Faisal had struggled against successive factions, in Egypt King Farouk had to contend with political dissension. Although it had been a sovereign state since 1922, Egypt had – like Iraq – been under the British yoke, and both countries were seen by Britain as strategic territories.

Nationalism had existed as a political force in Egypt for longer than in Iraq, but the war had brought the same sorts of tensions to the fore in both countries, with both Communists and pro-Nazi groups gaining ground. King Farouk's lavish lifestyle was a source of great contention, and his position was precarious. Egypt may have set out on the path of modernization and institution-building long before Iraq, but in both countries the growth of a native middle class – who could act as the backbone of a democratic society – was still under way. For many Egyptians the pace of change was too slow, even though the state had only existed for a couple of decades.

Yet for Hassan, Cairo was everything he had been searching for. He was captivated by its intellectual energy, its vibrancy despite the war, and the humour of the Egyptians. Life here was intense – the hotels, restaurants and cafés were packed, and music, champagne, bordellos and drama were in liberal supply. Larger-than-life Westerners such as Patrick Leigh Fermor, Max Mallowan and Evelyn Waugh made themselves at home in its high society.

Although Hassan was aware of these flamboyant figures, he did not mingle in their circles, but was content to enjoy a quieter existence. The one thing he did long for was a chance to meet his hero, Taha Hussein. A giant of Arabic literature, Taha Hussein had made a remarkable journey from humble beginnings as a blind boy in a small village in Egypt to the Sorbonne in Paris, stamping his homeland, philosophy and the Arabic language with his intellectual mark along the way. Hassan had been a schoolboy when Taha Hussein's autobiography *al-Ayyam* was published. The book's impact on him was tremendous, and it would guide his way through life, inspiring and motivating him whenever he lost courage. If Taha Hussein could overcome his blindness, Hassan could do the same.

Hassan's extracurricular activities in Cairo focused on the famous Groppi coffee house, and concerts held in the Azbakiya Gardens or at the concert hall of the American University. His favourite outing by far was to Tea Island at the Cairo Zoo, where he enjoyed sitting under the trees listening to the birdsong as he and his friends discussed politics. For the first time he felt he was the master of his own destiny, away from his overbearing family among whom his handicap set him apart.

While Hassan settled into university life, Rushdi was launching his political career. Although he was happy to move back under his father's roof, Rushdi's relationship with Hadi remained difficult. In spite of his relative inexperience, Rushdi felt that his time in Beirut and Europe had afforded him an awareness of the world that his father, who had left school at sixteen, lacked. He felt that Hadi's attachment to Kazimiya was symptomatic of his parochial outlook on life.

In particular, Rushdi disapproved of Hadi's continuing relationship with Weir & Co., who he believed were taking advantage of him. He took it upon himself to enlighten his father about his views, but Hadi was happy with the arrangement. He and the company had been working together for over a decade, and the relationship and trust built during that time were more important to him than merely increasing his own profits. He listened as patiently as he could to Rushdi's arguments, reminding himself that his son was still a young man.

Aware that Hadi was irritated by Rushdi's sense of his own superiority, Bibi tried to protect her favourite son from his father while simultaneously seeking to establish him as his father's heir both politically and financially. She was proud of the way Rushdi's university years had polished him, and dedicated herself to promoting his cause. In turn, she believed that she would be able to rely upon him in the years to come, and also privately hoped that she had found in him a travelling companion who had more time for leisure and entertainment than her husband.

In 1943 Bibi, Rushdi and Raifa decided to visit Cairo, both to catch up with Hassan and to enjoy a holiday. Raifa had recently finished secondary school, and was brooding over both her mother's veto on her wish to pursue her education at the Beirut College for Women and the loss of her sister Thamina to 'Salehbassam', as she would always call him with distaste. They travelled by Nairn bus to Damascus, and from there to Haifa, where they took the train to Cairo. On the train, much to her daughter's mortification, Bibi spent the entire journey admiring the smart young British soldiers who were travelling to join their battalions on the North African front.

At tea one afternoon in Cairo's Continental Hotel, Bibi was spotted by an Indian fortune-teller adorned in a colourful turban. He approached her and asked if he could read her palm. Bibi watched in fascination as he held her hand and traced the lines one by one. He closed his eyes and took several deep breaths. Bibi shuddered; perhaps this wasn't such a good idea after all. After a lengthy pause, with his eyes still tightly shut, he spoke: 'Yes, it is clear – you will have an odd number of children.' Bibi indignantly told him that she had eight offspring, and it was on the tip of her tongue to say that babies were the last thing on her mind at her age when the fortune-teller put his index finger to his lips to silence her. Bibi flushed, fidgeted in her seat and let him resume.

Next he told her that she would have a long life, but would suffer from a serious illness; and that soon the light of someone dear to her would be extinguished. As Bibi turned pale, he added that for every leaving of the world there was an arrival. Holding tightly onto her palm, he paused as if in deep thought for a moment, then said that her husband would continue his rise, but then there would be great change.

'Another house?'

'It is not clear ... somewhere else.' Confusion flickered across the fortune-teller's features. Bibi did not wish to tempt fate, so she took advantage of his momentary hesitation to pull some money quickly from her purse, hand it to him and leave. She had heard enough.

Unfortunately, this encounter poisoned the rest of Bibi's stay in Cairo: she was convinced she would lose one of her children, and would most likely die herself from a mystery illness. Not even the popular fancy-dress balls at the Continental could take her mind off these revelations. She wanted to go back to Baghdad as soon as possible and await the imminent catastrophe.

Soon after their return from Cairo, Raifa had a suitor. His name was Abdul Amir Allawi, a promising British-educated young doctor, and after the appropriate introductions he started visiting the Chalabi house, where he was offered sherbets of rosewater and cantaloupe juice, as was the custom for new suitors. Delighted with his wit and conversation, Bibi decreed that he was a suitable candidate for her middle daughter. When it came to the trousseau, she ordered many of the items from Cairo, having inspected the goods there a few months earlier. In this instance, Hadi respected her choice of husband for their daughter.

Shortly after Bibi's trip to Cairo, her mother Rumia died at the age of seventy-three. Najla felt bereft, as Rumia had always shown her love and affection, whereas her own mother often seemed too busy socializing to notice her youngest daughter. She shut herself in her room and cried for hours. Bibi's youngest boys, ten-year-old Talal and eight-year-old Hazem, remembered how their old-fashioned grandmother would entertain them with tales of her exotic travels, while Bibi and Saeeda recalled the miracles of Rumia's kitchen.

'There was the time she made the *fesanjoon* for the new King,' Bibi remembered quietly. 'She was a good woman. I wish I had been a better daughter.'

Saeeda nodded sadly. 'She was so kind; she was too good for this world.'

Not long after Rumia's death, Bibi discovered that she was pregnant for the ninth time. At the age of forty-four, she was shocked by her state. She had a two-year-old granddaughter, Leila, by Thamina, and another grandchild on the way. At first she did everything she could to hide her condition; she even claimed that she had no idea what could possibly have caused it. Nevertheless, she remembered that her ninth child had been predicted by the Indian fortune-teller, and in the summer of 1944 she exiled herself to Dhour el-Choueir, a delightful mountain town in recently liberated Vichy Lebanon, where she planned to remain until the child was born. In late October she gave birth to a son, whom she called Ahmad. Before Ahmad's first birthday he had two new nephews: Raifa gave birth to her first child, Ghazi, and Thamina had her second, Mahdi.

The end of the Second World War in August 1945 raised new questions about the future of Iraq. The country still did not benefit directly from its oil, which remained squarely in the hands of European consortiums. Moreover, one of the main employers during the war had been the British Army, which no longer needed to maintain such a strong presence in the region. The Allied victory had created a dearth of opportunities for the Iraqi people. It was a potentially incendiary situation.

Nuri Pasha al-Said's stock was high once more, as he had always championed the British. He was brought back as Prime Minister in November 1946. However, many of the army officers who had supported Rashid 'Ali in 1941 remained bitter, and looked for an opportunity to seek revenge. In addition, the tension between Nuri and the tenacious Prince Regent, Abdul Ilah, added more pressure to

A family outing from Baghdad. Ahmad as a toddler stands
on the far right, next to Bibi.

the situation, as they disagreed about the best way forward for Iraq.
Of the two, Nuri was the more able politician, but he lacked royal
authority. The Prince Regent could always exercise power over him,
although he had less ability. Abdul Ilah adopted a heavy-handed
approach to government affairs, curbing the freedoms the press had
enjoyed during the war.

The one issue that Nuri and Abdul Ilah agreed upon was the serious threat posed by Communism. The growth of the Communist
movement in Iraq held dire implications for them both, and threatened to draw the country into the Soviet orbit.

The public's desire for reform was strong and its patience low, with
increasing demands being made for political liberalization and changes
to electoral law. When some of the political parties that had been
banned during the war started to regroup, the Iraq Communist Party
emerged as the most organized of these. The far left had long had a
presence in Iraq. Marxist literature was freely available, and intellectuals from Iraq frequently met Communists from Lebanon, Syria and
Palestine. The Tudeh Communist Party in Iran also had a role to play
in influencing the growth of Communism in Iraq, as a substantial
number of left-wing Iranians lived in the Iraqi holy cities.

Communism attracted people from all backgrounds, especially Jews and Shi'a, and those from the southern rural areas. As members of a predominantly agricultural and impoverished society, the disenfranchised farmers were a natural audience for Communists. However, the intellectual backbone and the bulk of the party lay in the cities, not among the workers but among teachers and students in schools and universities, infusing an intellectual idealism into the movement which might have been absent had it developed in a factory environment. The Communist Party was the party of youth. Ironically, the Ministry of Education's own scholarship programme, which enabled many Iraqi students to continue their education abroad, meant that leftist influences from overseas made their way to Iraq.

Kazimiya took a particular shine to Communist ideals. Indeed, for all that Bibi criticized Hadi's cousin Salim for his beliefs, several of her own cousins and relatives joined the Communist Party. The mullahs' influence had been on the wane in the town since the revolt of 1920, with many Shi'a believing that the refusal of the clerics to negotiate with the British had been short-sighted, and had limited their community's access to power. Marx's vision filled the vacuum created by their disillusionment. Moreover, the Party represented resistance to the established order, and was deeply concerned with social justice, two major Shi'a preoccupations. Anti-British sentiment was another attraction. The old dependence on Islam and Britain became replaced by a dependence on Marx and the Soviet Union.

Belief in Communist ideology among the founders and intellectual members of the Iraqi Communist Party paralleled that of the zealots of early Islam on their proselytizing missions from Mecca. Pamphlets and leaflets were liberally distributed in Kazimiya. The Party's first public declaration – 'Manifesto of the Association Against Imperialism' – called for a total overhaul of the country.

22

Love in Strange Quarters

Of Marriage and Other Unions

(1946–1947)

'STOP, I'M GOING to die!' Bibi declared operatically, one jewelled hand on her chest and the other extended towards the driver of the horse-drawn carriage.

The man turned his head slightly, reins in hand, and replied dryly: 'Madam, must you die now? You really can't die in the *kallachiya*. Wait until I get to a more respectable neighbourhood and then you can die.'

Bibi was stunned. Najla snickered, and Thamina tried not to laugh even though she was appalled by the driver's familiar manner. They were on their way to a fitting with a seamstress on Rashid Street. None of the house cars had been available, so Bibi had asked one of the houseboys to get an *'arabana* to take them. The carriage had to pass through the old Midan Square, and Bibi and Najla had begun to bicker, so no one had noticed when the driver took a right turn into a narrow street lined by little huts with open doors in front of which women lounged listlessly. That was when Bibi realized they were in the *kallachiya*.

The mellifluously named *kallachiya* was an institution in Baghdad, a shabby maze of small brick houses, open courtyards, narrow alleyways and rooftops which were as busy as the rooms below in the summer. The lusty voices of *chanteuses* sang love songs, and locally brewed Farida beer and various flavours of *arak*, including date, watermelon and even meat, were always in ample supply.

Each house was a separate brothel, with its own madame or pimp and its own staff, including *daqbulis* – boys who carried water for the men to wash with after sex. It was said that these 'chicks', as they were known, were also for hire. They were notorious in Baghdad, and their testimonies were considered invalid in court. Besides the *daqbulis*, there were different ranks of prostitute: the more beautiful and sought-after they were, the quicker they left the muddy alleyways for more comfortable surroundings.

The *kallachiya* had acquired its character and traditions over many years. Tolerated if not legalized, it had continued to flourish under the Ottomans. An infirmary checked the workers regularly, and the gates to the area's entrance were guarded by policemen, who were empowered to search clients. Many of the girls who worked in the *kallachiya* had escaped from their families for one reason or another, and the possibility of honour killings and revenge was always there.

For many Baghdadi men, a visit to one of the *kallachiya*'s many brothels was a regular habit, even if it was just a fleeting stop during a lunch break. The clients came from many walks of life: musicians, shopkeepers, soldiers and politicians – everyone was welcome. A proud regular was the much-loved popular poet, Mulla Abboud al-Karkhi, who had been a Baghdadi landmark himself in the 1920s.

Karkhi had spent his life reciting the woes of the common man, and had been regarded as a barometer of public sentiment. He was also unabashedly interested in women, and devoted many verses, both lewd and refined, to them. One day he found inspiration in the funeral of Dawud al-Lampachi. Besides being the city's official lampman and lighting its gas streetlights, Dawud al-Lampachi was a famous pimp who was married to an equally famous prostitute, Fattuma Samachi. Karkhi created a satirical elegy of Lampachi's work among Baghdad's whores:

> Lampachi Dawud died and with him his arts.
> Come on people let us go and condole Fattuma ...
> Let us go and pay respects to Fattum and Zahiya
> And Najafi's daughter and Mariam al-Kurdiya ...
> You powerful mighty ram, every whore is crying for you ...

The poem quickly became legendary in the city, and metamorphosed into a song. However, Karkhi got his just deserts when Mariam al-Kurdiya, a beautiful and renowned prostitute, spotted him in the marketplace and attacked him with her shoe in revenge for some verses about her in the poem. Bibi remembered that when news of the incident reached the *dawakhana* at the Deer Palace, Abdul Hussein's guests had erupted into uncontrollable laughter which had echoed around the house. She also remembered how Karkhi had written a serious elegy for Abdul Rasul that had moved the family to tears.

She was woken from her reverie when she glimpsed a man walking out of an alleyway with his head bent down, adjusting his crotch. Even from a distance, he looked faintly familiar, and she prayed that he wasn't one of the household staff on his day off – the *kallachiya* was definitely no place for any self-respecting woman to be seen. Horrified by the prospect of discovery, Bibi leant forward and implored the driver: 'I really will die if you don't get me out of here right now!' She pressed herself back in her seat so that she could not be seen by anyone on the street, and urged her daughters to do the same. But Thamina and Najla looked around them in curiosity. They were shocked when they realized that some of the women seated on four-legged wooden frames were displaying their privates under thin gauze to passers-by, like cheap goods on show in a marketplace.

For her part, Bibi vacillated between the demands of social decorum and a profound sense of sadness. The whole idea of prostitution was deeply repugnant to her, but she was also depressed by the fate of these fallen women. She thanked God quietly for her family's good fortune, and muttered, 'Tff tff tff,' to ward off the devil. Then Najla nudged her accidentally. Bibi snapped at her to be careful, and another fight ensued.

Although Rushdi didn't resort to the *kallachiya*, he continued to dedicate as much time as he could to the pursuit of pleasure, which he combined when possible with his political and business interests. There were many more bars and clubs for men of Rushdi's social standing than there had been in Hadi's youth; several new hotels had also opened, including the Sindbad, the Semiramis, the Baghdad and the Khayyam. The last two had large terraces overlooking the river, where customers could enjoy a late-evening drink and watch the reflections of the stars dancing on the rippling water below. Rushdi particularly liked the Zia Hotel, where he often entertained friends and business associates.

Heading out for an evening in his black Cadillac, Rushdi and his companions would debate whether to head for the Roxy or for Abdullah's in the mixed Bataween neighbourhood. Both clubs staged famous cabaret shows from Europe. Abdullah's was Rushdi's preferred choice, as it served dinner at 8.30 p.m., followed by a show at 10.30. Many of his associates had special 'friends' among the European, Egyptian and Lebanese showgirls, who they occasionally entertained outside club hours.

In the early mornings, to soak up the spirits that coursed through their blood, the young men would often end up at one of the makeshift *patcha* restaurants in the Sheikh Umar neighbourhood. There they would buy *patcha* soup, a local delicacy, made from sheep's entrails, which was a renowned hangover cure. In a letter to Hassan, Rushdi told him how one of his friends had found an old shoe in his soup one day.

Rushdi greatly enjoyed his life as a single man-about-town, although he knew he would have to settle down at some point, if only to consolidate his political career. When he saw a photograph of Ilham Agha Jaafar, a young beauty from a well-known Basra family, he decided that the time had come.

Ilham had grown up in Beirut, and Rushdi knew that some of his friends were already very interested in her. After he enlisted the help of a pilot in the newly established Iraqi Airways to fly him to Beirut to meet her, she was duly impressed. Beating two other suitors, Rushdi celebrated his marriage to Ilham in 1947.

Rushdi's connections to his father, Hadi, and his grandfather Abdul Hussein helped him to develop a close rapport with Iraq's Prime Minister, Nuri Pasha al-Said. The two men had similar attitudes towards Iraq's future, and Nuri decided to keep an eye on young Rushdi, with a view to assisting his political career.

Rushdi also maintained the existing family ties with the royal family, although he could not muster much enthusiasm for the Prince Regent, who he felt was small-minded and politically incapable compared to

Nuri Said, far right, at a lunch party in the Sif garden, 1946.

Nuri. However, Abdul Ilah was a keen hunter, and Hadi owned some wonderful horses and saluki dogs, which he kept just outside Baghdad. So the Prince Regent was a frequent guest at the Chalabi house, and Rushdi organized hyena hunts in the surrounding wild lands. These started early in the morning, when the men would gather with their horses for a hot breakfast before the long day ahead.

For all his political aspirations, Rushdi lacked his father's ability to connect with the common man. He also lacked Hadi's humility and his commitment to philanthropic causes such as the Iraqi Red Crescent, children's education and health. Hadi remained unswervingly loyal to Kazimiya and many families in the town relied upon him for their well-being.

In spite of his ambivalent attitude towards the town, in 1947 Rushdi ran for the Kazimiya seat for parliament and won, while his father was appointed a senator. With the reins of power still in the hands of the Arab Sunni minority, their achievements were the few Shi'a exceptions.

* * *

Hassan, meanwhile, had completed his studies in Cairo, and was determined to take his doctorate at the Sorbonne in Paris. Eventually he secured the support of his parents, and arrived in the French capital in the summer of 1947.

He was alone. Having completed her law degree, Jamila had decided to return to Baghdad. Her mother had never been happy about her situation as a young Christian woman working for a Muslim man, and in spite of the depth of her feelings for Hassan, Jamila could not see any future for herself with him. It was simply not possible. They came from such different worlds.

His first few months in Paris proved very challenging for Hassan. He lived at the Hôtel Lutece, which had been the Gestapo's headquarters only a couple of years earlier. His only friend in Paris, Kazim, worked at the Iraqi Embassy. Kazim was an immense help to him, but nothing could compensate him for the loss of Jamila's care and attention. He had to find a way to convince her to join him in Paris, he told himself.

For several weeks, Hassan asked Kazim to write the letters to Jamila that he dictated to him. Sheltering from the evening drizzle in a café, cigarette in one hand and pen in the other, Kazim patiently wrote down Hassan's words. During the fifth letter, Hassan sighed and said: 'You're much missed, Jamila. I'm sure you'll love Paris.'

Kazim looked up at him intently and asked, 'Hassan, are you sure this is just business?' When Hassan did not immediately reply, he wondered cautiously if his friend had fallen in love with Jamila, and suggested that unless he planned to do something about it, Jamila's coming to Paris would only complicate matters. Flushing, Hassan insisted that Jamila was only his assistant, then cleared his throat and resumed his dictation. It took many pleading letters before Jamila was finally persuaded to join him in January 1948.

23

The Girl on the Bridge

Anger on the Streets

(1947–1949)

BY LATE 1947, Britain's presence in Iraq was regarded as problematic by all the country's political parties, regardless of their ideologies. The question was how to address the nation's strong anti-British sentiment while also safeguarding Iraq's interests, which necessitated a continuing good relationship with Britain. The 1930 Anglo-Iraqi Treaty had given Iraq its independence and a place in the League of Nations; in return it had meant that Britain could station her troops in Iraq. The terms of that treaty now needed to be revisited.

Anticipating public opposition to any attempt to ratify the treaty, Nuri and the Prince Regent initiated early and secret negotiations with the British in order to ensure that Iraq made valuable gains in any new deal that might be brokered. However, Nuri had been replaced as Prime Minister by Saleh Jabr, the first Shi'a to hold that position, and it fell to him to ratify the treaty with the British Foreign Secretary Ernest Bevin in Portsmouth in January 1948.

Under the terms of the new treaty, which was kept secret from the public, Britain would withdraw its troops from Iraq and hand over its two airbases, supply Iraq with arms and military training, and establish a joint common defence board. Saleh Jabr had also obtained guarantees from Bevin that under its Palestinian mandate, Britain would halt the creation of the state of Israel, which had been outlined by the United Nations in December 1947. This was a serious issue for all

Arab nations, and in securing this guarantee, Jabr believed he had achieved something of immense importance.

Several delegations from Iraq travelled to Britain to discuss aspects of the treaty. Hadi's brother Muhammad Ali took part in the economic negotiations, as he was in charge of the government-run Rafidain Bank. All the delegations shared the belief that Iraq needed to remain close to one of the world powers, as it was not strong enough to stand alone.

When news of the discussions was leaked to the Baghdadi press in early January 1948, it incited uproar among students, who took to the streets. Thirty-nine people were arrested, and an uneasy peace followed.

It was to be the quiet before the storm. A mass demonstration broke out in Baghdad a few days later, on 20 January. Lasting seven days, the protests became known as the *Wathbah* – the great uprising. They had been organized to a large extent by the Communists, uniting the diverse elements of society in their anger: there were displaced peasants from the south, students, railway workers, lawyers, doctors and artisans, as well as members of the growing Zionist element within the Jewish community, who disagreed vehemently with the clause which guaranteed that Britain would block a Zionist state in Palestine. All the individual grievances of the different divisions of society, urban and rural, Shi'a and Sunni, nationalist and Communist, became entangled and merged together into one enormous outburst.

During the week of the uprising, Baghdad resembled a battlefield as the demonstrators converged from both sides of the river towards the Ma'moun Bridge near the government district. Those on the east bank wanted to cross the river to join the protesters on the west bank; the air rang with their chants and cries as they distributed anti-government pamphlets, while ranks of police spread across the city, strategically placed on rooftops, in homes, cafés, mosques. Their orders were to stop the demonstrators and turn them back at any price.

Street warfare broke out. The police opened fire, and many were killed. The road was strewn with bodies, while others fell into the river. When police reinforcements were positioned on both sides of the bridge, a number of protesters found themselves trapped, with machine guns pointed at them. A young woman of nineteen, Adawiyah al-Falaki, a prostitute from the *kallachiya*, broke through the crowd.

Defying the hail of bullets and the armed police, she carried a protest banner across the bridge, encouraging everyone to follow her. She survived, and became known as the Girl on the Bridge; several poems were later written in honour of her courage.

During the week of the uprising, one hundred people died and three hundred were injured. Many of the dead were taken to the Shi'a shrine cities of Najaf and Karbala for burial, the funerals enflaming the sense of injustice and evoking the stories of the martyr saints. Among those who fell near the bridge was Jaafar al-Jawahiri, the brother of Iraq's Poet Laureate Muhammad Mahdi al-Jawahiri, and a Shi'a native of Najaf. Jaafar was immortalized by his brother in a poem that was read out by his graveside. The verses of *My Brother Jaafar* and *Day of the Martyr* came to symbolize the sentiment of revolution for many Iraqis:

> It is not fancy, what I say, my brother,
> For he who has to take revenge is always awake; he never dreams
> But is inspired by my patient endurance.
> For sometimes inspiration can reveal what is hidden in the future
> I see the heavens without stars, lit up with red blood ...

Jabr's government was toppled, and the Regent had to rescind the treaty. The left had succeeded in bringing down the government, but they had no plans beyond that. To appease the public, it was decided that another Shi'a should become the next Prime Minister. Abdul Hussein's old friend Sayyid Muhammad al-Sadr, who had participated in the 1920 revolt and then become an ally of Faisal, took office.

Fearing for his safety, former Prime Minister Jabr left Baghdad for his wife's tribe near Hilla. As an expression of solidarity at such a difficult time, Hadi decided to visit him there. Ni'mati, who had never previously expressed a political opinion in his life, stunned the entire household when he passionately declared his strong opposition to this trip. His son Muhammad, Hassan's school companion, had embraced Communism during the war, when he had worked for the British Army: no doubt he had influenced his father. In his poor Arabic, Ni'mati criticized Jabr to Hadi: 'She's a fool – a bad man who knows nothing of the people! How could you be dealing with this and his shameful ways?'

Hadi explained carefully that, as a fellow Shi'a, it was important for him to rally round Jabr, and not abandon him. Jabr had represented the government, and now he was being made the scapegoat. Ni'mati didn't agree, but as he was about to reply Rushdi entered the room. Such was Ni'mati's devotion to Hadi that he would never contradict him in front of his children. He closed his mouth, nodded to Rushdi and left them.

Saeeda, who was now Ahmad's devoted nanny, disagreed with Ni'mati's views and was highly disapproving of his son's political leanings. Because her relationship with Bibi went back years, she carried a certain authority in the house, and held court in her private quarters downstairs. Through her visitors from Kazimiya, she learned a great deal about the popularity of Communism in the town, what people were saying and what they were doing.

She also became an important conduit for many of those who sought assistance from Hadi or Rushdi. If she was convinced that they were deserving, Saeeda never hesitated to lobby their causes upstairs with Bibi or Rushdi, who was very fond of her. She never spoke to Hadi himself unless he spoke to her first: it was a self-imposed rule, which she upheld till the end of her life.

It soon became clear that the new Prime Minister, Sayyid Muhammad al-Sadr, could not bring about change. No sooner had he resigned and a new Cabinet been formed than the Arab–Israeli war – which Israelis call the War of Independence and Palestinians the Catastrophe (*Nakba*) – broke out in 1948, following the abrupt cession of the British mandate in Palestine.

As part of the Arab Legion, which included forces from Jordan, Syria, Egypt, Lebanon, Saudi Arabia and Yemen, Iraq initially sent 3,000 men to fight the Zionist troops, and later increased that number to 21,000. This turn of events affected the status of the Iraqi Jews dramatically, as they found themselves positioned precariously between their ancient homeland and a promised land. Many were divided in their loyalties, as they welcomed some form of reparation to their people for the horrors of World War II, but were appalled by the plight of the Palestinians who were driven out of their homes. Nevertheless, the conflict in Palestine drove a wedge between the Jews and the rest of the Iraqi population. A Zionist movement had taken root in Iraq following the *Farhud* of 1941, and many Iraqi Jews had been secretly migrating to Israel. Many non-Jewish Iraqis believed that the Zionists themselves had been responsible for some of the attacks against Jewish interests in Baghdad in order to convince Jews of the need to migrate, as Israel was in need of an able and affluent population.

In 1949 the Arab Legion withdrew from Palestine. There was great anger in Iraq at the defeat, leading to more protests and greater criticism of the government, as well as increased hostility towards Iraqi Jews. A decree was issued: those Jews who wished to remain were welcome to do so, but any who wanted to emigrate to Israel had to surrender their Iraqi nationality and dispose of their assets in Iraq. In the event, some 130,000 Jews left Iraq, where their ancestors had lived for millennia. While they lost their ancestral homeland, Iraq lost an entire community – only a few thousand Jews remained, the backbone of the country's first middle class.

Yet another Cabinet was formed, with Nuri Pasha al-Said back at the helm. Although the Iraqi component of the Arab Legion had fought well, the conflict had been a disaster for the Arab nations, many of which looked to allocate the blame, with Egypt accusing Iraq of losing the war. This accusation effectively ended Iraqi involvement in the campaign; Nuri withdrew his forces and proposed a solution that would have recognized Israel, but granted it much less territory.

The Arab–Israeli war marked the beginning of yet another new order in the region, one that would alter it irrevocably. It also highlighted the increasing rivalry between Iraq and Egypt as major players on the regional stage. While Iraq had arguably proved itself more successful in terms of nation-building in recent years, Egypt had the advantage of continuity: the monarchy there had existed since the mid-nineteenth century, after all.

Weary of war and of the tensions that continued to trouble Iraq, Bibi decided that the time was right for her first trip to Europe. She arranged to visit Hassan in Paris in 1949.

Saeeda was quietly proud of the Chalabis' international travels and their royal connections, which she described with pleasure to her visitors. She was Bibi's eyes and ears in the servants' quarters, and her rooms were constantly busy with guests. When her friend Batul complained to her about men, she would give her advice based on her own experience, although she preferred not to dwell on the failure of her marriage. When another friend, Umm Abdul Amir, complained that her husband beat her, she firmly advised her to leave him.

Having been snatched from her African village as a young girl, Saeeda craved security and order; she could not bear the violence and chaos that seemed to be connected to the Communist cause. Yet, for all that she disapproved of Ni'mati's son's Communist loyalties, she was as horrified as everyone else when the leader of the Iraqi Communist Party, Yusuf Salman Yusuf, was condemned to death in 1949 for treason. A Christian based in Nasiriya, southern Iraq, Yusuf Salman Yusuf had been imprisoned in 1946 and given a life sentence in 1947. His *nom de guerre*, Fahd, meant panther in Arabic.

At his re-sentencing in 1949, Fahd was given an open trial, which afforded people the opportunity to listen to his ideas at first hand. The government claimed that he had used his base in Kut prison to preach, recruit for and plot a Communist revolution to overthrow the established order and bring about – among other things – the expulsion of the British, the distribution of land to the peasants, social revolution and power for the workers and peasants.

The decision to execute Fahd and three of his comrades – Yehuda Siddiq, a Jew; Hussein Muhammad Shibibi, a Shi'a; and Zaki Bassim, a Sunni – was met with international condemnation. Apart from three officers who had been executed for treason at Abdul Ilah's insistence after the 1941 coup, these were the first public executions during the monarchy, and many people were left in shock. Ni'mati worried desperately for the safety of Muhammad and his Communist associates, although he knew that Hadi would always do his best to protect him and his family.

24

Precious Things

Towards a New World
(1950–1951)

OIL HAD BEEN discovered in the early 1900s at Kirkuk in the north of Iraq, then part of the Mosul province of the Ottoman Empire. A rivalry had developed between the British Anglo-Persian Oil Company, the German Deutsche Bank, Royal Dutch Shell and later the Turkish Petroleum Company for control of the precious resource.

The Iraqi Petroleum Company, a consortium of European and American oil companies, was formed in 1929, after many years of negotiation in which Calouste Gulbenkian – better known as Mr 5 Per Cent – played a leading role. At the time that the Kingdom of Iraq was created in 1921, control of and profit from oil were firmly in the hands of the oil companies, and the Iraqi state did not a benefit from its most important natural resource. It was only in the aftermath of World War II, when it was discovered that more oil lay in the south of the country than in the north, that Iraqi politicians began to campaign in earnest to obtain a share of the oil wealth in order to fund the country's development. Hadi and Rushdi were not directly involved in the negotiations at this stage, but both awaited the outcome with interest.

'What do you think, Father – could there be something in this for us?' Rushdi asked Hadi casually one afternoon, when the family had gathered for a large lunch in the house at A'zamiya. A calm had descended over the table; the silver platters had been emptied, the younger children were with their nannies. Rushdi and Ilham now had two little ones, Hussein and Nadia; Thamina had had her third, a baby girl called Kuku.

Left to right: Suham (Saleh's wife), Thamina, Raifa and Ilham
(Suham's sister and Rushdi's wife) at a party in the late 1940s.

'Perhaps, perhaps not.' Hadi reflected for a moment. 'I have enough
to occupy me with my farms and lands. Agriculture is what I know;
oil – well ...'

Rushdi nodded thoughtfully. He would wait to see what came of this
new business.

Under the terms of the new agreements Iraq finally began to receive
50 per cent of its oil revenue, of which it funnelled 70 per cent into the
Iraqi Development Board, founded in 1950, the aim of which was to
fund large-scale development, including agriculture, industrial, urban,
public housing and education projects. One of the more exciting projects undertaken was the modernizing of Baghdad.

Many Iraqis dreamed of a resurrection of the golden age of the
Caliphate, seven hundred years earlier, with Baghdad as the showcase
of the region and the torch-carrier of a new Arabic modernity. Iraq's
leaders increasingly viewed their country as a stronghold against
Communism in the region, and the United States began to strengthen
its links to it. Western-educated young Iraqis were returning home,
optimistic and full of new ideas for their country. Many among them,
especially the newly qualified engineers and architects, found the

Development Board fertile ground in which to sow their ambitions for Iraq.

Perhaps encouraged by rumours of the city's transformation, there was a high level of migration to Baghdad from the rest of the country, as people from all walks of life came looking for a livelihood. They had seen the success of relatives who had moved to the area before them, and wanted the same for themselves. *Ahl al Hawr* – the Marsh people – also moved north in search of water for their buffalo, and in the spring, when the floodwaters had receded, entire families travelled on foot up the river to pitch their shacks, called *sarifas*, in what grew to be a large shanty town by the dam to the east of Baghdad. The poorest of the population, they were often seen picking up buffalo dung to use as fuel.

The Marsh people were a formidable sight, the descendants of the Sumerians; they were fine-featured, dark-skinned, and many had pale eyes that shone like gems in their faces. Theirs was a different world, harsh and rooted, which intersected fleetingly and strangely with the arrogance of the metropolis. Amidst the hustle and bustle of Baghdad's busy streets, some of the Marsh women would cross the city early in the morning with their buffalo in tow, to the residential quarters where they sold milk directly by milking their buffalo outside the doorstep.

When Bibi's youngest child, five-year-old Ahmad, saw the Marsh women with their buffalo, he thought they looked like something out of a fairytale, and asked his brother Jawad, who wrote to Hassan regularly, to describe them in his next letter to him.

Jawad laughed at the request. 'He might be more interested to know that Najla is engaged at long last; we always thought she'd end up an old maid.' At the age of twenty-four, Najla was to marry Ilham's brother, Abdul Latif Agha Jaafar.

One cloudless morning in Paris in the spring of 1951, Hassan had woken up early to go to the bathroom when he collapsed in sudden pain. The thud of his fall woke Jamila, who ran to him and summoned help.

Diagnosed with an ulcer, Hassan underwent an emergency operation at the American Hospital in Neuilly. Almost two-thirds of his stomach was removed. It had been a very close call.

Jamila asked Hassan's friend Kazim to send a telegram to Bibi and Hadi with the news. Hassan spent ten days in hospital, but he would need another three months in which to convalesce. His young uncle Saleh came over from London, on leave from his hospital training, but he could only stay for a week, after which he entrusted Hassan entirely to Jamila's care.

Hassan's illness broke down a barrier between them; he had nearly died, and her love for him was made clear by her meticulous and dedicated care for him. Hassan didn't know what to do; he was extremely attached to her, but was it love? For Jamila's part, she was tortured by guilt, torn between her faith and her love. She would often walk to the church of Nôtre Dame des Champs, light a candle and pray for forgiveness, for resolution.

Bibi decided it was time to visit Hassan again, having enjoyed her previous trip two years earlier, during which she had discovered the delights of Parisian *haute couture*. Following that trip, she had decided that aeroplanes were not for her. The gravity-defying experience had brought her much too close to God for comfort, and she had arrived in Paris exhausted. This time Hadi booked a luxurious sea passage for the family across the Mediterranean.

Bibi and Hadi took with them their three youngest sons: nineteen-year-old Talal, already a *bon vivant*; the more austere seventeen-year-old Hazem; and six-year-old Ahmad. The trip was a huge success: they took in the sights of Alexandria, Brindisi, Venice and Milan, which Bibi discovered was a shopping paradise. For the two eldest boys, the highlight of the Italian leg of the trip was the beautiful young women, whom they went to seek out in bars after dinner. Despite Hazem's film-star good looks, Talal always had more success with the ladies. He was much more witty and good-natured than his younger brother, whose view of the world tended towards the sardonic and the absurd. Both of them practised the few words of Italian they had learned on the boat in the hope of charming the Italian girls.

One morning at breakfast in Geneva, Bibi demanded a jewel from Hadi. She told her sons earnestly how unkind their father had been to her, how he had spent all those years travelling to dangerous spots around the country, leaving her alone to raise the family and worry about him.

Hadi chose not to argue with Bibi and risk ruining the holiday. He took her to Adler, where he bought her an art-deco, princess-cut ruby ring which covered half her hand, and to Boucheron, where he treated her to a rectangular diamond brooch. Bibi was very particular about jewels: she never wore earrings, and only rarely necklaces, which she thought would draw attention to her neck, which she still believed was the feature that made her look so short. Her signature accessories were large rings, brooches, crocodile handbags and silk scarves.

By the time Hassan and Jamila greeted the family at the Gare de Lyon in Paris, Ahmad was nervous about meeting his elder brother. The last time they had been together had been back in 1947, when Ahmad had been barely three years old, and Hassan twenty-six. He had been told that Hassan was blind, but his dark glasses frightened him a little. But when Hassan grinned at him and reached out to ruffle his hair, he overcame his fear and gave him a hug. Bibi smiled at them; she still felt responsible for Hassan's condition, and continued to scour the papers for news of any possible cures for him. However, she had recently had a dream in which she had been told that Hassan would never see again, and was slowly beginning to come to terms with her son's fate.

That evening they all went out to dinner, during which Hazem insisted on doing the ordering, much to the irritation of his family and the maître d'. As they all waited patiently, he leafed through his anti-quated French dictionary, intent on ordering yoghurt for his father. Finally and triumphantly, he put the dictionary on the table and smugly asked for '*lait caillé*'.

Confused by these strange people and this unusual request, the waiter came back with a glass of cold milk. Everyone except Hadi and Hazem burst into laughter. Wiping a tear from her eye, Bibi asked Jamila to take over the ordering. 'It might not be as poetic, but at least we won't starve!'

Hassan was amused by his youngest brother Ahmad's precocious-ness, and decided to test him on his general knowledge. He had heard that Ahmad knew the names of many of the world's capitals. In fact, when Talal had friends staying over, he would often wake his little brother up as a joke and ask him to name the capital of some distant country. Ahmad always got it right.

'Tell me, Ahmad, what is the capital of Australia?' Hassan asked.

'Canberra.' Ahmad smiled up at his new big brother.

Hassan congratulated him, and asked him a string of other questions about a range of subjects, from the name of the King of England to the parentage of a mule. Ahmad got them all right. Finally, Hassan asked if he could recite the Fatiha prayer. Ahmad immediately launched into the verses: 'In the name of the merciful and the bountiful ...'

Over the course of that dinner, Hassan's heart opened fully to Ahmad, and he came to love the little boy almost like a son of his own.

Paris in the summer of 1951 was full of tourists, including many young Americans, both the moneyed and the penniless, who had come to inhale the city's chic and intellectual air. Bibi was very taken with the sophistication of the Parisians, and her first port of call was Dior's sumptuous *maison* on the avenue Montaigne.

Dior's New Look suits had been a hit since 1948, launching the designer as the darling of post-war fashion, but to Bibi's disappointment few of his clothes fitted her elegantly – she had always struggled

An oil portrait of Bibi
painted in the 1950s.

with her weight. She did, however, manage to find a few nice outfits, and bought her daughters those that she couldn't fit into herself. She also treated herself to an array of new accessories.

Ahmad hated accompanying Bibi on her shopping trips and began to miss Saeeda acutely. He began to write a letter to her, unaware that she couldn't read, to complain about his plight. When he got stuck in the middle, he asked Jamila to continue it for him as he dictated:

Dear Saeeda,
How are you? I miss you very much. I am not well. My mother
makes me wash every day. I like Hassan very much and also
Jamila. The boat was very big, bigger than the one you came in
when you came from Africa …

He smiled to himself as he remembered how Saeeda often complained about her false teeth; to his immense delight she sometimes took them out and chattered them at him, mumbling through wrinkled lips that her *babuj* – her slippers – were getting too big for her mouth. Best of all, he remembered how comforting it felt when she pulled him to her and cuddled him. With her, he was safe from the world.

On their return from Europe, the family moved into their newly built house. It was on the land that Hadi had bought when they had moved out of the Sif Palace, across the road in A'zamiya from the villa they had been renting. In the end, the house had been designed by Raifa's brother-in-law Jaafar Allawi, a talented architect who had been educated in Britain, but who had a distinct Bauhaus sensibility.

The construction had been supervised by Hadi, who loved building and design. The marble and pink granite were shipped in from Italy, along with furniture chosen by Hadi during the family's trip. He also purchased crystal chandeliers and other *objets d'art*, including Sèvres porcelain and sterling silver. The statue of the deer that his father Abdul Hussein had loved so much was allocated a place in the new front garden, between the two entrances to the house.

The household staff had changed over the years, expanding and contracting yet continuing to just about represent the ethnic composition of Iraq. For the many Iraqis among them, Baghdad represented the hub of the country, where they came to find employment and to improve their fortunes. Like their predecessors, the Iranian staff usually came as pilgrims to the Shi'a shrines and stayed on, finding the atmosphere in Baghdad agreeable. Saeeda and Ni'mati too remained; it would have been impossible to imagine the new house without them.

Bibi continued her tradition of being the worst house manager in the world, yet somehow she maintained sufficient respect among the staff for the house to run smoothly. Her daughters stepped in when there were large parties to host, writing the menus and supervising the staff.

Even on quiet days, the poor cook still had to produce enough food for an army, as there were endless members of staff to feed, besides the large lunches for the family who gathered around the thirty-six-seat dining table. Hadi and Bibi, their children, their sons- and daughters-in-law, brothers- and sisters-in-law and their spouses, grandchildren and cousins all came. It was an open house, and any member of the family could turn up unannounced to break bread. Bibi was oblivious to the fact that these lunches might undermine her married children's domestic arrangements, taking them away from their own homes.

Besides being the pivot around which her extended family revolved, Bibi played cards every day with her society friends. On particularly serious gambling days she would drop by Hadi's office in Baghdad, waiting in the car while the driver went in to collect cash. No words were spoken, and no refusal could be brokered. The office manager, Yusuf Zubaida, would duly appear with an envelope, Bibi would exchange a few words with him, then off she went. She never set foot inside her husband's office, and he never came out to see her when she called by.

Her card-playing companions were mostly from the Sunni elite, which raised strong objections among her children, particularly Hassan and Najla. Najla was outraged by what she regarded as her mother's decadence, and was also very aware of the sectarian inequality between Sunnis and Shi'a. Her disapproval irritated Bibi, but luckily Thamina and Raifa played cards too, so neither was in a position to criticize her. Nevertheless, she berated them for being unsupportive. In between sips of tea she would complain: 'With children like you, who needs enemies? You'd think I'd killed someone.'

Bibi's other activities were of a different kind altogether, concerned with charity and the life beyond the walls of the house. Although she didn't wear the *abaya* every day, she was still a firm believer. She didn't drink, she prayed daily, fasted during Ramadan, and performed the pilgrimage to Mecca with Hadi in early 1944. Her mother's influence ran deeper than she wanted to admit. She wasn't very keen on the festival of Ashura, because of its funereal aspects, but she took her visits to the shrines very seriously. Besides going to Kazimiya to make her wishes directly known to the Imam, she also went to Karbala a few times a year.

Bibi would put on a scented silk *abaya* whenever she entered a shrine. Following the greeting and recital ritual at the outer door, she often hired a reader to guide her through her visit. Each shrine had its own prayer, dedicated to the appropriate Imam and available in booklet form, but Bibi always wanted a direct experience. The reader, dressed in his traditional gown and fez, would read out verse after verse, stopping occasionally for Bibi and her daughters to repeat the words after him. Once he had recited the last verses they would step out into the courtyard, where, much to the dread of her daughters, Bibi would open her handbag to pay him. Stronger than a magnet, the handbag drew all the shrine's beggars and unfortunates, threatening to submerge Bibi as she handed out money to them by the handful. Thamina and Raifa always worried that she would be knocked over by the crowd, and wished that she would not subject them to this ritual every time they came with her on one of her pilgrimages. In the end, it always took someone from the shrine management to clear a path for Bibi so she could leave the shrine, outside which she would be met by yet more supplicants, young children holding their hands out. By the time she had finished her bag would be entirely empty, much to her satisfaction.

Walking away from a shrine with her on one occasion, Raifa said: 'Surely there's no need for you to hand out so much every time? One day you might get seriously crushed and hurt.'

Bibi gave a sharp little snort. 'What is the point of having all this money if we don't share it?'

Her philanthropic outbursts extended to her mother's family in Kazimiya, although these handouts were altogether more discreet. She often sent them money via Saeeda or other intermediaries, or slipped them a few notes if they dropped by the house for a brief visit. She made sure to hide this help from her husband. As many of her relatives were Communists who openly criticized him, she suspected that Hadi and her sons would disapprove.

There were also the social hangers-on, the ladies of modest means from Kazimiya and Baghdad who knew the family through an extremely convoluted set of connections. Several mornings a week, such ladies would come to visit Bibi, much to the consternation of the staff, who resented their very existence. They brought with them news and gossip, as a *quid pro quo* for the money they expected from her. Hadi's elderly aunts Munira, Amira and Shaouna were increasingly frail, and no longer came to the house for their monthly breakfasts as they once had, so Bibi appreciated their supply of tittle-tattle more than she might once have done. She always gave to them generously, and continued to adopt people as pet projects, as she had done with Zahra, the girls' nanny, many years earlier.

25

Storm Clouds Gathering

Family Feuds and Revolution

(1952–1956)

AS BAGHDAD'S NEW buildings gradually took shape and the arts flourished, it seemed that Iraq was about to enter another golden age.

Then, in June 1952, a military coup toppled the Egyptian King. Jamal Abdul Nasser, an army colonel, took the reins of power, declared a populist dawn and destroyed Egypt's cosmopolitanism in one fell swoop. Nasser was a charismatic orator whose voice shook the whole Arab world. To the common man, he spoke from the heart. He stood up to the West, and many felt he had restored pride to a people who for centuries had been either tutored or colonized by the West. He became a symbol of anti-colonialism, aggressive nationalism and an idiosyncratic brand of socialism.

Hassan was shocked by the news of the coup in Egypt. He had recently returned from Cairo, where he had been defending his PhD in law, and he had a deep love for the country of King Farouk, its institutions and cultural diversity. He knew that there was poverty in Egypt; that there was illiteracy and decadence; but he felt that the country had nevertheless been moving in the right direction – state institutions were in place, and there had been a real opportunity for debate and progress. In Hassan's view, Nasser was wrecking all that was good about the old state. He destroyed the governmental and administrative systems, inflated the bureaucracy and undermined the educated classes.

Following the coup, thousands of people were forced to leave Egypt. Many of them were Europeans, including Italians, Greeks, Jews and

French who had enjoyed a privileged position in the country for generations. Nasser introduced state censorship, and the existing parliamentary system was completely dismantled. To Hassan's mind, Nasser resembled a Pharaoh in his decrees. However, for the masses, he was a prophet resurrected. The Chalabi family were deeply disturbed by the events in Egypt. Bibi in particular developed a deep-seated fear and hatred of Nasser.

That summer, the family went on their annual retreat to the Lebanese mountains. One morning Bibi was enjoying a leisurely breakfast on the veranda when a member of the household staff approached her. 'Madam, there's a man with a boil on his forehead who says he wants to see Mr Chalabi. Should I kick him out?'

Bibi sighed. 'No, no, that must be Salim. Please show him in.'

Salim walked onto the terrace looking dishevelled and unshaved; Bibi knew that he was on the run from the Iraqi authorities because of his involvement with the Communists. She took one look at him, and told him she wouldn't talk to him until he had taken a shower and changed his clothes.

Washed and dressed in one of Hadi's suits, which was far too big for him, Salim returned to the terrace. Their conversation was brief. Bibi chided him about leaving his children behind in Iraq, and told him firmly that his wife was sick with worry for him. She added that his familial duties were far more important than the ideological battle he had committed his life to.

Salim listened to her as he heaped food onto his plate; he was starving. Finally, he cleared his throat and explained that he needed some money. Bibi agreed that by the looks of him he clearly did, and asked how much he wanted. He told her that 150 Lebanese pounds would probably do the trick, to which she replied that she would arrange for him to have two hundred monthly that he could collect in Beirut.

Salim shook his head. 'No, I only want 150. Seventy is for me, and sixty for the Party. If you want to give me more, the rest will go to the Party as well. It's up to you.'

Bibi nearly choked on her toast. 'Me, contribute to the Communist Party, to that murderer Stalin? You must be joking. Suit yourself, take 150. But at least stay here until you've rested. I'm sure your cousin will want to see you.'

Salim stayed for another day; neither he nor Hadi mentioned the past, or the dangerous rumours that had been circulated about Hadi. At lunch the next day there were some Lebanese guests. Salim feigned deafness every time anyone asked him about the Communist Party, or leading Communist figures. He clearly felt he was among ideological enemies, and the less he said the better.

One afternoon, as they were sitting on the veranda enjoying the cool mountain breeze, Hassan announced to his mother and his sister Raifa that he wanted to marry Jamila. He asked if Bibi would buy her a ring and ask for her hand on his behalf. Bibi was delighted.

Hassan, however, had reservations about marrying Jamila. Some months earlier, she had had a hysterectomy. A life with her would be a life without children. He also knew that she remained torn between her love for him, her Christian faith and her family. When they had spoken of marriage, she had asked him what she could offer him as a wife that she wasn't offering him now.

A week later, Hassan told Bibi not to worry about the ring, because he had changed his mind. He refused to explain why.

In May 1953 King Faisal II turned eighteen and took over the throne from his unpopular uncle, the Prince Regent, Abdul Ilah, who became the Crown Prince. Heads of state from around the world were invited to the coronation, but there was a lack of state-owned luxury accommodation for them. The royal family, who had always been modest in appearance and lifestyle, asked several leading families if they could accommodate visiting dignitaries. The Saudi royal family stayed with Thamina and her husband Saleh Bassam in their house on the river-front, and the government came to rely on Hadi's hospitality whenever a visiting head of state came to stay, asking him to throw many parties on its behalf over the years.

Hadi's relations with Andrew Weir & Co. were gradually winding down because of the company's decision to divest in Iraq in the late forties, as its trade had been affected by the dismantling of the British Empire. However, Hadi was a shrewd businessman, and by 1954 he was at the peak of his wealth. He shared his wife's attitude to money, and started to invest in large-scale philanthropic projects, such as a modern children's hospital in Kazimiya.

Aware of the rising value of land and the limited space available in Kazimiya, he purchased large plots on the outskirts of the town, which he then subdivided and sold at half their market value to people in need of new homes and shops. He also built a large mosque there that he bequeathed as *wakf*, religious trust. The area became known as Hadi City.

One day he was approached by Weir & Co. on behalf of its Chief Director, Lord Inverforth, who wanted to sell the company's largest property in Iraq, a 25,000-acre cotton plantation called Latifiyyah Estate which was situated between the Tigris and the Euphrates, bordering both rivers. It had been bought by the company three decades earlier, and although they had invested in it, using the latest farming techniques and technology, they had struggled to make a profit from it. Now they were offering to sell it to Hadi for £750,000 (approximately £30 million today).

At about the same time another proposal was put to him. A very different piece of property was on offer from Richard Costain Ltd, a British construction company that also operated in Iraq. The estate in question was Dolphin Square in London. It had been completed in 1937, and was thought to be the largest block of flats in Europe, with 1,250 units. Rushdi had had business dealings with Costain, and felt strongly that his father should buy up Dolphin Square without a moment's hesitation. Hadi went to London with his manager, Salim Tarzi, to see Dolphin Square for himself, and to meet the directors of the company who were selling it.

Rushdi couldn't contain his irritation at his father's caution, and vented his frustration to his mother. 'I can't believe he's done this! Can't he see beyond his own backyard? Doesn't he know a good business opportunity when he sees one?'

Bibi shook her head. 'Well, he's very fond of this country. What do we know? Perhaps he knows best.'

'Best?' exploded Rushdi. 'He's so narrow-minded it's beyond belief!'

Bibi tried to calm her favourite son down, promising him that she would have a word with Hadi when he came back. In the event, there seemed to be no need: Hadi's trip had gone well and he had decided to buy Dolphin Square after all, despite his preference for agricultural projects. However, when word reached Weir & Co. about his interest in the London property its directors appealed to Nuri Pasha to persuade him to buy Latifiyyah instead. At an informal meeting in his drawing room, Nuri told Hadi that it would be in everyone's interest to see Iraqi capital invested in such a large asset, and that he was the only person who could do it. Over cups of sugary tea, the fate of Latifiyyah was sealed.

Hadi bought the land, and received as a token of gratitude from Lord Inverforth a large engraved silver box from Aspreys. Rushdi was furious with his father for the rest of his days. He could never forgive him for missing such a golden opportunity. Yet for all his bluster, he would never match his father's business acumen.

For Hadi, Latifiyyah had possibilities that went beyond agriculture. Baghdad was expanding rapidly, and the property was only twenty miles south of the city. His plan was to keep the prime agricultural land for himself, and to divide the rest of the estate into residential plots and small farms which he would sell off.

Latifiyyah became Hadi's passion, and he spent as much time there as possible, overseeing the many aspects of its management. In addition to the farmers who worked for him, the agricultural land and its requirements, the irrigation canals and their continuous maintenance, there was the sale and delivery of the produce. Many people lived in Latifiyyah, and their welfare was very much a part of Hadi's responsibility. Besides all this, he needed to maintain good relations with the tribes in the area. As a consequence, many of the family's weekends were now spent in the main estate house at Latifiyyah.

Rushdi was distracted from his disapproval of his father's investment choices when he was appointed Deputy Minister of Agriculture in late 1954. Nuri Pasha wanted to test him, to see if he was able to handle responsibility before giving him more. Within a few months he became a fully fledged Minister. Rushdi recruited many young Shi'a men to the various ministries, especially from Kazimiya, which was now his parliamentary seat as well as his ancestral home. Not unlike his grandfather in the early days of the Iraqi state, he tried to redress historical imbalances by taking immediate action. He established a rapport with some unlikely people while in office. One of them was Hamed Qassim, a grain merchant and one of many small suppliers whom Hadi had dealt with over the years. Hamed took to visiting Rushdi at the Ministry, and over cups of tea he secured a job for his son through him.

In developing Latifiyyah, Hadi was following a trend in the country. Oil money was being channelled into building projects through the newly established Development Board, whose members were determined to fulfil their vision of Baghdad as a pioneer in the region, reclaiming its ancient prestige. With the support of American and British experts, the Board commissioned leading architects to redesign the capital city.

A master plan was drawn up, with plans for new bridges and civic buildings, and expanded avenues. Hadi's son Hazem, who was still at Cambridge University, enthusiastically followed the news of his home city's blossoming. His father's cousin Abdul Jabbar Chalabi was a member of the Development Board, and wrote to him of the various plans, including a campus designed by the leading Bauhaus architect Walter Gropius for the newly founded Baghdad University; a sports stadium by Le Corbusier; ministry buildings by Gio Ponti; a national art gallery by Alvaar Alto; and new parks and roads. The celebrated Greek urban planner Constantinos Doxiadis designed public housing to replace the shanty town that had grown to the east of Baghdad, and later created a plan for the city as a whole. Perhaps the most dramatic of all the commissions was given to Frank Lloyd Wright. Although Wright was in his early nineties, he was carried away by Iraq's rich history, and designed a vast opera house which was to be built on the empty islet in the middle of the Tigris.

When Hazem returned home in the summer of 1957, he was amazed by the changes he saw. The new roads were busy with Chevrolets, Pontiacs, Chryslers and Buicks, driven by Iraq's burgeoning middle classes. The horse-driven tram from Kazimiya to Baghdad had been replaced by the motorcar. Between Rashid and Ghazi Streets, Queen Alia Street had opened as a new commercial area, where fashionable young women mingled with the black *abayas* on the walkways. There was a new international railway station, a large cubist building with a high-domed central hallway and tall pillars that looked like a modern brick temple.

Alongside the innovations in architecture and town planning, intellectual and cultural activity was flourishing along the banks of the

Tigris. The new literary and artistic scene in Baghdad was pushing the limits of experimentation.

Iraqi culture had always been heavily reliant on the word; indeed, the first system of writing in the world had been devised by the Sumerians in southern Iraq 5,500 years earlier, and the oldest known epic in the world is the Mesopotamian story of Gilgamesh. Baghdad was in many ways a city dizzy with the glory of the word, with a rich poetic tradition. Whether written or spoken, from the golden-age verse of Haroun Rashid to the modernistic free verse of the contemporary poetess Nazik al-Malaika, poetry was part of the lifeblood of the people.

Besides poetry, there was a long oral tradition which manifested itself in the gossip of the cafés and literary salons, through the cries of the Ottoman drummers, the rhetoric of the mullahs, the lamentations of 'Ashura, lyrical *maqams* and love songs. Words were the key means

Ahmad (in hat) with, from left, his nephews Ghazi, Ali and Mahdi. Standing behind are Jassim Muhammad (Hadi's guard's son), Ni'mati's son Muhammad and Rijab.

through which people sought to express the complexity and variety of life in Iraq. A famous Arabic saying is: 'Cairo wrote, Beirut published and Baghdad read.' A great deal of translation was taking place from Russian, French and English into Arabic. For the first time many of the great European classics were made accessible to university students in search of the novel and subversive.

Even the younger generation, Ahmad and his eldest nephews Ghazi, Mahdi and Ali, enjoyed the literary flowering, becoming regulars at the Coronet bookshop and at McKenzie's on Rashid Street, where comic books captured their imagination – *Plastic Man*, *Superman*, *Batman*, *Archie*, *Donald Duck*, *The Adventures of Tintin* among others – as well as *Life Magazine*, *Time* and the *Saturday Evening Post*. The power of the word would soon also come for them in the form of the 45 rpm records that would begin to become available in Baghdad. The boys would acquire an extensive repertoire of pop trivia knowledge during their summers in Lebanon, where rock and roll was shortly to arrive in a big way. Together with Ahmad and his cousin Leila, Ghazi would scour the bookshops and the few record shops in Baghdad for the latest hits. His father had acquired a Telefunken gramophone with huge ivory knobs which Ghazi would turn up to high volume to listen to the songs of Paul Anka, Buddy Holly, Bill Haley and the Comets, and most importantly Elvis Presley, whose transliterated name 'Elbis Bridgley' meant 'I wear my shoes' in Iraqi. Universal favourites were Mambo Italiano and Shish Kebab.

Images were also important. In 1951 the sculptor and painter Jawad Salim had established the Baghdad Modern Art Group. Considered the father of modern Iraqi art, Salim had been educated in France and Italy, and later in England under Henry Moore. Inspired by Picasso, he revisited Iraq's diverse and ancient heritage in order to develop a fresh language with which to express the nation's artistic identity. His work became a continuous exploration of traditional motifs married to new forms, materials, colours and landscapes – a new art for a new country.

Baghdad Radio, the Arabic BBC which broadcast from Cyprus, had a large following. But one of Abdul Nasser's more effective exports from Egypt to Iraq was the radio station *Sawt al-Arab*, 'The Voice of the Arabs', launched in 1953, which devoted two programmes a day – one in the morning and one in the evening – to attacking the Iraqi monarchy and regime. These quickly became the most popular radio shows in the country, and *Sawt al-Arab* the most popular station, railing daily against King Faisal II and Nuri Pasha, inciting Iraqis to kill them and 'overthrow the colonial yoke of dictatorship'. Nuri's name became synonymous with 'British agent', while the King was portrayed as a meek British pet.

Expressing the anger and defiance of the masses, Nasser stood up to the French, the British and the Israelis by claiming back what was 'legitimately' Egypt's. The impact of his broadcasts became clear in 1953 when protests erupted in Baghdad against a backdrop of inflation and the government's continued pro-Western stance. The army had to be called in; however, many of the soldiers appeared to be sympathetic to the demonstrators' cause. The government succeeded in regaining control of the city, but as the protests had taken place so soon after Egypt's *coup d'état*, they sent shivers up many spines.

'Turn it off, turn it off!' Bibi would cry to Hassan whenever Nasser's voice came over the radio. 'Lies, lies, lies. He's just a rabble-rouser.'

Bibi's anti-Arabist views were shared by many other Iraqis. Although her privileged position inevitably helped to shape her political opinions, her stance was rooted in her sense of identity as a member of a religious minority. Many of the Shi'a viewed talk of Arab nationalism as thinly veiled Sunni propaganda aimed at controlling them, in much the same way that Kurds viewed it as Arab propaganda intended to subjugate them.

Hassan shared Bibi's dislike of Nasser. 'Egypt was such a beautiful place. No one can speak or criticize the establishment there now,' he said one day while he was sitting with her and Rushdi in the *andaroun*. With regret and nostalgia, he told them that he had heard the university's name in Cairo had even been changed in order to erase any trace of King Farouk. He felt that the country had become cut off from the world, all in the name of nationalism. All the same, he did wonder if Nuri Pasha had taken his Communist vendetta too far; he couldn't see why Iraq had to be at the forefront of the global battle against Communism when there were more important local matters to be dealt with.

Rushdi disagreed with him, arguing that Nuri's policies and the plans of the Development Board demonstrated the progress that was being made in Iraq. 'You can't build a new country overnight,' he insisted. Hassan countered that reform and social development had to go hand in hand. When Rushdi reminded him that half of the Iraqi Communists who supported Stalin and Nasser owed their education to the government, Hassan said that the new Ba'athists, with their talk of Arab rebirth and revolution, bothered him far more than the Communists.

Rushdi remained adamant that Iraq was leading the region in democratic progress. He argued that the NDP's opposition newspaper, *al-Ahali,* wouldn't be published every day if the government was as oppressive as Hassan claimed. Hassan retorted: 'Do you think what happened in Egypt can't happen here? Nuri might just save the day yet again, but he has to heed public opinion, even if it's at the expense of his foreign policy. I wish I could tell him myself.'

Bibi was a supporter of Nuri Pasha al-Said. She had often heard Hadi talk about his political insight and his achievements in the halls of government, but she had come to know him in a less formal setting. She had visited his home many times, playing cards there with his wife and other friends. If they played late into the evening, Nuri would often come in from work and chat with them. Bibi had developed a rapport with him that reminded her of the banter she had once enjoyed with her late father-in-law, Abdul Hussein, although she wasn't about to discuss politics with him.

Like the rest of her family, Bibi was terrified of Nasser's influence on the Iraqi streets, and feared that it would bring about a coup in Baghdad. Hadi decided that it would be prudent to send some of his money abroad, and worldly-wise Rushdi suggested that he transfer it in his name. Hadi also bought a building in Beirut as an investment, and started building a villa in the mountain resort of Bhamdoun.

As far as Hadi was concerned, the opinions of his sons carried no weight if they lacked practical application. While it may not have been in his power to change the political system, he poured as much energy and money as he could into private projects with a view to improving people's lives, creating jobs and homes for them.

26

Defiance

A Crisis and a Key
(1956)

ALTHOUGH NURI PASHA AL-SAID was not continuously in office as
Prime Minister during the 1950s, he came to represent the official face
of Iraqi politics for many. Despite his background as a privileged
member of the Sunni elite and a supporter of pan-Arabism, Nuri understood that Iraq's complexity and ethnic composition set it apart from
the rest of the predominantly Sunni Arab world. (After all, a third of
the Iraqi population were Kurds, who had continued their campaign
for autonomy ever since the creation of Iraq in 1921.) He also realized
that the Shi'a needed to be granted their share of political power. A
generation of educated and able men had already emerged among them
and served in the highest posts, including Prime Ministers such as Saleh
Jabr and Fadhil Jamali.

To achieve his goal and to secure protection from the growing
Communist threat from the Soviet Union, Nuri turned to Iraq's non-
Arab neighbours, first to Turkey, then Iran. Turkey was a member of
NATO and enjoyed a close relationship with the US, and Nuri made
positive inroads with the country's charismatic Prime Minister Adnan
Menderes, who had won Turkey's first free election in 1950. Together
they agreed on a joint cooperative aggression pact in the event of
attack.

Signed in 1955, the agreement became known as the Baghdad Pact.
In due course, with full British support but the more reserved blessing
of America, membership of the Pact was expanded to include Iran and

Pakistan, as well as Britain. The creation of a powerful political and economic bloc, which included Arabs and non-Arabs as well as Muslims of both sects, suggested a new way forward for the region and potentially a dynamic new political order, whose *raison d'être* was not solely defined by opposition to the West. The Central Treaty Organization (CENTO, as it became known) did not have a unified military command structure like NATO; however, it was connected with NATO via Turkey's and Britain's memberships, and to SEATO through Pakistan.

The Pact boosted morale amongst the political elite in Iraq, many of whom welcomed the rapprochement with Turkey and felt that the Pact offered protection from Nasser's continuing aggression. However, these sentiments did not trickle down to the people, who remained vehemently pro-Arab, and continued to identify strongly with Nasser. Moreover, some sections of the Iraqi population perceived the Pact as a last-ditch attempt by Britain to continue to exert its influence over the region.

As part of Menderes's liberalization policy, several delegations of Iraqi government officials were invited to Turkey. The Speaker of the

Hadi's reception in Turkey, while on an official visit as acting head of the Iraqi Senate, after the signing of the Baghdad Pact in 1955.

Parliament and Hadi, who was then the acting President of the Senate, led one group of ministers on a tour of Turkey's new industrial and commercial projects. Bibi was very disappointed that she could not join them, as she would have loved to have been fêted with official pomp and style. However, she was later relieved to have been spared an experience that nearly killed her husband and her son-in-law, Raifa's husband Abdul Amir. The delegation were on a flight from Ankara to some textile factories in Adana when their plane was caught in a violent storm that nearly caused it to crash. Menderes welcomed the rattled Iraqis back at Ankara, and told them they would only travel on trains from then on, or he would be accused of trying to get them assassinated.

The Suez Crisis of 1956 inflamed the political tensions that were already smouldering within the region. When the United States and Britain refused to finance his mammoth Aswân Dam project in Egypt, in retaliation Nasser declared the nationalization of the Suez Canal and the Suez Canal Company. Anthony Eden, the British Prime Minister, was outraged by this attack on a vital British economic asset. A few months later, having failed to reach an agreement directly with Nasser or through the UN, Britain attacked the Canal with the support of France and Israel.

American President Dwight D. Eisenhower condemned this action and demanded a ceasefire from the attacking forces. Humiliated, the French, British and Israelis retreated. The public showdown forced Eden's resignation, and represented a crushing demonstration of the decline of Britain's power.

Despite Nuri's support of Egypt in the crisis – he even broke off diplomatic relations with France – the outcome had serious ramifications domestically. In the eyes of his Iraqi detractors, Nuri was still seen to be in a security alliance with Britain and America, although it was not formulated as a proper treaty. There were demonstrations throughout the country, including in the shrine city of Najaf, which suggested that even the religious establishment was beginning to stir. As leading Shi'a figures, Hadi and his fellow politician Abdul Wahab Murjan went to the city to pacify the situation. Throughout his political career, Hadi had paid close attention to the Shi'a religious establishment, and he continued to be relied upon by the government and monarchy to help smooth the way for their own dealings with the clerics.

During his time in office, Rushdi attended very carefully to the rumours of growing discontent within the army; some units held strong Arab nationalist sentiments, were pro-Nasser and vehemently anti-monarchy. One small group known as the Free Officers had been formed by a man called Hajji Sirri, who belonged to the officers' engineering corps. They had begun to agitate for a mutiny similar to that which had propelled Nasser to power in Egypt, although by all accounts they weren't taken very seriously.

Rushdi had witnessed the polarization between Nuri and Crown Prince Abdul Ilah, with both men and their associates incessantly working to undermine each other. He wondered if they were aware of the potential danger, and the subject began to torment him. At a family lunch one day he was so emphatic about the threat posed by the military and their intense dislike – if not outright hatred – of the Crown

From left, King Faisal II, Crown Prince Abdul Ilah, Hadi and
Rushdi in the Sif garden, late 1950s.

Prince that his father decided to pay Abdul Ilah an informal visit at Qasr al-Rihab Palace.

Hadi relayed to the Crown Prince the rumours and his own opinions of them. As he was preparing to take his leave, he hinted that perhaps His Royal Highness might consider making some provisions in case of a possible attempted coup. The Prince laughed and jovially replied, 'Don't you worry, old fellow! There's nothing in the least to be concerned about – at least 60 per cent of the military is behind me, or I'd be the first to pack my bags and hotfoot it!' He laid his hand on Hadi's shoulder as he escorted him to the door. 'That said, never can be too careful. Always a good idea to have a bit put by overseas, just in case. Could you or Rushdi fix that, d'you reckon?'

A few days later, a soldier arrived at the house and presented Hadi with a large metal key. He was followed shortly by another soldier, who delivered a wooden box. Inside it was the equivalent of £100,000, for Hadi to deposit offshore.

27

Revolution

Slaughter of a Family

(1958)

'INQILAB, INQILAB, wake up, *inqilab*!' Jawad cried.

'What?'

'Wake up, get up! Come on!'

'What? Who died?'

'There's been a coup. The army's stormed the Palace.'

Bibi was instantly shaken from her deep sleep into a state of absolute terror.

'Where's Rushdi?'

'He's at home.'

Bibi knew that, as Ministers for the Economy and Health respectively, Rushdi and his brother-in-law Abdul Amir were meant to be going to the airport to see off King Faisal II. The King was travelling to Istanbul to meet the Muslim leaders of the Baghdad Pact and also to formalize his engagement to Fazileh, a princess of the Ottoman royal family. Rushdi had just returned from London where, at Nuri's behest, he had led a delegation to negotiate an increase in Iraq's oil share from the Iraqi Petroleum Company to 80 per cent. Hassan's immediate thoughts were of his father, who was out of the country on official business as head of the Senate. He was relieved by Hadi's absence, as it meant there was no immediate concern for his physical safety.

The phone soon started to ring. The first to call was Thamina, startled by the sound of bullets raining over her house from the opposite bank of the river. Nuri Pasha al-Said's house was two houses down from hers. Across the river lay Qasr al-Rihab, the Royal Palace. Having spent the hot night on the roof, Thamina and her family had been woken by the shots and had come down into the house.

Worried about her friend Princess Badiya, the Crown Prince's sister, who lived near the Palace, Thamina asked her husband, Saleh Bassam, to phone her house. When he got through, Princess Badiya's husband, Sharif Hussein, told them they were worried about their children, and wanted help in taking them to a safer place. Saleh Bassam froze. Thamina took the receiver from his hand and asked the Sharif to tell her what was going on. They had barely exchanged any words before it was clear to Thamina that they had to help rescue the Princess's children. She ran into her dressing room and threw on a light summer dress.

Her children's nanny, Fahima, was preparing milk for Thamina's youngest, still a baby. Fahima was a trustworthy Chaldean woman who had been with Thamina for over ten years and was devoted to her children. Thamina told her that she had to take the children straight away to their paternal uncle Sadiq's house down the road, where they would be safe. There was no time for talk.

Next, she put out her husband's clothes for him. Snapping out of his fear, Saleh rushed to get dressed. Then Thamina remembered that

Princess Badiya (centre), King Faisal's aunt and the only surviving member of the royal family, flanked by Thamina and Raifa in England, 1961.

their eldest son, Mahdi, was sleeping over at her sister Raifa's house. She called Raifa and told her that she was sending their driver to pick him up.

She and her husband drove in the direction of Princess Badiya's house, which lay in the Mansour district on the other side of the river. The city already looked and smelt different; there was a heaviness in the air. Except for a few fishermen, who drifted on the glittering water in their little boats, seemingly impervious to life beyond the river's cycles, the Tigris was empty.

As Thamina and Saleh Bassam approached a roundabout they were confronted by a huge mob of shouting men. It was clear that most of them were from the *sarifas*, the shanty town that lay to the east of the city. Migrants from the south, these men lived in makeshift reed huts without water or electricity; they were poor and angry; their legitimate grievances had been channelled into frenzied action by pro-Nasser slogans and speeches that preached class hatred, and anti-monarchy and anti-British sentiments. A cry went up among them: *Allahu akbar, allahu akbar. Al mawt lil malik, al mawt li 'Abdul Ilah!* 'Death to the King, Death to Abdul Ilah!' There were soldiers from the mutinous army units amongst the mob. They were still wearing their khaki uniforms. Although many of them weren't wearing their black berets, they brandished their rifles in their hands.

Thamina saw a body being dragged along the street by a rope. One of its arms was missing. She looked again. It was Abdul Ilah, the former Regent and Crown Prince. She was sure of it.

Suddenly, she found herself circled by men thumping on the car's roof and windows, jumping on the bonnet. Faces pressed up against the glass, shouting, spitting, blocking out the light. She and her husband were trapped. Cowering with her arms over her head, she was afraid the windshield would shatter over her.

Through the car's open windows, dark arms started to reach in, catching hold of her and pulling her out. Tugging at her sleeves, the men tore her dress, exposing her chest. Although she was desperate for air, all Thamina could think about was covering her body with her free hand. She cried out for her husband, who was still in the car.

Someone started screaming: 'This is the daughter of Abdul Hadi Chalabi the traitor! This is his daughter!'

As she struggled to cover herself, Thamina heard a shout above the other voices: 'Leave them alone! Move away!' A man grabbed her, then opened the car door, pushed Saleh Bassam out of the driver's seat and took the steering wheel. With Thamina beside him, he reversed and drove away from the crowd, back towards the river.

As they drove through empty streets, the man explained that although he was dressed in civilian clothing, he was an army officer. Wiping the sweat from his brow, he explained that he had recognized Dr Saleh Bassam through the windscreen. The doctor had treated his wife a few years earlier, saving her pregnancy, and he had not forgotten.

A random twist of fate had spared Thamina and her husband from death that morning.

In A'zamiya, the family was in uproar. As Hadi's house overlooked the main road and was relatively vulnerable to attack, the family had decamped to Jawad's home, on a quieter stretch of Taha Street. Within the house, news and hearsay were rapidly exchanged as hysteria mounted. The army had broken into the Royal Palace; the cabinet ministers would surely be their next target.

The family's fears were confirmed by a radio broadcast in which Abdul Salam Arif, the officer in charge of the army's 3rd Division and a staunch supporter of Nasser, announced that all ministers were to be shot.

Ear-splitting martial music followed this broadcast, punctuated by lurid declarations of war and death. Bibi listened, her throat dry, as a popular Egyptian song was played again and again:

> Allahu akbar fawqa kaydi al mu'tadi,
> Wallahi lil mathloum khairu mu'ayidi ...
> ana bil yakin wa bil silah sawfa aftadi.
> – ya arab qulu, qulu ma'ai
> allahu allahu allahu akbar,
> allahau fawqa kul mu'tadi.

> God is Greater above every aggressor.
> To the mistreated God is a support ...
> I will fight with arms and the truth.

Oh Arabs, say with me, say with me,
God is, God is above every aggressor.

Mahdi was on his way home from his aunt Raifa's house. He hadn't
wanted to wait for his father's car to arrive, so Raifa had put him in a
car with one of her husband's drivers. It was a mistake.

At the corner of Rashid Street, Mahdi saw an enormous crowd of
uniformed soldiers, traditionally-robed men and women in *abayas*.
The driver stopped the car and got out to watch. The mob were dragging a corpse. Mahdi recognized the Prince's battered face. The driver
cheered. People were swarming over Abdul Ilah's body, hacking at it
with knives. One man bit off a piece of flesh and tossed it to the crowd.
The look of frenzied delight on their faces was unlike anything Mahdi
had ever seen.

Some of the mob broke away from the rest, and were heading
towards the car. The driver called out: 'There is a Chalabi boy here! I
have a Chalabi boy here, do you want him?'

Seeing the knives shining in the mob's hands, Mahdi quickly slipped
from view, then carefully opened the car door and crawled out in the
opposite direction, towards the bridge. No one noticed him. By the
time the group had arrived at the car there was no Chalabi boy to be
had. Angry, they turned on the driver and started to beat him for lying
to them; he shook them off and ran away.

The Salhiya Bridge had disappeared beneath the sea of people sur-
ging across it. Mahdi ran down to the riverbank and jumped into the
water, allowing the current to carry him south, away from hell.

Nuri Pasha al-Said was in a deep sleep when the baking woman
knocked on his bedroom door. She came to the house early every morning to bake fresh bread, which he enjoyed for breakfast, but this time
she was not alone.

Under orders from the leaders of the coup, Nuri's former assistant
had come to arrest Nuri in person. He convinced his unit to wait
outside while he delivered the news, thinking that in this way he would
be able to safeguard himself against the Pasha's reprisals in the event
of the coup's failure. He ordered the frightened baker woman to wake
her master up. Breathlessly, she knocked on Nuri's bedroom door,

woke him and told him what had happened. Nuri immediately took two pistols from his bedside cabinet, told the woman to try to delay the soldiers, then, still dressed in his pyjamas and slippers, slipped out of the house to the levee below.

At the waterside, Nuri frantically waved a nearby fisherman over, jumped into the boat and asked the man to take him to the opposite bank as quickly as possible. The fisherman was a humble man, and it didn't occur to him to refuse Nuri's request. Nuri lay flat on his stomach, pressed against the bottom of the boat and hidden under a thick net. Just before they reached the opposite shore, they heard the sounds of a radio coming from someone's house. The voice of Abdul Salam Arif announced the new regime's Proclamation No. 1:

> Noble people of Iraq: trusting in God and with the aid of the loyal sons of the people and the national armed forces, we have undertaken to liberate our beloved homeland from the corrupt crew that imperialism installed ...
>
> Brethren, the army is of you and for you and has carried out what you desired ... Your duty is to support it ... in the wrath that is pouring on the Rihab Palace and the house of Nuri Said ... We appeal to you, therefore, to report to the authorities all offenders, traitors, and corrupt people so that they can be uprooted, and we ask you to be one hand in purging those individuals and eliminating their evil.
>
> We have taken an oath to sacrifice our blood and everything we hold dear for your sake.

Panic-stricken, Nuri asked the fisherman to turn the boat around. He couldn't risk walking through the streets; he was a wanted man. Unsure of where to go next, he remembered that his neighbour on the riverfront, Thamina, was the daughter of Abdul Hadi Chalabi, his friend and ally. He told the fisherman to row towards her house.

When he arrived, he found the house empty apart from a couple of servants. He stole into the drawing room and waited anxiously for Thamina. Before long she entered, alone and dishevelled in her torn dress. Her surprise at seeing him was compounded by his strange appearance, in pyjamas and with his upper set of false teeth missing.

Nuri's first question was whether King Faisal II was still alive. Although Thamina could not answer this, she told him that she was certain Abdul Ilah was dead; she had seen his corpse being dragged through the streets.

After a pause, Nuri attempted to reassure her: 'The members of the Baghdad Pact will retaliate against this outrage quickly; our allies Turkey and Iran won't stay on the sidelines.'

Thamina nodded numbly at him. 'I need to change my clothes. Here –' she switched the radio on – 'Maybe we can find out what's happening.'

When Mahdi arrived home, dripping wet from crossing the river, he found Nuri and Thamina listening to the Egyptian revolutionary songs that blared from the *Sawt al-Arab* radio station. His mother threw her arms around him, hugging him tight as he shivered in her arms.

Nuri decided that he would leave Baghdad for Kazimiya, where he could hide with members of Bibi's family until it was safe for him to cross the border into Iran or Jordan, where he hoped to obtain support to quash the coup. It was decided that Murtada Bassam, Thamina's brother-in-law, and Fahima, her children's nanny, would take him there. And so, hidden in the back of the family car, Nuri left the burning city behind him.

Once he had left, Thamina suddenly became afraid of her own servants. The radio broadcasters had started to announce that there was a 10,000-dinar prize for anyone who delivered Nuri Said to the revolutionaries.

They turned off the radio, but the images on television were even more harrowing. There was footage of the bloodthirsty mob dragging the corpse of Abdul Ilah to the gates of the Ministry of the Interior, where they raised up what was left of him, as if he were a cow's carcass. A butcher started to fillet his body, throwing pieces of it towards the people. Raifa, who had been hiding at her neighbours' house that afternoon, watched in horror as the television showed an old woman clad in an *abaya* carrying a bag filled with meat. She gleefully told the camera: 'Extra special meat for sale! It's the most delicious meat in the world – prime cuts of Abdul Ilah, the Crown Prince and former Prince Regent!'

Later that afternoon, Jawad took Rushdi and his brother-in-law Abdul Amir Allawi and their fellow politician Dr Fadhil Jamali to hide in the Sheikh Jamil farm, which lay three quarters of an hour's drive north-west of Baghdad, near the Tigris River. The farm belonged to

Hadi, and they hoped it would be safe. Bibi took Hassan and Talal, Raifa and her children, Rushdi's wife Ilham and her children to the nearby home of their old family friend Sadiq Istrabadi, whose release from jail Hadi had helped to secure in 1935. Saeeda was taken to Kazimiya. She looked the most confused of them all as she collected a few of her belongings to take with her. When she said goodbye to Ahmad, she squeezed him tightly to her and whispered some prayers in his ear. Nothing was said of Nuri Said, although they were all convinced that he must have made his way to one of the friendly tribes around Baghdad, from where he would escape from the country.

Everyone was waiting for some form of intervention, be it divine or from abroad. The expectation was that the British would come to the aid of their allies in Iraq, or – as King Hussein was a cousin of King Faisal II – would use Jordan as a means of offering support. The men of the revolution stood for everything that Iraq's foreign allies – Britain, the US, Turkey and Iran – were supposedly against. Surely someone would do something to stop this?

The 19th and 20th brigades of the 3rd Armoured Division of the Iraqi Army, the Free Officers that few had paid attention to earlier, were responsible for the coup of 14 July 1958. Their leaders were the pro-Communist Abdul Karim Qassim and the pro-Nasser nationalist Abdul Salam Arif.

Hassan later learned that it must have been 5 or 6 a.m. when the 3rd Battalion of the 20th Infantry Brigade stormed the Royal Palace. Three battalions of the brigade had already broken camp that night and were en route to Jordan from their headquarters in the west. Abdul Salam Arif, one of the battalion commanders, succeeded in avoiding suspicion when he ordered his battalion and two others to proceed to Baghdad instead. There they dispersed and seized the main strategic bases such as the Ministry of Defence and the television and radio stations.

Whatever resistance the guards at the main gate of the Royal Palace put up was swiftly overwhelmed; they were completely outnumbered. Thabet Yunis, Prince Abdul Ilah's personal guard, informed him that there had been a coup, and the palace was surrounded. He gave the Crown Prince the impression that the royal family were to be taken to some other location.

Princess Badiya, the King's aunt, had heard the sounds of fighting from her house a few streets away, and rang the palace to find out what was happening. Twenty-three-year-old King Faisal II spoke to her calmly and told her to listen to the radio.

As the family came down the stairs they could hear loud voices outside. The first to step outside into the pale dawn air was Queen Nafisa, the King's grandmother on his mother's side. She held a Quran in her hand. Some officers were standing by the fountain in the court-yard in front of the house. Queen Nafisa walked over to them and asked them to swear on the Holy Book that no harm would befall her grandson and her cousin. They did so.

The family lined up in a semi-circle in front of the fountain: the Crown Prince, Faisal, two orphaned girls who had been adopted by the family, Princess Abdiya and Queen Nafisa. The Crown Prince's wife, Hiyam, stood slightly behind them, and his guard, Thabet, next to him. Silent and solemn, they waited to learn their fate.

Abdul Sattar 'Abussi, an officer in the battalion, walked up to the Crown Prince and started to abuse him. Thabet interrupted, crying out that he had no shame to be using such language. 'Abussi raised his gun and shot Thabet dead in mid-sentence. Abdul Ilah started scream-ing in horror. Another shot was heard nearby. 'Abussi turned round towards the noise, then swung back to face the family and started firing, empting his machine gun into their bodies.

Faisal, the young King, didn't die instantly. After he fell, he asked for some water. 'Abussi stepped over to him and fired more shots into his head, which split in two, spilling his brains onto the ground. Queen Nafisa held out the Quran and called to the soldiers, 'May God curse you!' as they shot her down. Several members of the palace staff were also killed. The water in the small fountain ran red as the bodies lay like slaughtered sheep on the ground.

'Abussi, who committed suicide a few years later, would say that he hadn't wanted to let any members of the royal family live, in case they came back to power again, as they had done after the 1941 pro-Nazi coup. The only person to survive the massacre was Hiyam, the Crown Prince's wife. She fainted, and was rescued from the pile of corpses by a soldier who pretended she was a palace maid and his cousin. He helped to smuggle her to her tribe, the Rabi'a near Mosul.

The officers argued among themselves about whether they should bring the bodies of both Abdul Ilah and the King out into the streets. They decided to leave Faisal behind, for fear of turning the public against them, because he was well-liked. They tied a rope to Abdul Ilah's body and proceeded to drag it around the capital, followed by an ever-growing, insatiable mob.

Early in the morning of 15 July, the day after the coup, Nuri Pasha was still hiding in Kazimiya. He had learned that his only child, Sabah Said, the Director of Civil Aviation, had been killed. His family was away for the summer, so Sabah Said had spent the night at his club. No sooner had the mob started to rampage through the streets than a member of staff at the club had told them where Nuri's son was staying. Part of the mob broke off and stormed the building, dragged him out onto the street and hacked him to pieces.

Nuri knew he was not safe. Suspecting that his whereabouts were already known, he asked to be taken to Bibi Istrabadi's house in Baghdad. Bibi Istrabadi was Bibi's friend from childhood; her father, the Pivot, had brought Saeeda to Iraq. Her family was not involved in politics, and Nuri thought it unlikely that they would be suspected of harbouring him.

The Chalabi family had slept terribly that night. Bibi and Ahmad had been given the main bedroom in the house of Sadiq Istrabadi, an old friend and former Mayor of Kazimiya, who had been arrested with Hadi back in 1935. Bibi had lain awake, worrying about what had happened to Rushdi. Early in the afternoon her dear friend Bibi Istrabadi, who was a relative of their host, ran into the house still wearing her nightgown, over which she had hurriedly thrown her *abaya*. She was very distressed.

Before Bibi had time to greet her and commiserate about the catastrophe that had befallen them, Bibi Istrabadi started to beat her chest. She told Bibi that Nuri Pasha had come to her home, seeking shelter, and that she had taken him to her stepdaughter's house, where she thought he would be safer. Now she didn't know what to do.

Bibi thought carefully. 'May God help us. He must leave Baghdad. We'll have to wait for my son Jawad to come back. Nuri Pasha must go to Abdul Razzaq Ali Salman of the Duleim tribe in Anbar – they're our friends, and they'll smuggle him to Jordan. It's the only way.'

As Bibi Istrabadi left, Bibi called out to her: '*Fiman illah*, go in the safety of God.'

While Bibi Istrabadi had been out of the house, Nuri Pasha had taken a short nap. He had been on the run for over twenty-four hours. He was exhausted.

Before sleeping, he made sure to count the number of people in the house. When he woke up after Bibi Istrabadi's return, he immediately noticed that one person was missing. He became anxious, and told Bibi Istrabadi that he ought to leave straight away. Unable to wait until Jawad had returned, she decided to take him to the Uraybi house after all.

She and Nuri put *abayas* on and walked out of the front door. They had barely taken a step when they saw that the house was entirely surrounded by soldiers, their guns pointed at them. One of the officers shouted out to Nuri that his son was dead, and now it was his turn. Standing behind Bibi Istrabadi, Nuri cocked his gun, ready to shoot himself. Suddenly an officer fired. It was all over.

Although Nuri Said was buried, his body was dug up two days later and dragged through the streets by the mob. He was hacked to pieces as the baying crowd spat on his remains.

The person Nuri had noticed was missing from Bibi Istrabadi's house was Omar Jaafar, an in-law of Bibi Istrabadi's stepdaughter. He had betrayed Nuri's whereabouts and collected the 10,000 dinars.

Bibi collapsed when she heard the news of Nuri's death on the radio a few hours later. She knew that her dear friend had been killed with him. Bibi was inconsolable. 'How could they kill a woman like that?' she sobbed.

It was clear that the Chalabi family could not stay at Sadiq Istrabadi's house any longer: they had become refugees in their own city. They decided that they would join Rushdi at the Sheikh Jamil farm. At dawn the next morning the cars set out, carrying Bibi, Hassan and Jamila, Talal and Ahmad, Hadi's sister Shamsa and her son Issam, Rushdi's wife Ilham and their three children.

Taking the Kazimiya road towards Samarra in the north, they passed many farms. Whenever they spotted a farmer or shepherd they were filled with panic, wondering whether it was obvious that they were on the run. Bibi was terrified whenever they approached one of the palm orchards by the roadside, imagining that soldiers would ambush them.

After an hour, however, they reached the farmhouse. From the outside the building resembled a fort, and its interior was equally austere. It was sparsely furnished, with a long hall that extended right through the house to the garden and the river at the back.

The immediate task was to find water and food for everyone. Rushdi's children were tired; his youngest, Muhammad, was only two years old. While the women attempted to rustle up something to eat, Jawad went to check on Rushdi and Fadhil Jamali, who were hiding in the pump houses, three small wooden shacks that were just about big enough for a man to fit inside.

About twenty minutes later, Ahmad heard the rumble of engines and looked out of the front door to see three armoured trucks approaching the house. In them were perhaps sixty soldiers, fully dressed for battle, wearing helmets and with fixed bayonets on their rifles. Once the trucks had come to a standstill, the commanding officer shouted out orders and the soldiers took up positions around the house, setting up machine-gun tripods.

The commanding officer, a gruff man, entered the house. His hand lay on his Webley revolver, tied by a cord to his belt. He stared at Bibi, Hassan, Talal, Ahmad, Shamsa and her son Issam, then turned to Talal and snarled at him: 'Are you Rushdi?'

'No.'

'Where is Abdul Hadi Chalabi?'

'He's out of the country.'

'Where is the Minister for Reconstruction, Dhia Jaafar?'

'Out of the country.'

The officer turned to Bibi, took out his revolver and pressed it hard against her chest. Ahmad tried to step in front of his mother, but the officer pushed him back with one hand and demanded that Bibi tell him Rushdi's whereabouts.

'If I knew where he was I'd tell you, to save my three other sons in this room,' Bibi replied, her voice shaking.

Unsatisfied, the officer turned to Hassan, but snorted in disgust when he realized he was blind. Ahmad screamed at him to leave Hassan alone, and that he would go with him.

Both Ahmad and Talal walked out of the farmhouse ahead of the officer, who ordered them into the front of one of the trucks. They bumped along the unpaved track to the pump houses, where the officer ordered Talal and Ahmad to get out. Pointing the gun at their backs, he commanded, 'Take me to your brother now.'

Talal knew the fugitives were hiding in the pump houses, but he didn't know which. He couldn't focus; his throat was burning and he had to drink something. He bent down to scoop up a handful of water from the runnel that led to the pumps, oblivious to the worms swimming in it. He looked at his younger brother, who looked back at him. What if he pointed to the shack where his brother was hiding and caused his death?

'Come on, what're you waiting for?' shouted the officer. 'Move it!'

Talal walked slowly towards the pump houses, unsure of what to do. Impatient, the officer overtook him, went to the middle pump house and swung open its door. No one was in there. He went to the right-hand one; again, no one was there. Turning sharply to face the boys, the officer ordered them back to the house.

If he had opened the door to the left-hand pump house, he would have discovered Dr Fadhil Jamali. Unknown to the brothers, Rushdi had left the pump houses earlier and gone to the other side of the farm, looking for a way for them to flee to Jordan through the desert.

Both Rushdi and Fadhil Jamali were dressed in Bedouin clothes, long *dishdashas* robes and *zuboun* jackets. A farmer in the neighbourhood had spotted Jamali early that morning, but had then heard a report of his death on the radio. It had been read out by Henry Cabot Lodge, the US Ambassador to the United Nations, who decried the savage behaviour of the coup leaders. The farmer decided to inform the authorities, but they had accepted the news that Jamali was dead.

Back at the house, Ahmad and Talal's cousin Issam invited the officer and some other soldiers to join them for lunch. Sitting around the edge of the room, the soldiers ate greedily. In between mouthfuls, the officer turned to one of them and said: 'They're not so bad after all, these people.'

After lunch, Bibi got up to go to the toilet, which was situated down the hallway. Ahmad followed her, and stood guard outside the rickety door. The officer approached, shoved him aside and made to push the door open. Gripping the man's wrist, Ahmad stammered, 'She's a woman, it's … it's impolite!' With a mock threatening gesture, raising his free hand as if to slap him, the officer scowled at Ahmad, but then turned away, saying that they would not leave without Rushdi. Behind the door, Bibi thought she might be about to faint.

A few moments later, Rushdi walked calmly into the house with Jawad. Although they had initially been unaware of the soldiers' presence on the farm, as they were approaching the house the situation had become clear to them, but by then it had been too late to do anything about it.

Rushdi barely had time to kiss everyone goodbye. Bibi and his wife Ilham were in tears. A few words were exchanged; Bibi told him to take care of himself, and as he left, called out to him, '*Allah wiyak aini*' – God be with you, my darling.

Jawad asked for permission to follow the soldiers back to Baghdad; he wanted to make sure that his brother and Fadhil Jamali were taken to the Ministry of Defence, as all the other arrested men had been. When Rushdi was taken into the building, Jawad embraced him and repeated what Bibi had said: *Allah wiyak*. Rushdi barely reacted. It was almost as if he had abandoned himself to his fate.

From the Ministry of Defence, Rusdhi and Fadhil Jamali were taken to the Abu Ghraib prison, as all the other political prisoners arrested during the first two days of the coup had been. At Abu Ghraib, Rushdi was told in person by Brigadier Abdul Karim Qassim, the leader of the new military regime and now Prime Minister and Acting Minister of Defence, that as a member of the deposed Cabinet, he had to await justice. However, as an afterthought, he added, 'You'll be fine.' On Qassim's desk Rushdi glimpsed a dirty copy of yesterday's newspaper, with a headline announcing the news of the government's oil renegotiation in London. He felt nauseous. How much could change in a few hours.

A few hours later, another proclamation was declared by the revolutionary government: all former Cabinet Ministers and high-ranking officials under the monarchy were to be executed. The soldiers on guard duty in the prison went out of their way to taunt the arrested officials, delighted to have such eminent men at their mercy.

The same day Rushdi was arrested, a mob broke into Thamina's house to arrest Saleh Bassam for helping Nuri Said. Upon entering the house they saw Thamina's sixteen-year-old daughter Leila in the hallway. A soldier put a rifle to her chest and asked her where her father was. Leila's nose started to bleed, but the soldier kept his rifle pressed against her. The other soldiers ransacked the bedrooms, destroying the children's toys and anything else they came across. They found Saleh Bassam in one of the rooms and took him away. They also arrested his brother Murtada.

Despite the condemnation of the coup by the US representative at the UN, forty-eight hours after it had taken place there was still no sign of foreign intervention. Jordan, which had recently formed a United Arab Union with Iraq, did nothing. It was obvious that help wasn't going to come.

Whether it was as a direct result of the Suez débâcle which Britain had suffered a few years earlier, or a calculated plan to sell out their friends, it was unclear what lay behind the apparent indifference of the British. There had been a subtle shift of attitude amongst the British in Baghdad, who were clearly aware that propaganda against the monarchy had reached immense proportions within Arab circles.

The British Embassy was attacked by the mob on the day of the coup, and the residence was burned down. One British staff member was shot dead. Several other buildings were also attacked, and the statue of General Maude, like that of King Faisal I, was destroyed. Yet the only action the British government took was to lodge a protest with the new military government through its Ambassador. With few exceptions, like the Minister of Finance, Muhammad Hadid, a British-educated economist who had been a founder of the former opposition National Democratic Party, the entire Cabinet was now in uniform.

Although bound to Iraq by a treaty of friendship, the British did not intervene to stop the coup. Sam Falle, an Embassy official in Baghdad, wrote of the new regime as having 'some quite decent people known to us'. It was he who would unwittingly confirm Britain's attitude towards the new government in the week after the coup. Waving to a neighbour who happened to be Hadi's brother-in-law, Falle called out in Persian: '*Halla khob shud*' – Now things are good.

The British were not averse to finding new friends, or building upon their well-established relationship with Brigadier Qassim, who despite his closeness to the Communists promised that British interests in Iraq and its oil would not be affected by the change in regime. Britain officially recognized the Qassim government two weeks after the coup, on 1 August 1958.

The extent of Britain's commitment to the Baghdad Treaty proved to be a tribute made to King Faisal II in the House of Commons by Prime Minister Harold Macmillan, followed a few days later by a memorial service at the Queen's Chapel for the dead King and his family. Faisal, Abdul Ilah and Nuri were made honorary Knights of the Grand Cross.

When the revolutionaries had seized Baghdad, Hadi had been in Tehran. There were strong cultural, social and commercial ties between Iran and Iraq: a few years earlier, one of the Shah's sisters had been considered as a bride for King Faisal II. Like the Turks, the Iranians wanted to counteract Nasser's interference in Iraq, but they were clearly waiting for US approval to act. Everyone's hands seemed to be tied.

Initial reports of events in Baghdad were inaccurate and chaotic. In the Iraqi Embassy in Tehran Hadi listened to the local radio station, expecting to hear that Iraq's friends were mobilizing to retaliate against the renegade officers. Instead, he heard declarations of support for Abdul Nasser, and worried how his family would survive. As long as the coup's leaders were intent on arresting supporters of the old government, he could not return to Iraq. Even in Tehran he was not safe: pro-coup staff at the Iraqi Embassy had alerted Baghdad to his presence there, forcing him to seek lodgings elsewhere.

As the days progressed, a covert channel of communication was opened between Hadi and his family, with the Iranian Ambassador in Baghdad, a friend of Thamina's husband Saleh Bassam, transmitting coded messages. The most important message explained that Hadi had to keep quiet lest he jeopardize Rushdi's life.

In Iraq, the entire family was under surveillance, and were banned from travelling. At the Ministry of Defence, Talal was warned that should his father be convicted of conspiring with the Shah's government against the new authorities, the Chalabi family would be eradicated to the last person.

Over 120 individuals, among them Rushdi, remained behind bars awaiting trial. The charges against them were not disclosed. In the weeks that followed the coup they were moved to the jail in Bab al-Mu'adham. They were forbidden visitors, although they were allowed letters from home, and in some cases food. Ni'mati delivered Rushdi's food to the prison, but was not allowed to see him. The family had absolutely no news of him.

In the Chalabi household, Hassan assumed the role of strategist, looking for ways to save his incarcerated older brother, and keeping everyone's morale up while his younger brother Jawad implemented their plans.

Meanwhile, Thamina still lived in terror of her servants turning on her. Several had helped themselves to her valuables, taking some silver pieces and valuable porcelain vases; they justified their thefts on the grounds that they were more deserving than the mob, and then they disappeared. A few servants remained who she was too afraid to fire, although she did not have the means to keep them either. Throughout it all, Fahima stood firmly by her side, helping her to cope amidst the chaos.

When the schools finally reopened at the end of the summer, the changes were palpable. Martial, drum-based Egyptian songs were played over and over as a background to lessons. There was an unspoken acknowledgement that something irreversible had taken place – with the street scenes of death and mutilation, sacred taboos had been broken – and the situation was balanced on a knife's edge. The killing of the royal family had emboldened many people who, regardless of their personal ideologies, embraced the new order as a means to survive. Even schoolteachers were not immune, singling out children whose families were associated with the old regime. The first thing the children had to do when they returned to school was to rip out pictures of King Faisal II from every textbook. For Ahmad and his nephews Ghazi and Mahdi, the playground became a battleground: they were 'traitors, sons of traitors'. One boy accused Ghazi's mother, Raifa, of being so decadent that she washed in milk. Another day, Mahdi, whose father Saleh was in jail, was cornered by a group of older boys and had a toenail pulled out as the group prodded him and chanted abuse.

The younger children, Ali, Hussein, Nadia and Kuku, were enrolled at Madam Adel's school because their English school, Ta'sisiyah, had been closed down temporarily owing to the revolution, as there were fears for the safety of its British teachers. At Madam Adel's the teachers taunted the children.

Even home was no longer safe. Gangs of young boys started appearing in A'zamiya, standing outside the various family houses and shouting out insults and accusations: 'You traitors, you foreign agents, go to hell!' Men whose military uniforms gave them licence to enter any house they pleased carried out searches which were designed to provoke and humiliate the family. It seemed to be almost a type of tourism for them, as they stomped through the elegant houses, hating the inhabitants even more for living in such luxurious surroundings.

Ahmad could not bear it. Although he was not yet fourteen, he attacked the soldiers when they rummaged through his family's belongings, and openly carried a large photograph of the dead King with him. He had been especially upset by the killing of Faisal II, for only a few months earlier he had escorted him on his visit to the Jesuit-run Baghdad College. When he carried his photograph with him on a train to Basra, accompanying his sister Najla to her husband's house there, soldiers yanked it out of his hand and slapped him across the face.

One morning, a month after the coup, Thamina woke to the sound of soldiers knocking on her front door. She pleaded with God to save her, but they had come for Fahima. Thamina pleaded with them to let Fahima be. What could they possibly want with a poor woman like her? she asked. She was just a nanny. Ignoring her, they took Fahima and shoved her in the back of their jeep. Unable to sleep, Thamina maintained a vigil of prayer for Fahima, begging and supplicating God to save her.

Three days later the telephone rang. It was Fahima. She was at her family home in the old town quarter of Baghdad. She had just been released from jail, and would come back to Thamina's house the next day. Too afraid to talk over the telephone, she saved her story until she saw Thamina in person.

The women hugged, and Fahima told her that the soldiers had offered her 500 dinars to tell them about her mistress's involvement in Nuri's escape. Nothing had passed her lips, even though the soldiers had roughed her up. Fahima had persuaded them that she hadn't even been at the house, as it had been her day off.

The safety of the children became an even more serious issue following the government's announcement that a new 'people's court' would be set up to try the imprisoned 'traitors and imperial agents'. Colonel Mahdawi, a cousin of Brigadier Qassim's, was appointed president of the court, although he had no judicial or legal experience.

The first trial was that of Major General Ghazi Daghistani, a high-ranking military officer loyal to the royal family. The standards of interrogation were extremely poor, and the witnesses unreliable. The court quickly turned into a platform of incompetence, providing the masses with entertainment as former figures of the regime were insulted and berated by the court's president. It was labelled a kangaroo court by the Western news agencies, which reported on the irregularity of the procedures, the erratic and verbose judge and the silencing of defence lawyers. The children of those who appeared on trial were taunted daily by their schoolmates. One boy was so traumatized by the abuse that had been hurled at his father, General Arif, during his trial, that he dropped dead from a heart attack at the age of thirteen.

Finally it was announced that the charge against Rushdi and two other ministers was standing against the 'freedom of the people'. In despair, Bibi called on all of her ancestors who were descended from the Prophet for divine intervention, although deep down she wondered whether they would heed her prayers, given what had happened to the King and his uncle. They had been from *Ahl al-Bayt* too, and they had been killed, as Hussein had been at Karbala a thousand years earlier. No one had stopped it. She could not believe the abruptness with which the coup had turned her entire life, and those of her family, upside down. They had taken a terrible fall, personally, politically, socially and economically.

Abdul Karim Qassim declared Iraq a republic and appointed himself President of a three-man Revolutionary Council, whose other members were 'Aref and Talib. Qassim was to be called *al za'im*, the leader. The constitution that had been created in 1925, along with its executive, legislative and judicial bodies, was scrapped. Martial law was declared. There was to be no parliament, senate or any other representative body, elected or otherwise.

FEBRUARY 2005, SADR CITY

I'm standing on an unkempt, sandy football pitch in the middle of Sadr City in east Baghdad. This is the reclaimed land that the city planner Doxiadis designated for public housing in the 1950s. There is no sign of public housing, not even a proper sewerage system in place. Situated in Iraq's capital and housing over two million people, this impenetrable shanty town is the largest urban village I have ever seen, with animals of all shapes and sizes roaming the unpaved streets. It is difficult to believe that I am in a country with huge oil reserves in the twenty-first century.

It's the tenth day of Ashura, and I've come to see a re-enactment of the ancient battle of Karbala. With me is my friend Abu Muhammad, an Iraqi Turkmen from Kirkuk who has worked for my father for many years. Although I never met Ni'mati, Abu Muhammad is the closest person I can imagine to him, as he is both an employee and a true companion. He is also a very devout Shi'a. Even before the ceremony begins his eyes well up with tears, in anticipation of Imam Hussein's eventual martyrdom in this morning's passion play. Abu Muhammad is one of the former regime's many victims: his brother was executed by Saddam in the early 1980s, forcing Abu Muhammad (who was a grocer at the time), his wife and five young children to flee to Iran. Like many other Iraqis, they struggled to survive, and found themselves in a shanty town called Karaj.

Today the football pitch is packed with thousands of people, all of whom live in Sadr City, many below the poverty line. Besides featuring in the news as the headquarters of the anti-American rebel Mahdi army, Sadr City was previously known as 'Saddam City', and before that as 'Revolution City'. Its inhabitants for the most part are descended from the tribesmen who migrated from Iraq's southern marshlands fifty years earlier, attracted by the bright lights of Baghdad. They are still struggling to get education, economic empowerment and political rights. As I look at the little children playing in the mountains of garbage which must have been piling up for years, I reflect for a moment on my enthusiasm as a student for Baghdad's modernizing projects of the 1950s. I can understand now why God and the martyred Imams are a welcome distraction amidst such squalor.

The play starts with the narrator's jarring voice lamenting the tragedy of Karbala over rusty loudspeakers. The main characters in the battle re-enactment ride Arabian horses that have been 'liberated' from one of Saddam's palaces. Nearby I can see their stables, converted rusting municipal buses. Glancing at Abu Muhammad beside me, I notice that tears stream down his face as the lamentation grows louder and the first victim falls on the battlefield.

I remember my uncle Hassan's story of going to the Ashura ceremony at the Kazimiya shrine as a blind boy seventy years earlier, and wonder whether the cumulative pain people felt then would measure up to that shared here today.

BOOK FOUR

Fields of
Wilderness

DECEMBER 2007

I'm sitting with a friend, Muhammad, who has recently returned to Iraq for the first time since fleeing the country after the 1991 uprising against Saddam's regime. He has been living the life of an exiled, penniless refugee in the West. He's sombre and pensive as he tells me about meeting his mother and siblings a few days earlier for the first time in fifteen years. His father died while he was away, and he feels his absence today more deeply than ever.

An intellectual, Muhammad is the son of a mullah who instilled in him a smattering of Communist ideals as well as religion, giving him Marxist literature to read alongside the staple Shi'a texts. He likes to quote Gorky, but his references are still quite Islamic. I tease him, telling him that he proselytizes even when he thinks he doesn't. He always retorts by adopting Marxist terminology, using Arabized versions of class labels, such as *al-aristoqratiya* and *al-bourgoisia*. He loves to wind me up by mentioning the 'expired epoch' in reference to Iraq pre-1958. I ask him to tell me of any other period in Iraq's modern history that was as peaceful, relatively democratic and progressive. He doesn't answer me. Instead he laments the evils of Saddam's regime and the criminal destruction of Iraq's soul. I agree. The unpaved roads across the country can be tarmacked fairly quickly, but how do you rebuild a human being, revive his spirit after such abuse and humiliation?

Like many Iraqis who welcomed Saddam's fall, Muhammad had great hopes for positive change. Now that he is back, he finds himself struggling to balance his expectations with the realities. His bewilderment extends to his own family who stayed behind, who look different to him, dress differently and sound different. This causes him grief as he wonders where he belongs.

During his exile he devoted a great deal of energy to the Iraqi opposition, hoping for the day when Saddam would be gone. Now he tells me that he doubts whether he can live in Baghdad any more. I remind him that in London he lived on the margins, having little interaction with mainstream British society. He nods, but then he tells me that he doesn't expect me to understand what he means, because I come from *al-aristoqratiya*. I tell him that I doubt exile as a personal sentiment has class consciousness. It's something deeper. It becomes another layer of oneself.

'You're very argumentative today, aren't you?' he remarks.

He turns to the subject of the current ruling elite of Iraq, *al-tabaqa al-hakima*, and their incompetence. He feels disdain for many of them – among them those who were once his companions in exile – especially those with Islamic political affiliations. Perhaps I am right, he tells me, and the 'expired epoch' really was the best period the country has known since it was created, even though everything he was taught at school told him otherwise. I laugh and tell him he should really stop struggling to make sense of this 'expired epoch' – there have been several other expired epochs since, including Saddam's, which pulled Iraq deep into a dark abyss and the current uncertainty. I tell him he should revisit his class labels, as they have expired too.

28

Lost Lands

Seeking Shelter

(1958)

HADI WATCHED THROUGH the small window as the plane circled over the green and gold English countryside. His thoughts went back to his first, terrible trip to Britain in 1930, with his dying brother Abdul Rasul. Over the following twenty-eight years he had returned to London many times, on business trips to the offices of Andrew Weir & Co., and on holiday. When he came on business he always travelled with an assistant who could translate for him, and often with his manager, Salim Tarzi. This time he was arriving as a very different kind of visitor. All he had with him was the single suitcase he had taken to Iran six weeks earlier.

For the first time, he had experienced great difficulty in obtaining a British visa. His diplomatic passport proved to be an obstacle now that he was the ex-deputy head of a defunct Senate and a former official of a deposed monarchy. He understood the delay in granting him his visa to be a sign of Britain's support for the leaders of the coup: the British did not want to send out the wrong messages to the new government in Iraq.

Nevertheless, London seemed the logical place in which to seek refuge. His son Hazem was here, training at a merchant bank, and in London he would be far enough from Iraq not to jeopardize the safety of his imprisoned eldest child, Rushdi, by inflaming the situation with his presence. He also had some Iraqi acquaintances there, including friends who had been stranded in Britain because, like him, they could not return home.

As he walked out of customs with Shamsa's son Issam, he spotted Hazem waiting by the barrier for him. His son embraced him. 'It's good to have you here,' he said, his voice wavering with emotion. Hadi could see that there were many questions Hazem wanted to ask, but he suspected that he would not have the answers he wanted to hear. He smiled wearily; he felt so very tired. But at least he could now focus on the present and establish some order in his life.

Travelling into the city from the airport, London seemed different to Hadi; perhaps because so much else had changed in his life. At first he thought the taxi was taking them by the wrong route, for he had forgotten that this time he would not be staying in his usual quarters, the Athenaeum Club near Hyde Park Corner. Instead he would be living with Hazem in his small rented flat. Watching the pedestrians in the streets, he wondered whether he would ever really understand the British and what made them tick; he had believed they were his country's friends. He compared the importance they placed on order, discipline and manners in their own society with, as he saw it, their outright double-dealing over the coup in Iraq. They had let their friends down badly. *They let us be slaughtered*, he thought, *without so much as the bat of an eye*.

When the taxi finally pulled up to the kerb, Hadi looked at the enormous complex of flats and sighed. As they walked towards Hazem's building he noticed that on the wall outside each block was a wooden board that listed all the flats beside the words IN and OUT, which were covered by a small slide made of brass depending on whether the occupants were in residence or not. Hazem explained that Dolphin Square was a base for many British families who lived abroad. The flats were serviced, but to Hadi's eye there was an air of impermanence about them. Going up in the old brass lift to the sixth floor, he reflected that he could have owned all this.

Each of the many blocks in Dolphin Square was named after a British naval hero: Hazem's flat was in the Hood block. Hood 202 was small, dark and drab, its floors carpeted in a nondescript taupe. Hadi inspected the place without betraying his emotions: there were two untidy bedrooms, a bathroom, a small kitchen and a bland sitting room with slightly tired red-and-white sofas.

Hadi shook his head brusquely, dismissing his thoughts; he could not afford the luxury of self-pity. He turned to his son disapprovingly and told him that the place needed a proper clean. Finding that the fridge was empty, he announced that they should go and get some groceries before the shops closed. His mood lightened fractionally when he learned that Dolphin Square had its own small supermarket, a novelty for the Chalabi family.

Six weeks after Proclamation No. 1, the first law passed by the new Qassim government, came Law No. 30. Under it, all the large land-holdings in Iraq were to be sequestered by the state. Being among the largest, Latifiyyah was the first to go. With a sweep of an anonymous official's pen, Hadi lost his most prized possession. In fact the government impounded all of Hadi's properties, which included more than 400 square kilometres of prime real estate near and around Baghdad. There was no financial compensation, nor was any of the land redistributed to the peasants. Instead, Iraq's agriculture was left to die.

The news that he had lost Latifiyyah came as a terrible blow to Hadi. As he sat in the flat in London with his nephew Issam, he indulged himself in a rare outburst during which he recalled the many years of hard work he had put into his career. He told Issam stories about his early adventures with desert marauders, when he had ridden out into the countryside with Ni'mati. He described every piece of land he had cultivated, every project he had undertaken, every enterprise and business venture, until he came to the subject of Latifiyyah. Then he stopped.

'With Latifiyyah, my entire life's work is gone,' he said numbly. He did not reveal that he had lost the equivalent of US$1 billion. Instead, he rubbed his brow and continued, 'But it's not the end of everything. And I thank God every day that your cousins weren't physically harmed in what happened.' He told his nephew that they had to trust to the infinite wisdom of God that all would be well.

Centuries of family roots were being ripped up from the country Hadi loved. He knew he could never live in Iraq again. That had become clear to him with the murders of the royal family and Nuri Said. The thought of never again setting foot in Kazimiya – of never entering the shrine, visiting his grandfather Ali's grave, greeting his old friends – pained him greatly. He remembered how his father Abdul Hussein had suffered when, in 1922, Sheikh Mahdi had prohibited him from entering the shrine and effectively ostracized him from Kazimiya.

As an exile in Britain, Hadi would be less able to effect change and help his country. Moreover, Iraq was his identity; the country meant everything to him. He didn't know who he would be without it. He thought of his family, and of how proudly he had helped his country set out on the path of progress. Was it really all gone?

29

Migration

Precious Cargo

(1958)

HASSAN WAS SUMMONED to the Law Faculty two weeks after the coup had taken place. Never having served in politics, he did not face the same risk of arrest and imprisonment as his brother Rushdi. Even so, his situation was extremely precarious.

There had been serious opposition to Hassan's return to the university, particularly from the Dean, Abdul Rahman Bazzaz. Hassan's family was closely associated with the old regime – his brother, father and grandfather had all served in the Cabinet – and many people could not forgive these connections. However, several faculty members were his friends, despite their political differences, and they lobbied for Hassan when he risked losing his position. What ultimately saved him was that he was a well-regarded professor with a good teaching record. Now more than ever, Hassan needed to keep his position in the faculty and earn his living. It was a question of survival.

On his return, Hassan had to attend a lecture given by the Dean that emphasized the need for professors to educate students about the revolution, and to promote an understanding of revolutions in general, including the concepts of nationalism, anti-colonialism and freedom. Most members of staff already supported the coup, and were supporters of either Nasser's Arabism or the Communist Party. Although fundamentally different in their beliefs, for the time being the two groups were united in their anti-British and anti-monarchy sentiments. The memory of Iraq under the Hashemite monarchy was already tarnished, and had acquired the label of the 'expired epoch'. A new language was developing, inspired by the spirit of the revolution, in which the most popular words included the labels 'agent', 'traitor' and 'thief'.

In the immediate aftermath of the revolution, the Iraqi Communist Party (ICP) had emerged as the most organized of the political groups, with followers nationwide. There was a mood of revolutionary euphoria amongst them. Many left-wing intellectuals and artists celebrated the new era, and the women's movement, which had started humbly in the 1940s in the field of education, was adopted by the ICP (indeed, in 1959 the first female minister in the Arab world was appointed in Baghdad). A campaign was relaunched to eradicate illiteracy, particularly among rural women.

It seemed that there was still a surfeit of energy available for demonstrations. When Qassim officially withdrew from the union with Jordan and the Baghdad Pact, street demonstrations followed in support of his decisions, and the statues of the old regime were destroyed amidst large celebrations. Everything belonging to the old order was deemed evil and repulsive. The names of bridges, roads and public buildings that related to the old regime, such as the Royal Hospital, were changed to commemorate the revolution.

Marches and protests frequently impinged on the campus, and several of Hassan's students started to agitate about their right to join them. During term time, Jamila fearlessly protected Hassan from potential student attacks and faculty antipathy. By now the precise nature of their relationship had become increasingly unclear even to the couple themselves. Jamila was much more than Hassan's secretary, yet she was not his wife. She was resolute in the face of the hostility towards him, ignoring verbal assaults and dismissing students who disrupted the rest of the class. Her protective efforts were much needed: Communist students had formed groups aimed at 'protecting the revolution', and outperformed the police in reporting on individuals' political and revolutionary transgressions.

Throughout this time Rushdi remained in prison, where he was surrounded by former ministers such as Tawfiq Suwaidi, Fadhil Jamali, Abdul Rasul Khalisi and Abdul Wahab Murjan. Conversations among the inmates were held in low voices for fear that the guards might be eavesdropping, but they continued to receive food and letters from home.

Rushdi never wrote to his family, preferring to pass on messages via Ni'mati, who delivered his food to him, rather than risk the guards opening his letters. Ni'mati was very worried about Rushdi. When he was allowed to see him he seemed pensive and withdrawn, resigned to his fate, and he grew paler and more emaciated with each passing day. Rushdi told him that he hated the toilet facilities, and found washing a nightmare as the bathrooms were so dirty. But small victories such as when Ni'mati managed to smuggle in some clean towels distracted him from his circumstances temporarily, and he always asked Ni'mati about his children.

Ni'mati tried to keep Rushdi up to date about what was going on in the government, but there was little that he wanted to know apart from whether he and his fellow prisoners were going to be charged, tried and executed in the near future. He didn't know what crime the authorities would concoct to charge him with, but he was aware that the newspapers delighted in slandering members of the former government without levelling any specific accusations against them.

Meanwhile, the family was struggling to cope with diminished funds and fewer contacts. The new secret police had been instructed to keep a close eye on members of the old regime and their families. They were forbidden from travelling, and risked being searched at home without prior notice if the police felt like it. They didn't know who they could trust. While most of the country embraced the new era with revolutionary zeal, the Chalabis and many others like them lived in terror.

Bibi continued her visits to the shrine, praying for godly intervention in the family's circumstances. Perhaps her prayers were heeded. In spite of the authorities' overt hostility towards the Chalabi family, Hadi's office in Samaou'al Street remained open as Jawad attempted to keep the firm afloat and salvage what he could of the family's business interests.

One morning, about a month after the coup had taken place, an innocuous man with deep-set eyes walked into the office. He asked about the family and about a flour mill that Hadi owned, and explained that he had grain for sale. It took Jawad a few moments to realize that his visitor was none other than Hamed Qassim, the brother of the leader of the coup, President Abdul Karim Qassim; he was a grain supplier from whom Hadi had bought stock in the past, and with whom Rushdi had enjoyed a good rapport.

Hamed was a simple man whose brother's sudden rise to power had left him unsure of how to react to his own newly elevated status. Having re-established contact, he developed the habit of dropping in at the office during the afternoon, when Jawad and Hassan were often to be found there. Making himself comfortable in one of the leather chairs, Hamed would praise his brother, reading out fawning articles in the press about him. He began to push to sell ever larger quantities of grain to the mill, even though it was no longer in the family's hands, as Hadi's assets had been frozen. Hamed clearly wished to benefit from his new status, and to make as much profit as he could from his grain, but it became apparent to Jawad that he also wanted to assist the family, and that he might even be able to help Rushdi.

By October, Jawad had succeeded in getting the travel ban lifted from his thirteen-year-old brother Ahmad with Hamed's help. It was no longer safe for Ahmad to attend Baghdad College, where he was being viciously bullied by other students. Getting him out of Iraq had become a priority for the family, and as soon as the ban was lifted arrangements were made for him to join his father in London.

A week before Ahmad was due to leave, Ghazi, the son of his older sister Raifa, got wind of the arrangements. The same age as Ahmad, and having also endured months of taunting at school, he was completely distraught. He ran into the kitchen and found a sharp knife, which he put to his uncle Jawad's throat. 'Ahmad can't leave without me!' he screamed. 'If I don't leave with him, I'll kill you! I have to leave with him!'

Motioning to Bibi and the other women to stay back, Raifa, her voice shaking, told Ghazi to let her brother go, and said that they had enough enemies without fighting amongst themselves. From the other side of the room, Bibi added, 'If Ahmad can go, then there may be a way for you too.' Ghazi collapsed into a chair, sobbing. Nursing his throat, Jawad agreed to help him. Through Jawad's efforts, Ghazi's travel ban was also lifted so he could travel with Ahmad.

Before setting out for England, Ahmad went to say goodbye to Saeeda, who was living with friends in Kazimiya. Her angina was wearing her down. She looked at Ahmad for a long time and then hugged him to her as tears flowed down their faces. Theirs had been a very strong bond ever since Saeeda had first set eyes on Ahmad on the day he was born. They both knew it was the last time they would see each other.

When the time finally came for Ahmad and Ghazi to leave, everything was rushed and fraught. They were lectured on the qualities of fortitude, courage, good behaviour and hard work, first by Hassan and then by Jawad. Bibi, in tears, whispered travel prayers in their ears and circled their heads three times with the Quran, then made them kiss the book, hoping that in doing so she would be shielding them from the dangers of their forthcoming flight. She was also frightened that the guards would stop Ahmad at the airport and confiscate any messages or valuable items that were being given to him to carry, although in the event this did not happen. She did not know when they would see each other again.

The boys flew from Baghdad via Istanbul and Frankfurt to London. Ahmad carried with him on the plane a Leica camera that had been given to him as a birthday present. Sitting next to them was a well-known poet from Najaf, whose face they recognized from the newspapers. When they landed in Istanbul, Ahmad noticed his camera was missing. He and Ghazi started searching for it, until Ghazi spotted the distinctive camera string dangling from the bag of the poet, who had already passed through Turkish immigration. The two boys watched in outrage as the man walked out of the airport with the stolen camera. The theft was like a continuation of their recent experiences in Baghdad, where people seemed to think they could take what they wanted from the family.

On the day Ahmad and Ghazi landed at London's Heathrow Airport the sky was dark, and dense fog stripped the colour and life from the landscape. The contrast with the golden warmth of Baghdad was stark. It was like landing in Limbo. On their previous few trips to Britain on holiday, they had never noticed the shabbiness of the airport. The terminal was a shack-like building where they were welcomed by two large signs: 'UK Citizens' and 'Aliens'. It was obvious which one they had to pass under.

After they had passed through customs and immigration they were greeted by Hadi, who hugged his youngest son close with tears in his eyes. On the way to Dolphin Square Hadi asked the boys about their flight, and for news of the family in Baghdad. Only when they entered the flat did Ahmad and Ghazi realize how much things had changed. The contrast with the large spaces they were used to in Baghdad could not have been more striking. They were equally surprised when Hadi opened their suitcases and started unpacking for them. They had never seen him perform a domestic chore before. Until that moment he had always been the very symbol of everything powerful and strong in their lives; he had been, in Ni'mati's words, Little God.

Despite his surroundings, Hadi seemed cheerful. He had already prepared them a meal, which he reheated in the tiny kitchen and served to them. By now Ahmad and Ghazi were on the verge of panicky anger; having witnessed the mighty head of their family prepare their supper, they couldn't imagine what other humiliations might lie in store for them. Seemingly oblivious to their mood, Hadi warned them that the rice wasn't quite as tasty as the basmati rice they were used to back home. He explained that an old acquaintance of his, Naji Ibrahim, an Iraqi Jew who had served in an RAF bomber squad during World War II, had introduced him to the Hellenic Stores, a Greek shop in Charlotte Street near Soho, where he could buy Middle Eastern food. Such provisions were hard to come by in London, so the shop was an important find. However, the ingredients were not quite the same, and many were not available. Although he had never cooked for his family before, Hadi was determined to do his best and to try to recreate the simpler dishes they loved.

With Hamed Qassim's help, it became possible for the family to send more of the children to Britain on visitors' visas. A month after Ghazi had left, his brother Ali found himself in London the guardian of his young cousin Hussein. Ali tried to take his comic books with him, but there were too many for him to carry. He had to content himself with two albums of his stamp collection instead. Ali had the added burden of having to take a long message to Hadi, which he was made to memorize verbatim by his uncle Hassan:

1. *Don't speak to X, Y and Z. We know that they've already made contact with the new government and might harm us if you've said anything.*

2. *Stay in London; don't go anywhere else.*
3. *The money in Switzerland has to cover the costs of all the schools as well as your stay.*
4. *Don't start any new business venture; just stay put.*
5. *Bibi's still forbidden to travel, but we're working on it.*
6. *We're still trying to get Rushdi released from jail, even if under house arrest.*

Rushdi's daughter Nadia and her two Agha Jaafar cousins, the children of Najla, followed next. All were under the age of ten. They were sent with an acquaintance of the family who they did not know well, and who didn't pay much attention to them. During their stopover in Rome she left them alone in their hotel room while she went out. Nadia was so frightened she cried herself to sleep.

By the time Thamina's daughters Leila and Kuku were ready to go to Britain a few weeks later, a bolder plan had been concocted. Now that the initial goal of staying alive had been achieved, the family started to think of how best to safeguard any valuables they had left, since it had become clear that leaving the country was an inevitability for all of them. Bibi's precious jewels had to be smuggled out somehow, but she could not be the one to do it, because that would have been too obvious. Instead, Bibi and her three daughters devised a plan to smuggle them out in Leila's thick blue woollen coat.

The coat was unstitched, the shoulderpads taken out and the hemline unpicked. Piece by piece, Bibi ran her hands over her jewellery before handing it to her daughters to sew into place. Each item had its own special story: she lingered over her large solitaire, her topaz from Istanbul, her rubies and emeralds from Geneva, the long strings of delicate Bahraini pearls that Hadi had bought her several years earlier and the diamond watches she had once collected.

Finally, the coat was restitched, and under her mother's watchful eye Leila wriggled into it with its heavy invisible cargo. Terrified that she would be searched, she repeated the prayers Bibi had taught her for protection. Convinced that she would be separated from her sister at the airport, she stormed ahead of Kuku, and passed through the security checks unhindered.

30

Hunger Pangs

Yearning for Home

(1958)

IN LONDON, HADI dedicated himself to organizing his grandchildren's lives, sorting out their schools and establishing a daily routine for everyone living in the apartment at Dolphin Square. It was his way of compensating the children for what had happened to them. Organization was something he had always been good at, and he preferred having something to focus on to dwelling on his family's predicament. No task was too menial for him. Eight-year-old Kuku would stand obediently in the bath while he scrubbed her back for her. She was so confused by the situation and her new circumstances that she stiffened like a statue.

The older children found it very disconcerting to see their grandfather in this strange domestic setting, away from their familiar world. Hadi had always been a remote figure in their lives, a commanding, totemic presence who exuded power and confidence. Here, in the dingy little flat, he seemed disconcertingly smaller than he had appeared at home in Iraq.

The darkness and dampness of London seeped into the children's bones from the moment they arrived. The city seemed a forlorn place to them. It had not yet recovered fully from the hardships of the Second World War, although rationing had ended seven years earlier. They could not help comparing grey London to golden Baghdad, which for them represented space, safety, warmth and plenty. They knew why they were there, but somehow they could not square the circle. The leap from the palm-lined roads of Baghdad to the grey streets of London was too big. Food was another sore point. Used to crates full of fresh fruits and vegetables, they found the idea of buying a single apple or banana from a shop depressing.

The boys escaped from their straitened circumstances by going as often as they could to Leicester Square to see the latest films. The cinema had played an important part in their lives in Baghdad, and it continued to do so. But they were disillusioned with London itself. It was not the modern metropolis they had dreamed it would be.

The children also felt starved emotionally, for Hadi could only provide them with so much care and affection. They initially felt resentment towards the people they encountered in shops and Dolphin Square, and later, at their boarding schools, they disliked being called by their surnames and bristled when yelled at for walking with their hands in their pockets. At school their feelings were compounded by constant hunger, owing to the tiny portions they were served and the inedible nature of the soggy dishes.

The reserved manner of the British equated in their minds to coldness and emotional meanness. Several times a day there would be subtle reminders of their foreignness when they couldn't make themselves understood, or a passer-by would scrutinize them in the street. Unlike other immigrants who were able to join existing expatriate circles in the city, the children became their own community, isolated and separate. But London was their home now.

31

Arrivals and Departures

The Importance of Contacts

(1958–1959)

THE NEW PRESIDENT'S brother Hamed Qassim soon became the most important contact that the family had, as he was able to keep them informed about Rushdi's situation in prison. Bibi arranged to be at the office one afternoon when she knew that Hamed would be visiting, although she usually stayed away from it. When she heard Hamed coming through the front door, she bustled into the hallway as if she were just on her way out. She exclaimed brightly at the sight of Hamed; he smiled politely, they exchanged pleasantries and Hamed proceeded to share what news he had of Rushdi. As Bibi turned to leave, she paused as though a thought had just occurred to her. 'Perhaps it might be possible to speak to the President, to explain Rushdi's unique circumstances?' She left the question dangling in the air.

Hamed reflected for a moment, before saying that such a meeting would be very difficult to arrange. He suggested that an alternative might be to call on his wife, Umm Adnan. She was his first cousin, and as a blood relative she was highly trusted by Abdul Karim. Bibi knew of her, although she had never socialized with her. That was the way things worked in Iraq: people knew of each other even if they moved in different circles. There was always a brother-in-law of a cousin who was the husband of the sister of someone … In short, connections were always to be found. Hadi's high standing in Iraqi society had meant that the Chalabi family had an extensive social network, yet there had never been a reason for Bibi's path to cross Umm Adnan's before. Had

298

it done so, Umm Adnan would have been the one paying a visit to Bibi, as the more socially eminent of the two; but now the tables had been turned.

Bibi went to visit Umm Adnan the next day at her house in Karadat Mariam. She listened respectfully while Umm Adnan sang Abdul Karim's praises, telling her how much the new President relied on her, how hers was the only food he ate, for fear of being poisoned, and how he relied on her two sons as trusted guards. Bibi was pleased to hear all this, as it suggested that Abdul Karim Qassim would listen to Umm Adnan if she were to put in a word on Rushdi's behalf.

Bibi told her, 'I can see how dear you are to the *Za'im*, and I'm confident he wouldn't refuse you anything. You're a mother, and you wouldn't want your son to be sitting in jail for no reason ...'

Umm Adnan took a deep breath, then nodded her head. She got up and walked to the hallway, where Bibi saw two telephones on a narrow table. She picked up the receiver of the nearest phone and asked to speak to the *Za'im*. She was connected immediately, and repeated to him what Bibi had said to her. Bibi watched her, clutching her handbag as her heart beat very loudly in her ears.

'Fine. I'll see you later.' Umm Adnan put down the receiver and came back through to join Bibi. She smiled. '*Inshallah*, it's good news. The *Za'im* promises your son will come to no harm. He'll do his best for him. You have his word.'

Holding back her tears, Bibi thanked her profusely, then excused herself and left. This was the best she could do for Rushdi: a promise wrung out of a tenuous connection with the new leader's cousin. Such an arbitrary event symbolized the situation in Iraq, where powerful individuals dominated and the rule of law went only so far.

Umm Adnan was as good as her word. A week later Rushdi was allowed to come home and live under house arrest. He moved into his parents' house in A'zamiya to be near the rest of the family, but his eldest two children, Hussein and Nadia, were at boarding school in England. Rushdi's situation remained uncertain, as he was still awaiting trial. Four soldiers were posted outside the house around the clock, and Rushdi had to report each day to the nearby Farouk police station, until another intervention from Hamed spared him that ordeal. His wife Ilham revealed that she had made a *nidir*, a plea, to one of the saints to release him, in return for which she had promised to wear black during Muharram and the following month.

Inside the house, Rushdi could do as he pleased. His family were allowed to visit him, as well as a limited number of friends. He could use the telephone, but it was bugged. The guards wrote a report on him every day which was delivered to the secret police at the Ministry of the Interior. They were not abusive, but their presence about the house was stifling, particularly when they strode in to collect their meals from the kitchen, and made themselves at home.

Bibi had expended much of her energy in her efforts to have Rushdi released. Resorting to comfort eating, she had put on a lot of weight. She was still under great emotional strain, and was fearful that she might be arrested next, as she had always freely expressed her opinions about Nasser at social gatherings and during her ladies' card games. Fearing that someone might report her to the authorities, she made plans to join Hadi, Ahmad and her grandchildren in London.

Bibi's permit to leave Iraq was finally obtained with Hamed's help. She was loath to leave her older children, but they urged her to go. Her goodbye was tearful: she was leaving everything behind, including her past. Yet a large part of her felt such revulsion at the place, and many of the people, that she couldn't wait to obliterate the last six months from her memory. These days she felt like an outsider in her own home. Her existence was steeped in terror and apprehension as each new proclamation chipped away a little further at her old way of life, with more lands being taken, more laws implemented. She would never have believed that her mother's Communist relatives could have been so gleeful about what had happened to the country, or to her, and was deeply hurt by their comments on her situation. Her memories of the violence of the early days of the coup – the regicide and Nuri Said's murder, as well as that of her friend Umm Abdul Amir – were indelible. A very deep root had been severed, and the damage was irreparable. Her life in Iraq would never be the same again. She felt deeply betrayed by what had happened to those she loved and to her country.

This time, Bibi had no choice but to fly, despite her fears. She went first to Beirut, and from there to London, arriving in January 1959.

Talal (third from left) in military uniform with friends
during his military service in 1959.

She cried when she saw Hadi at the airport, her emotions a chaotic mixture of joy and sorrow. But she became downright depressed when she set foot in the flat in Dolphin Square. When she saw her grand-daughter Leila, she burst out uncontrollably, 'Your father's in jail – he might die and you'll become orphans. What are we going to do?' It hadn't occurred to her to put on a brave face in front of the children. Her sense of release at having left Baghdad was soon replaced by an overwhelming horror of being stuck in London. She was a nobody there, another foreigner drifting through the city. She was fifty-eight years old, and had fifteen grandchildren. She was too old for this.

Soon after Bibi arrived, most of the family moved out of Dolphin Square into a slightly larger flat near Regent's Park, leaving Hazem to enjoy the restored tranquillity of his modest apartment. A tall, modern building, the White House looked more like a hotel than a home. The park was within walking distance, but Bibi missed her old life and her house, especially the conservatory on the first floor where she used to sit in the winter sun. Even her interest in clothes and materials waned, despite the many choices that were available in London. She didn't interact positively with the city, feeling rejected by its very bricks and stones, which seemed to speak a different language to her. For all the time she spent there, she never warmed to the city. It was as if she were allergic to it.

Within the confines of their characterless new home, she prayed out loud, 'Please, God, get me out of here,' and waited for divine intervention as she worried ceaselessly about her children back in Baghdad. Her days were slow and dull, punctuated by prayers and the meals she threw together, her afternoon siestas and ineffectual attempts to do the household chores.

When her daughter Raifa followed her to London soon after, she listened patiently to her mother's woes, managed the house and ensured that Bibi and the family were well looked after. She attempted to exert some control over the wayward Ghazi, but was relieved when her younger brother Hazem stepped in to help.

Food remained an important focus for all the household, particularly the matter of where to acquire it, and they often resorted to the Greek shops around Charlotte Street. The ease of the family's old life in Baghdad was in dramatic contrast to the laboriousness of their existence in London. Bibi repeatedly lamented her fate, asking time and time again: 'Whatever happened to us?'

Schools were found for the children through the recommendations of friends and acquaintances; the fees were to be paid from the funds Rushdi had transferred out of Iraq on Hadi's behalf a couple of years earlier. Interviews were arranged for Ghazi, Mahdi and Ahmad at Seaford College in Sussex. Positively revolted by the place, Ghazi and Mahdi were irreverent to the headmaster during their interviews, and were duly denied admission. Ahmad, on the other hand, displayed all the required respect. As far as he was concerned, the situation was so awful that it made no difference what school he attended. He was told that he could start immediately.

The girls went to Huntington House in Surrey, where they had to wear an orange uniform, except for Najla's daughters, who had inherited their father's family superstition about the evils of that colour and were allowed to wear yellow instead. All the girls suffered unbearable homesickness. They hated the cold and the rain, and felt bewildered at having being yanked from their lives to come to this austere, loveless place. Nadia took to staring out of the classroom window. She concocted a daring escape plan with her cousin Kuku, but they were caught in the act of creeping out of the school. Their punishment was to clean the dining-hall tables for several weeks.

While the younger boys were sent to St Leonard's Forest School in Horsham, Sussex, where they were overwhelmed by an atmosphere of Protestant religiosity, the two rebels, Ghazi and Mahdi, were the last to find a suitable institution that would take them. After the Seaford fiasco, Hazem decided to send them to Lysses School in Hampshire. Within seventy-two hours of their arrival, Mahdi had stolen a map of the area, noted down the local train times and drawn up an escape plan, which Ghazi codenamed 'the X plan'. They sneaked out of school one evening, walked through the woods to the station and caught a train to London, where they made their way back to Dolphin Square.

Hazem was not pleased to see them, although he didn't send them straight back to school. Instead, as a punishment they had to endure a harsh regime of their uncle's devising: they were allowed no heating and little bath water, Ghazi had to sleep on the dining table and Mahdi in the bathtub. Several weeks later, seething from the punishment Hazem had inflicted upon them, the boys bought some red ink which they added to his underwear when it was being washed at the launderette. Unfazed, Hazem wore the pink underwear until it decomposed many years later. Eventually the boys were accepted by Millfield College in Somerset, where they remained despite their antics.

The first three months at Seaford College were gruelling for Ahmad as he slowly adapted to the dampness and the cold, the constant feeling of hunger and the shock of sharing bathwater which had taken on a sickly tinge of brown by the time the third boy had stepped out of it. In time, he found the solution to his bath dilemma: once he had proven his academic mettle, he did the prefects' prep for them in return for getting cleaner water.

Although he was the only Iraqi at his new school, he wasn't the only foreigner – there were a handful of boys who came from overseas – but the majority of the pupils were from the English middle classes. As he had joined the school late in the term, he stuck out from the others, and in the classroom he adopted the survival techniques he had perfected at school in Baghdad, shrugging off the teasing until it ceased.

Fortunately, there was a friendlier atmosphere at the school house where his dorm was, and where his sympathetic house master, John Allerton, a former RAF pilot, reached out to him through literature. Later Mr Allerton encouraged Ahmad to sign up to the air force cadets, and he joined other cadets on flights all over the British isles, testing radar.

At first his hunger followed him to school, where there was a scarcity of food. This was eventually remedied by Najla, who set up a standing order for him from a shop in nearby East Mosley. Each week the shop sent him a package that included six boiled eggs, two loaves of bread, cheese, two cans of tuna, six oranges, four Cadbury sandwiches, two tins of corned beef, a packet of raisins and a packet of McVities digestive biscuits.

Ahmad in 1961 at Seaford College receiving a prize from
Douglas Bader, a famous WWII RAF pilot.

As a foreigner, Ahmad was put in the lowest class in the fourth form. Within a term he had risen to the highest class, acquired a whole new vocabulary, learned to play rugby and managed to avoid the practice of 'fagging' by the skin of his teeth. There was now another aspect to his life beyond mere survival. He threw himself into his studies, and his newfound passion was mathematics.

Yet he remained obsessed with Iraq, as his new English friends soon discovered. He missed his life there and the people he had left behind, as well as the comfort of a home whose security he had never questioned until it had been lost. He could not reconcile himself to the injustices that had been meted out to him and his family. He wrote often to his older brothers, asking for news, and read the *Daily Telegraph* religiously. His more carefree fellow pupils found his attachment to the news a little too serious for their own tastes. Whereas they contented themselves with Radio Luxembourg and the tunes of the Everly Brothers and Cliff Richard, Ahmad spent hours fiddling with his pink-and-white Akkord radio, trying to tune in to the insufferable *Sawt al-Arab* station in order to get more news from home.

From the radio and his family, he learned that Nasserite Arab nationalists such as Abdul Salam Arif were pushing for a union between Iraq and Egypt, but that Abdul Karim Qassim was not in favour of this move. Arif had fallen foul of Qassim, who imprisoned him some months after the July coup. Ahmad surmised from various conversations he overheard when he visited his parents in London that they believed Qassim's attempt to distance himself from the nationalists was proof of his strong ties to Britain, and that the British were advising him. Similarly, the dominance of the Communist Party in Iraq's politics had become a cause for American concern, as several articles in the US broadsheets reflected. Qassim had been told to distance himself from Moscow. By playing the Communists off against the nationalists, he could focus his efforts on strengthening the state.

When Ahmad tried to interest his schoolfriends in these developments, they looked at him askance. Nor could they understand his growing determination to get into the Massachusetts Institute of Technology in the United States. There were constant debates among them late at night about the merits of America versus Britain, with Ahmad arguing with adolescent certainty that America was the future, as after the Suez Crisis Britain's moment had passed. The anger he felt towards Britain's treatment of Iraq was palpable.

One morning at breakfast Ahmad had a disturbing premonition. He was sure that Saeeda had died. A few days later a letter from Baghdad confirmed the news. Ahmad was inconsolable; he walked outside and sat staring at the South Downs, unable to explain his loss in any language that would have been comprehensible to his schoolmates. Saeeda had been the linchpin in his life. He learned that she had been buried in Najaf, and that, not having any descendants of her own, she had left him her only valuable possession: a small plot of land in Kazimiya that Hadi had given her many years before.

After six months of living together in London, Hadi and Bibi's attitudes to the city could not have been more diametrically opposed. Bibi could not reconcile herself to life there, and took the British reserve as a personal affront. Her rejection of her surroundings extended to the activities of her own family members. One morning while she was having a cup of tea in the kitchen, she noticed that Hadi was attempting to make baklava. Bibi was so scandalized by the sight that she angrily barked at him: 'I didn't know I'd married a confectioner. Have you gone mad?' Hadi shook his head wearily and got on with the task in hand.

Hadi's in-built serenity served him well in London. He was able to retreat inside himself and to find simple pleasures, such as feeding the pigeons in the park nearby. He didn't mind being a foreigner, and believed that the city was as good a place as any in which to find work, even though he spoke little English and was not well-versed in British ways. He realized that he would need to rely on his grown-up sons, who spoke English much more fluently than he did.

He spent time with a few Iraqis he knew who were in the same predicament, often meeting them for afternoon tea at the Athenaeum Club on Pall Mall, where they would discuss the possibilities for changing the situation in Iraq. He had always been a great walker, and took to exploring the streets, closely examining the Georgian and Edwardian façades. Instead of the *faisaliya* he had worn in Baghdad, he now wore a tweed hat to keep his head warm. Trying to recapture his old life, he sometimes visited the antique shops where he had bought favourite items in the past. He remembered buying two large silver pieces that had belonged to an Austrian émigrée who had fled her country during the Second World War. The larger of the two, a sculpted bucolic scene depicting two young lovers in a wood, had touched him deeply. There was more resonance to the émigrée's story for him now. He thanked God for his life, aware that he could have been killed in the coup.

To calm Bibi down, he told her to be grateful, that money came and went. Besides the jewels that had been smuggled out in her grand-daughter Leila's coat, some of her clothes had been shipped over from Baghdad, but receiving them upset her further, as they carried so many memories. Throwing them in a heap on her bed, she looked at them in despair. They struck her as unfit for use in this life.

In early March 1959, news from Baghdad further clouded Bibi's horizon. A special military court launched a trial of the men of the 'expired epoch'. Another of Karim's cousins, Mahdawi, was the head of this court; during the course of the trials he was to become better known for his unmeasured outbursts and rough language than for any legal expertise, which he lacked entirely. All of the former Cabinet members as well as many other politicians of the old regime were to be charged, including Rushdi. Along with Abdul Rasul al-Khalisi and Abdul Wahab Murjan, he was accused of violations of the constitution and human rights. He had been present at a meeting several years earlier when the Cabinet had voted in favour of expelling a renowned Communist, as at the time it had been illegal to belong to the Communist Party in Iraq. The Communists now had powerful friends, and wanted to settle the score.

32

Escape to Nowhere

The Threat of the Clown Court

(1959)

RUSHDI HAD BEEN under house arrest for four months by the time the trials were announced. He spent his time listening to the radio and watching the Mahdawi hearings on television, witnessing the ways in which his friends and former colleagues were derided and belittled one after the other. Rushdi lived and breathed the court, imagining what it would be like to endure the endless humiliation, foul language, insults and jeering crowds. Although he was still banned from leaving the house, he became obsessed with fleeing the country, and started to talk to his siblings about his plan. At the age of forty, he was prepared to create a new life for himself and his family, leaving everything behind.

One day Saadoun, a friend of Hadi's from the Dulaimi tribe near Falluja, came to visit him. It occurred to Rushdi that he could travel by car with Saadoun to Falluja, then with his help and knowledge of the local terrain, cross the western desert from there to Jordan. He and Jawad spent days making plans and mapping routes out of Iraq. Their younger brother Talal warmed to the plan as well, as did Najla, who was desperate to leave, although like all of them she was banned from travelling. Baghdad had changed for them. Their social lives had vanished, their assets had been frozen and their freedoms curtailed. They had to be careful who they talked to and what they said, as they were under constant scrutiny. Hassan was the only one who refused to become involved in the scheme. He thought it foolish and extremely risky, as failure could cost them their lives.

At dawn one morning in March Rushdi's wife Ilham and Najla quietly packed a few belongings, as well as milk, biscuits, water and canned food for the small children. They whispered as they got into the cars, one of which would be driven by Jawad, the other by Talal. The children worried them most, in case one of them started crying and woke the neighbours. Nannu, the houseboy, realized what was going on, but he could be trusted; and Ni'mati's loyalty was never in question.

The two cars drove slowly out of Taha Street and along the riverside, then crossed the A'zamiya bridge towards Karkh, on the right bank. Rushdi held his breath as the metal bridge rattled under the wheels. The shadows of night were gradually thinning as ribbons of light crossed the sky. Further down the river a lone fisherman waited silently for the day's catch.

They passed 'Atayfiya, then the Mansour district as they headed west towards Falluja, just over an hour's drive from Baghdad. The dirt roads were lined with green fields and palm trees, with the occasional hut in the distance. Beyond the greenery, the desert beckoned. For the first half hour the only noise was from the vehicles themselves, bumping along as the children slept in the back and the other passengers, apart from Rushdi, who could not rest, dozed.

Suddenly, military planes appeared in the sky and there was a loud bombing noise. Jawad and Talal braked hard, and Rushdi's heart leapt to his mouth. His two-year-old son Muhammad woke with a start and began to cry. For a long minute they sat in the cars, expecting to be bombed at any moment. Najla recited a prayer, waiting to die. Then the bombing stopped and the planes banked away towards what the family assumed must be the Habbaniya air base.

What they didn't know was that the day of their escape coincided with an attempted military coup aimed at overthrowing Qassim. It was led by an Arab nationalist Free Officer based in Mosul, and Qassim's planes were bombing the Abu Ghraib military camp, which lay on the route to Falluja, in retaliation. A few days earlier a large Communist rally in Mosul had provoked the disapproval of the Free Officers there, marking the beginning of a power struggle between the Communists and the nationalists throughout the region. Qassim, who sympathized with the Communists, had started to purge the armed forces of pro-Nasser nationalists in order to strengthen his grip on them.

As the cars entered the outskirts of Falluja, they approached some military trucks that had been readied for action in the event of a coup. The last truck in the convoy stopped, an officer got out and waved at them to pull over. Everyone froze. This was surely the end. They would certainly be imprisoned, and any clemency that might have been shown to Rushdi would evaporate. The officer walked to the car, and asked Jawad why they were driving so early in the morning. Jawad told him they had come to spend some time on their farm, which lay further along the road, and that they had brought the children with them for a change of air.

The officer peered into the back, and saw a row of little eyes staring back at him. He looked puzzled, then frowned. 'I'll need to take you to the *Qa'immaqam*, the district director of Falluja. Follow me.' Trembling, Jawad restarted the car. The drive into Falluja took ten minutes, but it felt more like an hour. When Ilham started to wonder out loud what would happen to them, Rushdi snapped at her to be quiet.

At the old district building, they were told to get out of the cars and were led into the offices. The *Qa'immaqam* had not arrived; it was not yet seven o'clock. Najla gave the children some biscuits to distract them. She kept her face lowered; women rarely entered such buildings, and certainly not in provincial towns. Luckily, as it was so early in the morning only a few employees were there to stare at them.

The *Qa'immaqam* finally arrived, and loudly ordered tea as he walked towards his office. The family sat in suspense as he was briefed by the officer. Jawad, Rushdi and Talal were called in first. Having instructed them to take a seat, the *Qa'immaqam* asked them to explain the reason for their journey. He already knew who they were, and was aware that Hadi owned land in the district. He also knew that Rushdi was awaiting trial.

The men repeated their story: they had come to check on the farm, to see the land manager and to give the children a change of scenery. The *Qa'immaqam* looked at them suspiciously. Why should he believe them? But he had no proof that they were up to anything. He left the room, glanced at the women and children, then went into the toilet along the corridor. Waiting in his office, Rushdi didn't utter a word, even when Talal said, '*Akalnaha*, we've had it.'

The *Qa'immaqam* came back in and sat at his desk. He explained to them that it was a dangerous day to be on the road, because a military coup had just been quashed. Jawad seized on this piece of information and started to ask him about the coup. The *Qa'immaqam* paused, then said that he recommended they return to Baghdad, as he expected there to be more military activity in the area. With that they got up, thanked him and left. Whether the *Qa'immaqam* believed their story or not, they could not tell. Talal wondered whether he had changed his mind when he saw the women and children, realizing they were harmless. Deflated, they returned to the cars and headed back to Baghdad.

When Hassan came home from university at the end of the day, he was surprised and relieved when Talal greeted him at the door, sombre and shaken after the day's adventure.

Hassan had had enough. His family was falling to pieces, and he couldn't stand idly by. Overcoming his reservations about his family's growing debt to Hamed Qassim, he rang him one Friday and asked for an appointment. Jamila took him to the meeting, and sat with him as he made the case for his brother's innocence.

Hassan told Hamed that there was no basis for the case against Rushdi, as the Cabinet had taken a collective decision to act against the individual in question, and that decision had not been in violation of the law, since the Communist Party had been banned at that time. To single out individuals with trumped-up charges spoke of revenge, not justice. He also pointed out that the court was regarded as a joke, and had acquired the nickname 'the clown court'. Hassan suggested that the hearing be delayed, secretly hoping that the court would run out of steam after seven months of mockery. He left Hamed's house with assurances that he would do his best to spare Rushdi the trial. A week later, the court announced that Rushdi's hearing had been delayed.

33

A Temporary Home

Visits to the Park

(1959)

IN JULY 1959, a year after the revolution had taken place, Rushdi's travel ban was finally lifted, again with Hamed's help. It had become clear that the government didn't have much use for Rushdi as a scapegoat: he had never posed a serious threat to them, and their appetite for revenge had been partially sated. The Mahdawi court fizzled out after sentencing thirty members of the old regime to death, four of whom were hanged. The executions took place because the court needed to validate its purpose and authority in the eyes of the public. Several military officers, the Arab nationalists who had participated in the revolution, were also executed for their roles in the attempt to overthrow Qassim.

The most intense trial was that of Said Qazzaz, the former Interior Minister. He never wavered, even when the judge insulted him in the foulest language. When the time came for him to speak in his defence, he stood up and with a firm voice told the court that it was little more than a circus of prejudice and abuse. He declared: 'This situation has convinced me that my fate was decided before the beginning of this trial. Yet as long as I remain alive, so long my fate is known, and since I don't fear death or the noose, I will make this statement so that my voice will be heard outside this court by all my Iraqi brethren. I confirm to them that I have served them with all my loyalty and honesty for more than thirty years. If I have committed any mistakes, let them know that I was denied the right to defend myself, and that I was brought to this place only to be insulted and cursed by officials. I have listened to the witnesses you have brought. There isn't a single specific event in which I have committed a legal violation.' He concluded, 'As a Muslim who sees only God's justice, as an Iraqi who has served for thirty-three years for national unity, I declare that I am proud for what I have given to my dear country.'

The news that Rushdi would be spared a trial and that his travel ban had been lifted raised the spirits of everyone in the Chalabi household in London immeasurably. Even Bibi was moved to suspend her ritual critique of life in the city (temporarily) as she counted the days and hours until Rushdi arrived.

When he landed at Heathrow with Ilham and his youngest son Muhammad, he was greeted by Hadi, Hazem, Raifa and Najla's husband Abdul Latif Agha Jaafar. Although both men rarely displayed physical affection, Hadi and Rushdi embraced and held each other tight as Hadi's eyes welled with tears. Ilham stood next to them, weeping quietly with relief and exhaustion. Meanwhile, Bibi waited impatiently at the flat near Regent's Park.

Now that his eldest son was safe, Hadi was keen to attend to a pending matter. It concerned the £100,000 that Abdul Ilah had asked him to deposit abroad on his behalf three years earlier. The money was still there, secure in Rushdi's bank account, but he and Hadi were the only people who knew of its existence now that the Crown Prince and his family were dead. Hadi decided to give the money to the royal family's next of kin, their cousin King Hussein of Jordan, who was now the head of the Hashemite family. He still hoped against hope that Hussein would help restore the monarchy in Iraq. It was decided that Najla's husband the Agha, who knew the King personally, would go to Amman and inform him of the amount to be transferred to him. The King accepted it with delight, asking a legal notary to validate the transaction.

Morale improved in London once Rushdi joined them. His arrival led to another migration a month later, from Regent's Park to a spacious flat in Oakwood Court, a large red-brick Victorian mansion block in Kensington. The entire family rejoiced in their more comfortable new surroundings; it felt as if the frayed and confusing first phase of their flight out of Iraq had finally come to an end. The apartment was bright and airy, with two large drawing rooms and a kitchen that was big enough to cater for the entire family. They employed a Spanish maid, who could never come to terms with the number of different family members who kept appearing in the apartment at all hours of the day.

Bibi became quite popular with the driver of the daily fruit and vegetable van that came to Oakwood Court to collect orders from the residents, as hers was always the largest order. Instead of the pound of onions ordered by other residents, Bibi ordered eight, and instead of three apples, she ordered four pounds of fruit. There were many mouths to feed, and however hard the van man struggled to fulfil her order, for Bibi the quantities always seemed a compromise when she compared them to the enormous crates of deliveries she had been used to in Baghdad.

There was less of a feeling of transience as the family settled into Oakwood Court. The upmarket neighbourhood, with department stores nearby in High Street Kensington, was convivial, and Hadi was grateful for the proximity of the parks, taking daily walks through their wooded areas before lingering over the many different plants and colours of the formal gardens, remembering the elaborate flowerbeds of the Deer Palace. Before heading home he would often sit on a bench, watching the proud peacocks strut past.

Rushdi was very grateful to be in London, and felt reassured by the city's atmosphere of order and safety. Baghdad now seemed to him a place of evil, danger and chaos. As he considered the events of the past year, he cursed the day 900 years earlier when Haroun al-Rashid had moved his court near Baghdad, marking the rise of the city. Then he would curse his sixteenth-century forefathers for taking part in the Ottoman conquest of Baghdad, rather than staying at home in Turkey. 'Then at the very least,' he would say, 'we'd have been Turks today, and still have a country.'

Outside the family circle he flinched whenever he heard an Iraqi accent, because he now feared his fellow countrymen. Yet the option of integration into the upper levels of British society was not open to him, as it was a self-contained world with little interaction with foreigners. The main quality the family noted in the British was a combination of reserve and politeness. Few of them seemed to be aware of what had happened in Iraq, let alone of the Chalabis' involvement in the country's story. Like many newcomers in London, Rushdi and his family had to create their own little world. Their status put them on a par with other affluent exiles, such as ex-Maharajas from India, Persian Qajar grandees and the Egyptian aristocrats who had left following Nasser's revolution in 1952. Rushdi began to explore possible business opportunities, and also took the time each week to visit his favourite restaurant, the Rib Room at the Carlton Tower Hotel in Chelsea, as well his favourite shop, Gardners', a new supermarket on Kensington High Street.

It was not long before the old topic of Dolphin Square was raised once more, with Rushdi arguing that his father had made the worst mistake of his life by not buying the development when he had had the chance. He elaborated at length on this point, enlisting his mother's support. Hadi's was not a confrontational character; he was more interested in keeping the peace, as his own father Abdul Hussein had been. He attempted to ignore the criticism directed at him by his son and wife, and became increasingly withdrawn.

London was regarded by the Chalabis as merely a temporary home; none of them was interested in putting down roots there. The family was gathering its energies to determine what its next step should be on the road to survival.

34

Return to the Shrine

A Life by the Sea

(1959–1963)

LEBANON WAS ENJOYING a golden age, most particularly Beirut. By the early sixties, the growth of its banking industry and the influx of money into the country meant that the city had a stronger claim than ever to be 'the Paris of the East'. It boasted *haute cuisine* French restaurants and elegant nightclubs, as well as rustic gourmet Lebanese restaurants with captivating views of the sea. The Lebanese felt, perhaps rightly, that they outdid the rest of the world in their hospitality.

Beirut had long been a place of refuge: the mountains behind it had for centuries been home to religious minorities such as Maronite, Druze and Shi'a communities fleeing persecution elsewhere. The Levantines had been expelled from Egypt by Nasser, and had been followed to Lebanon by Syrians and Palestinians. Following the revolution of 1958, there was an influx of Iraqis of the old regime.

Bibi loved Beirut, with which she had been familiar for many years, and as soon as her grandchildren were settled in British schools, she began to agitate to leave London. There seemed to be no good reason for her and Hadi to stay on in the cold city, whereas it made complete sense to her to move to Lebanon. She could speak the language, and she knew many people there. Hadi agreed, and they flew to Beirut in October 1959. Bibi assumed that her daughters and Rushdi would see sense, and would follow soon afterwards.

In the event, it was Hassan who was the first to arrive, taking a sabbatical in late 1959 and travelling to Beirut with Jamila. He had not seen his father since the revolution, and they had an emotionally charged reunion. Both had survived so much in the intervening year that it was difficult to know where to begin.

Hadi and Bibi found an apartment on the top floor of the Kazan building, in the elegant quarter of Verdun. There were many exiled Iraqis like themselves in Beirut, as well as their old Lebanese friends and acquaintances. Owing to the local pronunciation, Hadi and Bibi soon became Mr and Mrs Shalabi, or Jalabi.

Hadi's brother Muhammad Ali, a successful banker, was living in the city with his family, and he and Hadi started exploring the possibility of founding a new bank with Lebanese partners. While Muhammad Ali had been one of the founders of the Iraq state bank al-Rafidain, and was well-versed in international banking, Hadi's social and diplomatic expertise was to prove essential in smoothing the way and resolving disagreements with their local partners.

Bibi with her sons in 1960s Beirut: left to right, Jawad, Hazem and Talal.

Bibi established a rapport with the charismatic but often ill-tempered doorman of the Kazan building, Khalil, a 1948 Palestinian refugee from Nahariya in Israel. He began to run errands for her, and in time several members of his family came to be in her employ. His unique style of swearing caused great amusement among the younger generation, as he freely mixed the religious with the profane, mentioning prophets, shoes and vaginas in the same breath to tremendous dramatic effect. He had certain allergies, including an intolerance of chewing gum which was so extreme that he once stopped the lift to escape the torture of a woman chewing it.

Bibi ensconced herself so quickly in Beirut society that in the early 1960s her sons Hazem and Talal married women from prominent Shi'a

From left to right: Raifa, Thamina, Bibi and Najla at a function in Beirut in the 1960s.

Lebanese families, the Beydouns and the Khalils respectively. The substantial Lebanese Shi'a community originated mostly in the south of the country, and were predominantly impoverished and disempowered. They had suffered from similar prejudices as other Arab Shi'a, labelled ill-educated and backward. The Shi'a elite, however, were mostly well-educated and based in Beirut. They were generally more Westernized and open than their Iraqi counterparts, by virtue of living in a more Westernized society.

The rest of the family moved in stages. Thamina came to Beirut after her husband was released from jail in 1960, while Rushdi moved his family in 1962. Bibi and Hadi came to represent the nucleus of the family, a substitute homeland. The sense of exile had become so entrenched in the whole family's world view that it formed an integral part of their identity. Bibi developed a hierarchy of trust within the family: at the centre were her children, with whom she could express herself uncensored, whereas most non-blood relatives were viewed with caution. She became even more dependent on her daughters, insisting that one or other of them accompanied her wherever she went.

Try as he might, Hassan could not find work in Beirut. He didn't want to be a financial burden on the household, and he certainly didn't want to lose the only career open to him, so he decided to return to Baghdad with Jamila and take up his old post at the university. As an academic, he was able to avoid politics to some degree and dedicate himself to his work. Within a year of his return to the university he was made head of the Law Department.

Among his many students he encountered a few exceptional individuals, such as Jalal Talabani, a future leading Kurdish politician who would rise to become the President of Iraq and the first non-Arab leader of an Arab country. Another notable student and future President was Saddam Hussein, who at that time was rising through the ranks of the Ba'ath Party, which espoused an even more extreme version of Arab nationalism than Nasser did in Egypt. Saddam was academically unexceptional, but he had already established a reputation on campus as a thug. When Hassan failed him for his poor work, he became the target of Saddam's insults. He would hear him growling at the back of the room: 'Only a blind man would wear a ridiculous tie like that – or a member of the bourgeoisie.' Hassan refused to be drawn into an argument with the young bully.

Soon after Hassan returned to Baghdad, Jawad followed him. Like his older brother, he had not held a political position prior to 1958, and so was not in danger of being arrested.

The brothers lived in Jawad's house, a few doors down from the old family home, which had been seized by the government. Tormented by this daily reminder of the family's misfortunes, Jawad dedicated himself to securing the release of his father's frozen assets. Although Latifiyyah had been sequestered by the government and was gone forever, the mill was still functioning and generating some income, and there was a hope that some of Hadi's other lands might be reclaimed, as they had not been legally confiscated. However, there was little that the brothers could achieve in the absence of Hadi, the owner of the lands. Hassan decided that the only way forward was for an agreement to be reached with the Qassim government that would permit Hadi to return to Iraq for a short visit, on the understanding that he would not resume his political activities.

Authorization was cautiously given for Hadi to return to Baghdad in late March 1962. He knew that he was taking a great risk by going back, as he was inextricably linked with the old regime. His journey home was extremely poignant. He had been abroad at the time of the revolution, and had never had an opportunity to say his goodbyes. Now, at last, he would be able to bid his home farewell properly.

Hadi's grain-trading protégé from the old days, President Qassim's brother Hamed, met him at Baghdad airport, where an unexpected reception was held for him. And when he reached A'zamiya a large group of people were waiting outside his old house to welcome him back. As a public figure he had known people from all walks of life – tribal sheikhs and small landowners, bureaucrats and merchants – and many of these old acquaintances were now nostalgic for the old days, as the revolution had not delivered all that it had promised them.

Ni'mati was also waiting for him, delight lighting up his tired face. He had been worried for days that something might happen to Hadi at the airport should the authorities change their mind about his visit. The two childhood companions embraced each other clumsily in the manner of old men, the lack of words between them compensated for by the many pats they gave each other on the back. Ni'mati's life had changed drastically after the family's departure. He had aged quickly, retreating to the little house Hadi had given him near Kazimiya, close to his grown-up children. He still came to visit Hassan each week, and received a monthly salary, as did several other former household employees.

While in A'zamiya, Hadi stayed with his sons, but he did visit his house nearby. When he first entered it everything seemed reassuringly familiar, yet on closer inspection it was all quite different from how it had been when he left it four years earlier. The house was dark. Most of the furniture was covered with sheets that had collected dust and a fine covering of the sand that had seeped into the rooms. There was no life. However, he was so happy to be back that he didn't dwell on these differences. Instead, he immersed himself in the sights, smells and tastes of his beloved home and country. He felt alive again. Being in Iraq once more reinforced his sense of how inextricably linked to and defined by his country he was, just as being away from it made him feel powerless and redundant.

He was overwhelmed by the warmth of his reception, especially when a large banquet was held in his honour by Kazimiya's leading families, the Jawahiris, Istrabadis, Hasouni, Kazimi, Ugaili and Kanaan. He caught up with many old friends and acquaintances, and momentarily forgot all that had passed. Yet when he entered the court-yard to the Kazimiya shrine he gave in to his tears as he stood by the grille of the tomb in the inner room. Kazimiya lay at the very core of his being, and since 1958 he had always kept a picture of the shrine with him.

The Communists' power in the town was waning, now that Qassim had been pressurized by the West to curb Communist activities in the region as a whole. The Nasserists and increasingly the Ba'athists were on the prowl. There were some loud mutterings among them about Hadi's presence in the area as he seemingly moved freely among his old network of friends and erstwhile colleagues. However, he was not allowed outside Baghdad, and his requests to visit the shrines of Najaf and Karbala were rejected.

The state security forces became increasingly suspicious as to the purpose of his visit, and articles by Yunis al Tai in the official newspaper, *al-Thawra*, criticized the government for allowing such an anti-revolutionary figurehead to enter Iraq. When his sons urged him to leave the country promptly, Hadi put up as much resistance as he could, but his sons' arguments were stronger. He was so distressed on the day of his departure that he was unable to speak.

In 1963, less than a year after Hadi's visit, another *coup d'état* took place in Iraq. The Arab nationalist wing within the military, which had supported Abdul Salam Arif in the 1958 coup, challenged Qassim's rule on the grounds that he did not favour pan-Arab policies, and had given the Communists too much leeway. Qassim had become politically isolated over the last year as he attempted to strengthen his grip on the country through his security services and by undermining any opposition.

A growing number of army officers had joined the Ba'ath (or Renaissance) Party, which had existed since the late 1940s. It was Arab nationalist in orientation, and espoused some socialist ideology, mostly linked to economic matters. Originally inspired by the conservative doctrines of a group of nineteenth-century German thinkers, the party operated on a cell-based system. The bloody coup of February 1963 was led by the Ba'athists, who had put a lot of energy into rallying support for their cause. Qassim, like Abdul Ilah and Nuri before him, had refused to believe the rumours about growing dissent within the armed forces. He was arrested and executed by firing squad outside the Ministry of Defence, where only five years earlier the mutilated remains of Crown Prince Abdul Ilah had hung.

Hassan and Jawad found themselves in yet another nightmarish situation; there was talk that the new regime might arrest them because of their links with Qassim's brother Hamed. Fortunately, the rumours came to nothing. Back in Beirut, Hadi tried to come to terms with the fact that his children were in danger once more, and that his prospects of returning to Iraq were now non-existent. Yet he still held out hope that the monarchy might be restored with the help of King Hussein of Jordan and the Shah of Iran. In late 1959, when Hadi had first moved to Lebanon, the Shah had sent the head of his secret services, Teymour Bakhtiar, to meet several prominent Iraqi exiles for discussions about overthrowing Qassim's regime and restoring the monarchy. Hadi had hosted several of these meetings. Subsequently, King Hussein had sent various messages to Iraqi exiles in Beirut through the Jordanian Embassy, exploring the possibility of taking action against the Iraqi government. The plan was always the same: to contact officers in the army who remained loyal to the old regime and convince them to stage a coup with the support of Iran and later Jordan. A combination of factors prevented these schemes from taking shape, and many Iraqi exiles later felt that the Shah had never really intended to act for fear of incurring Western wrath.

At this time, in 1963, the relationship between the Iraqi exiles and the Iraqi Army was undoubtedly poor. The new government officially requested the Lebanese government to extradite several leading exiles, Hadi among them, to Baghdad, accusing them of conspiracy. Hassan received threats in Baghdad which he relayed to his father in Beirut. Rather than comply with the Iraqi government's demands, the Lebanese authorities suggested to the exiles that they should leave Lebanon temporarily until the situation had calmed down. Thus, Hadi and the former Iraqi Prime Ministers Tawfiq Suwaidi, Ali Jawdat Ayubi and Ahmad Mukhtar Baban departed for Europe. As the charges against the exiles were based on hearsay, the Iraqi government could not apply pressure on European governments to extradite its citizens. Hadi sat out two months in London waiting for the storm to pass.

The Ba'athists' takeover in 1963 proved short-lived. The party was disorganized, and could not sustain its grip on power; it was, moreover, ideologically divided and indecisive about where the future of Iraq lay. This lack of cohesion was reflected within the army and the National Guard. Once more Iraq's troubled relationship with Nasser of Egypt came to the surface, and factional fighting spilled out onto the streets. Abdul Salim 'Arif, a Nasserite but not a Ba'athist, took charge of the military and established order. He ruled for the next three years by decree as Supreme Commander of the army and President, perpetuating the military rule that Qassim had established. Under his reign the Republican Guards were established, an elite fighting unit that would play a brutal repressive role in the future. During 'Arif's dictatorial rule, civil liberties were suspended and clan and tribal loyalties superseded those of state bureaucracy.

* * *

Once things had quietened down in Lebanon, Hadi returned to Beirut, where he found himself forced into premature retirement. His absence from Iraq meant that his son Jawad was responsible for overseeing his properties in Baghdad, with which he provided an income for the family in Beirut, while Rushdi remained in charge of the family's overseas assets. Hadi yielded grudgingly to the new generation. He knew he was fighting a losing battle, and that he had lost much of his standing as the all-powerful head of the family.

Neither Rushdi nor Bibi would let him forget his Dolphin Square mistake, and they dismissed any plans he might have with respect to real estate or agriculture. Stock markets and speculation were the way forward, Rushdi argued – things his old-fashioned father didn't know anything about. Bibi's comments were equally disparaging: 'Hadi, you're too old, what do you know? The children are more capable than you are; Iraq is gone.' She resolutely confined his achievements to the past.

Bibi remained angry, bitter and resentful about what had happened to her family. She wanted to forget about her life in Iraq, unlike some of her sons and her husband, who yearned for what they had lost. It took the close encounter with death which the turbaned Indian clairvoyant in Cairo had predicted many years earlier to make her long for home. One day in late 1963 she collapsed at home and had to be rushed to hospital. A perforated ulcer was causing severe internal bleeding, but the doctors could not immediately locate the bleeding artery, and Bibi fell into a critical state as her blood pressure plummeted. For a time she was on the brink of death, and Hadi, Rushdi, Thamina and Talal prayed frantically for her. When she recovered, they appreciated just how precious she was to them, just as she realized how important her family and her homeland were to her.

35

Of Carpets and New Blood

The Emergence of New Patterns

(1967)

IN 1967 THE family moved to a building in the neighbourhood of Bir Hassan, in south Beirut, which enjoyed a sparkling view of the sea. Their new home was a pristine modernist four-storey building, which they called simply 'the Building'. Made of stone and concrete, it was understated on the outside and spacious on the inside. The exterior stone was an unusual reddish-brown colour, reflecting the red of the surrounding earth. Two large palm trees flanked the entrance, an oblique reference to Iraq as they were not native to Lebanon.

Although the family still retained its Iraqi identity, Lebanon had pervaded it permanently. Bibi and Hadi now had several half-Lebanese grandchildren – Ali, Reem, Peri, Sarah and Bashar – the offspring of Talal and Hazem. These children's francophone and Francophile affectations and interests, owing to Lebanon's historical and cultural links with France, contrasted with the distinctly anglophone Baghdad-born older grandchildren. Even their Arabic dialects were different, with the older grandchildren speaking in more guttural Iraqi tones, whereas the younger ones had a softer Lebanese speech pattern.

But it was the law of Bibi that dominated the Building, and for all her love of Lebanon she still hailed from another land. Dogs were not allowed, because she didn't like them, and could always use the excuse that they were considered dirty animals in Islam. Cats were disdainfully tolerated, as Rushdi's children Hussein, Nadia and Muhammad were obsessed with them, so long as Bibi didn't have to see them.

Following his grandmother's dictum to a 'T', Salem, Jawad's youngest son, bought a goat as a substitute for a dog. For the children who were born in Lebanon, Iraq and the family's life there acquired a mythical status. This was personified by Hadi's mysterious and dignified aura. He never talked about Iraq, especially not to his son Rushdi. His was a silent lament, expressed through small sighs and muted inhalations between sips of tea.

Not long after the family had settled into the Building, Khalil, the Palestinian doorman of the Kazan building who Bibi was very fond of, moved his family to Bir Hassan, followed by two brothers and a cousin. The Eritrean maids Hidat, Betahon, Pitcher and Rishan dominated the corridors and stairwells in their starched white aprons as they chattered in high-pitched Amharic. Exiles from their war-torn homeland, they were fiercely honest and loyal, yet also stubborn and volatile, often shouting back when aggrieved and locking themselves in their rooms in protest. Some of them had crucifix tattoos on their faces, a common practice among Ethiopian and Eritrean women. They were supported by a team of local dailies. Finally there were the Egyptian and Sudanese cooks, who ruled the kitchens and found great favour with the family by recreating the dishes of their homeland, in particular *kubbat hamud* (meatball and rice stew with turnips), *sabzi* (green stew with fenugreek) and of course the much-loved *fesanjoon*.

Each floor of the building vibrated with a different variety of music. The pop songs of the Beatles and the Doobie Brothers competed with Bibi and Hassan's classical Arabic music from their quarters on the third floor, while Sudanese music reverberated from the kitchen and Eritrean folk tunes wafted with the fragrance of incense from the maids' rooms.

Bibi still adhered to a strict social code when it came to the women in her family. By the mid-1960s she was terrorizing her adolescent granddaughters Nadia, Kuku and Zina in Beirut with her judgements on their behaviour. It seemed as if her social points of reference had become stuck in the Baghdad she had left years earlier. She often used antiquated Iraqi terms which sounded faintly comical to her granddaughters, although they would never have dared to laugh to her face. When miniskirts came into fashion in 1965, the girls embraced them enthusiastically, shopping in fashionable Hamra Street with its growing selection of *prêt-a-porter* boutiques. However, without fail Bibi would reprimand them whenever she saw them in short skirts. She didn't exactly forbid them from wearing them, but would simply tell them they looked like prostitutes. She was equally horrified to discover that her granddaughters paraded along the seafront in swimming suits – or, she described it, 'naked for the world to see'. She would loudly lament her misfortune at having such slovenly relatives, conveniently forgetting the disapproval she herself had incurred when she had decided to throw off her *abaya* and wear short sleeves decades earlier. For their part, Raifa's daughter Zina and her cousin Nadia felt a combination of outrage and shame towards their grandmother. Nadia dealt with Bibi's criticisms with humour, concocting images of her in a miniskirt – which would not have been a pretty sight.

Conservative by nature, Bibi adhered to a self-formulated set of rules and customs. She held herself to be above reproach, and expected

Bibi at a party in Beirut in the 1960s.

compliance from her tribe. Her view of women had not changed since the days when her own daughters had been young, and she may have felt pressured to preserve her granddaughters' virtue in a society that wasn't her own. Regardless of the freedoms the younger generation of women was increasingly able to enjoy, including access to education and careers, Bibi understood a woman's role to be primarily that of wife and mother. For her, a single woman beyond a certain age was not only socially undesirable but a serious burden.

As Hadi's mother Khadja had done so many years earlier, Bibi held court in her sitting room, an idiosyncratic and ageing woman with a penchant for silk dressing gowns and high-heeled mules. She seemed unaware that her criticisms might be destroying her granddaughters' self-confidence. Although she could not force them to change their clothes when they visited her, she managed to instil in them a sense of difference, marking them apart from their friends, who didn't have to face such harsh judgements from their families. The natural challenges of adolescence were complicated for the girls by a deeper sense of alienation, that of exile and of having lost their original identity. In time, some of these girls grew up with conflicting identities, a part of them remaining deeply entrenched in their family and its experience of loss, another part rebelling against this legacy as they tried to make sense of the world.

In the wake of the 1958 revolution, the family came to view itself increasingly not just as exiles, but as being at odds with the most popular political trend in the region: Nasser and Arab nationalism. Nevertheless, they remained intently engaged with the latest developments.

Their most immediate access to the political opinions of the common people was provided by their Sunni Beiruti drivers. Bibi's chauffeur Shehab al-Din al-Arab may not have been as grand as his name (there was a murky halo of suspicion around his character and his past), but that did not diminish her fondness for him. Shehab came every day after her siesta and drove her to the seafront, where she would go for a walk along the promenade, the Corniche. His younger brother Nabih became Rushdi's chauffeur, and a popular hit with the grandchildren. He had a great fondness for Johnny Walker Black Label, and the gift of a bottle or two always went a long way in securing his good will. A colourful figure in his early thirties, he had a shameless appetite for sex and women. He often shocked the girls by showing them pornographic photographs while he was driving. They found his lewdness disgusting, but were secretly thrilled that he seemed to be treating them like adults (unlike Bibi). Nadia rather enjoyed these incidents; they were her way of resisting her overbearing father and grandmother.

The radio in Rushdi's midnight-blue Oldsmobile became a political battleground. In the morning, Nabih drove Rushdi and his son Muhammad to work and to school. Father and son listened to the BBC World Service, the Voice of America or Israel Radio's Arabic-language station. The tone of the reports was restrained and poised. In the afternoons, the ambience in the car was entirely transformed when Nabih picked up Muhammad alone. By then the radio would have been retuned to *Sawt al-Arab* from Cairo, Nabih's favourite station and the complete opposite of the BBC, noisy and unstructured. It was the station that Muhammad's young uncle Ahmad had listened to at school when he wanted news of home.

When the 1967 Arab–Israeli six-day war broke out, the presenters of *Sawt al-Arab* strove to outdo each other with their exaggerations. At the precise moment that the Israeli air force had destroyed its Egyptian counterpart, *Sawt al-Arab* was declaring victory for Egypt, claiming that the Egyptian air force had destroyed the Israeli planes amongst a host of other fictional triumphs. Two days later the Israeli military captured the Golan Heights in Syria, the West Bank in Jordan, the Sinai desert, the Suez Canal, and was within close proximity of Cairo. Egypt had lost nearly all of its military equipment, and thousands lay dead or wounded. The war represented an overwhelming humiliation for the Arabs. Nasser's lies to the people were exposed, and a feeling of dismay grew among the millions of listeners to *Sawt al-Arab*. It was the beginning of the end of Nasser's influence and the dominance of Arab nationalism.

The Chalabis kept an open house, and their most colourful visitor was undoubtedly *Hadji* Abbas Faili, the most successful carpet merchant in Baghdad. *Hadji* Abbas had taken over the family business from his equally brilliant uncle several years earlier, and both nephew and uncle had enjoyed a longstanding relationship with Hadi that dated back to the thirties. They had supplied a large percentage of the precious Persian carpets he had collected over the years. On moving to Lebanon Hadi had arranged for many of these to be smuggled out of Iraq overland, through the Syrian plain and the mountains to Beirut. A selection of other valuables, such as his collection of silver sculptures, crystal and porcelain, had been moved in a similar way, in small batches discreetly over time, but everything else remained in Iraq.

Hadji Abbas was a member of the Faili Shi'a Kurdish community of Baghdad, who were distinctly woven into the fabric of the city, dominating the markets in the old quarter, and who would be among Saddam's worst victims in the years to come. A small, dark, wiry man with a large nose and big ears, *Hadji* Abbas was invariably dressed in a beige or grey suit without a tie. He would always arrive at the Building without warning, causing a great commotion as he pulled up outside the entrance. Reaching the third floor, he would call out loudly: 'Chalabi, I have new carpets to show you!' This announcement would be followed by an extended lunch, with all the family gathering around to listen to his stories.

Hadji Abbas had travelled extensively in the pursuit of his business. He had visited Malta before the outbreak of the Second World War, seeking out British admirals as potential buyers for his luxurious goods, but had been unimpressed by their parsimony. On one occasion, one of a pair of sixteenth-century Shah Abbas gold-embroidered carpets fell into his hands. It was regal beyond imagination, and inspired him to write a letter to the Queen of England herself, proposing that she buy it. *Hadji* Abbas was scandalized when he received a letter from Buckingham Palace stating that there was no budget to buy such an expensive item at that moment. It shattered all his preconceptions about the British monarchy.

He told a story of how he had once had an ulcer operation in London while he was visiting a relative. Afterwards, he asked his nephew to get him a tasty, oily dish from an Iraqi restaurant on the Fulham Road. The nephew smuggled it onto the ward under his coat, and when the plump matron found out and started berating him, *Hadji* Abbas just screamed at her: 'Get out, you buffalo!' These were four of the fifteen words he knew in English.

The family elders would hungrily ask him for news from Baghdad and he obliged in his street Baghdadi, which they hadn't heard spoken by anyone else in years. Rushdi was particularly fascinated by *Hadji* Abbas's tales, although he was horrified by his table manners: he dug directly into the dishes with his fork, ignoring the interim step of his plate.

After lunch *Hadji* Abbas would walk out onto the balcony and call down to his workers to start unloading his latest goods. Carpet after precious carpet was unfolded in the drawing room before the haggling started. Bibi was entirely ignorant about carpets, but was nevertheless prepared to chip in with her opinions, prompting *Hadji* Abbas to turn from Hadi to her and ask, 'Madam, when are you going out today, so I can do my job properly?' Once his business was concluded he would often leave his carpets at the Building for months, giving Hadi the opportunity to ponder them before he appeared again.

Hadi had always loved carpets. He had grown up in a culture that was heavily influenced by Persia, where carpets were greatly valued. Carpets created spaces, not simply as objects on the floor, but around which people lived and upon which they prayed, ate and slept. For Hadi, they were also a manifestation of a spiritual beauty inspired by nature, which he loved, whether they had ornate floral sequences or medallions that spread out from the centre; whether they were Isfahanis, Tabrizis or Serafians with their intricate pictorial depictions of fables in many colours. Carpets were an expression of creativity that Hadi cherished; unable to create them himself, the next best thing was to collect them, discovering their origins and assessing their delicateness according to the number of stitches per square centimetre. Besides being an integral part of his culture, they were also a measure of his success, as he collected valuable and rare specimens.

There were other, less agreeable, aspects of life in the Building which brought back memories of Iraq. In 1966 Hadi's Lebanese friend Kamil Muruwa was assassinated. Muruwa was a dynamic Shi'a writer from south Lebanon who had worked his way up through the ranks of Lebanese journalism and founded his own newspaper, *al-Hayat*, in 1946. A forthright critic of Jamal Abdul Nasser, he was shot dead while sitting at his desk. All fingers pointed to Nasser, who had resented

Muruwa's censure. There was much anger and rage in the Building at the murder, which stirred up painful memories of the 1958 coup and Nasser's role in it.

There was other tragic news in 1966. A few months after returning from the *Hajj* and visiting Bibi and Hadi in Beirut, Ni'mati died of kidney failure. Hassan and Jawad buried him in Najaf, near other members of the family. When Hassan broke the news to him over the phone, Hadi was devastated. He had known Ni'mati since he had been fifteen and Ni'mati eleven, and remembered how Ni'mati had become separated from his parents at the shrine. Now Hadi was the one who was inconsolably lost.

36

The Ruins of Kufa

A Coup and a Birth
(1968–1972)

THE BA'ATH COUP of 1968 followed a similar pattern to the previous ones, with the use of the military to seize power in Baghdad at the end of July 1968. In the past five years the Ba'ath Party had undergone restructuring, recruiting high-ranking Iraqi Army officers and building up its intelligence wing, which under Saddam would develop into a brutal, repressive organization. The use of violence under this regime would reach an unprecedented level, with access to power limited to an ever-shrinking circle of patronage that was based on mistrust and exclusivity, limited to members of Saddam's tribe in Tikrit. A wall rose around the country, making Iraq darker and even more inaccessible to the rest of the world, although because of its oil wealth it could continue to exert its weight. After Jawad, unable to tolerate the pressures of the new regime, finally left Iraq, Hassan had prided himself on the fact that, as the only family member remaining in Baghdad, he was representing the Chalabis and perpetuating their presence in their homeland. But within a few months it was clear that he too could no longer stay.

In 1967 Hassan had joined a group of intellectuals who wanted to found a modern secular university in the ancient city of Kufa. An important centre for Arabic literature, theology and philology in the ninth and tenth centuries, Kufa was also the birthplace of the renowned Kufic script. There was much enthusiasm for the project, and its supporters included prominent Shi'a figures from afar such as Sayyid

Musa al-Sadr, a liberal Iranian cleric of Lebanese origins, and a considerable number of the Iraqi intelligentsia. When the Ba'ath Party took over the government in 1968, the society came under attack and one of its members, Kadhim Shubbar, was arrested. It transpired that the secret police had been closely following its activities, convinced that it was merely a front for Shi'a political activity. In early 1969 the Ministry of the Interior issued an order closing down the society completely, condemning it as a Shi'a Iranian project run by agents and spies. The same old theme had resurfaced, with the Sunni ruling elite accusing the Shi'a of unpatriotic conspiracies. News spread that Shubbar had been tortured, but had refused to denounce his colleagues.

Hassan lived in a permanent state of fear, expecting to be arrested at any moment. Rumours of the new regime's human rights abuses were rampant, and some of his colleagues had already been interrogated roughly by the authorities. However, he dared not risk attracting unwelcome attention to himself by attempting to leave the country in the middle of the academic term, despite a desperate telephone call from his father urging him to do so. Travel was restricted, and the telephone lines were bugged. Hassan and Jamila started staying with friends in order to avoid being alone at home. They were refugees in their own city, living in a state of uncertainty.

Hassan decided to risk travelling when the summer term ended in June; for him, leaving represented a major defeat, as there would be no one left to carry on the family name, which had existed for generations in Iraq. He was also devastated at the prospect of abandoning his academic career, which he knew from experience could not be easily resumed elsewhere. He and Jamila knew there would be no going back as long as Saddam Hussein held the reins of power.

Unable to find a teaching post in Beirut straight away, Hassan threw his energies into contacting Iraqi political dissidents. He also found a captive audience in his youngest brother Ahmad, now a postgraduate student in America, with whom he maintained a regular correspondence. As a sixteen-year-old Ahmad, desperate to leave his East Sussex boarding school, had applied to the Massachusetts Institute of Technology, and his talent for maths meant that he had been accepted. America seemed to him to promise more opportunity, freedom and energy than England, and his time there coincided with a period of great social upheaval, including the nationwide anti-Vietnam War protests, Martin Luther King's Civil Rights movement and the dramatic outbreak of civil disobedience at the Democratic Convention in Chicago in 1968.

In 1969 Ahmad completed his PhD in Mathematics at the University of Chicago. Before joining the American University of Beirut as a mathematics professor, he travelled with Hassan to Tehran, where they met representatives of the Iranian government to discuss the possibility of a counter-coup to the Ba'athist regime in Baghdad.

During the trip the brothers met and forged a lifelong friendship with Mulla Mustafa Barzani, the charismatic leader of the Kurds who was fighting for autonomy for his region in northern Iraq, and had sought refuge in Iran. Any effort to foment a coup against the Ba'ath regime was of great importance to Ahmad, and he enthusiastically assisted Mulla Mustafa in every way he could, particularly through his growing political and media contacts in Beirut. In 1972 he would arrange a meeting with Saleh Samerai, an Iraqi officer and an associate of Mulla Mustafa's who was plotting to overthrow Saddam. Samerai's body was found dumped in a ditch two days later. That same year, an assassination attempt narrowly failed against Abdel Razzaq al-Nayif, who had led the 1968 Ba'ath coup before being ousted thirteen days later while in London. In 1979 he would be shot dead in the lobby of the Intercontinental Hotel on Park Lane. The message was clear: opponents of Saddam's regime could be punished by death, wherever they were.

As in 1968, it seemed that once again the Shah had another agenda, and was using his dealings with the Iraqi dissidents as bait to force through a treaty with Saddam. The Persian Gulf Treaty was eventually signed under US auspices in Algiers in 1975. As part of the bargaining process, autonomy for Iraqi Kurdistan was reneged on, and Mulla Mustafa was expelled from Iran to the Iraqi side of the border, where he was a wanted man.

When Sayyid Musa al-Sadr appeared in Lebanon, the family welcomed him warmly and he became a regular visitor to the Building. The Chalabis were close to his relatives, the Sadrs of Iraq. Rushdi gave him a car, and Bibi often sat with him discussing the finer points of Islam.

Sayyid Musa was a powerful advocate for the rights of the poor and disenfranchised Shi'a.

Ahmad had become captivated by the demure beauty of a young woman he had spotted at several social functions. Her name was Leila, and she came from a leading political Shi'a family. Her father, Adel Osseiran, was one of Lebanon's 'men of independence', having fought the French occupation in the late 1930s and early forties. Later he served several times as Speaker of Parliament, the highest political position constitutionally permitted to a Shi'a in Lebanon, and held several posts as a Cabinet Minister.

Ahmad decided to pursue Leila, but he had to take into account the conservative customs of the time and seek permission from her strict father. After a suitable period of official courting, the couple married in early 1972, with Sayyid Musa al-Sadr officiating. They rented an apartment in a building near the rest of the family with stunning views of the sea, where I was born – their first child. Both sets of grand-

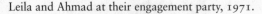

Leila and Ahmad at their engagement party, 1971.

Tamara and Ahmad in late 1970s Lebanon.

parents were delighted by my arrival, and I enjoyed both Bibi and my grandfather Adel's undivided attention for four years before my sister Mariam's birth, followed by those of my brothers Hahsim and much later Hadi. Many of my childhood memories are happy and warm, with my time divided between urbane Beirut, pleasant strolls with my father on the American University campus and running wild in my maternal grandparents' orange groves by the sea.

37

Civil War

A Shattered Sanctuary

(1975–1982)

IN THE EARLY 1970s Lebanon began its descent into civil war. The growing unrest was linked to a variety of complicated factors, including the composition of the country's population, its political system and even its geography; and it was exacerbated by the large influx of Palestinian refugees after 1967. Beirut had become the headquarters for the Palestine Liberation Organization (PLO), which had been established to represent the rights of the disenfranchised Palestinians following the creation of the state of Israel of 1948. It was headed by Yassir Arafat, whose indifference to Lebanon's sovereignty would become an issue of serious contention with the Lebanese Phalange Christians.

The civil war in Lebanon was a melting pot that included every possible Middle Eastern ingredient, from Nasser and Arab unity to the liberation of Palestine, Christian revivalism, Nehru-style socialism and the Shi'a awakening, through to a Greek Orthodox brand of Communism. Every country in the region had a stake in the outcome, as citizens were chased out of their homes, forbidden to enter certain neighbourhoods and denied basic services, with many being forced to leave the country altogether.

The war broke out in 1975. My earliest memories include days spent in underground shelters and seeking cover in corridors, as well as an occasion when I narrowly avoided death when a bullet hit the wall near my bed. In those early days, Beirut was still a habitable city with rich cultural diversity. Growing up there during a civil war only seemed unusual in retrospect. As a child, I felt that the outbreaks of violence that plagued Beirut were normal, because I didn't have any other experiences to compare them to. That doesn't mean I didn't feel fear at the sounds of bullets and explosions, but they became a part of life.

Even the parades of militias in their makeshift uniforms, with machine guns welded onto pickup trucks, did not disrupt our everyday routines. We children still went to school, shopkeepers to their shops, employees to their jobs, teenagers to their parties and housewives to their social gatherings. However dramatic the outbreaks of fighting, life resumed almost instantly afterwards. There was more concern about the dwindling supply of water, the power cuts and the diminishing availability of goods. People's energy was gradually consumed by the difficulty of accomplishing the most mundane tasks. Coming home with a packet of bread was often an achievement worthy of celebration. Yet, as daunting as everyday life came to be, it could still be full of pleasure.

As a small child I was unaware of the fact that I was the product of two cultures, and I enjoyed spending time with both sides of my extended family. My mother's family owned lands in the south which were a source of great adventure and fun for me as I played under the loving gaze of my grandfather Adel. The colourful and exotic residents of the Building spoke differently from everyone else, but they didn't appear particularly foreign to me. Even my father, whose Iraqi accent never faded, seemed to me to belong in Beirut. I knew my grandparents came from Iraq, but I did not know much about the country other than that the residents of the Building had important ties to it.

The civil war in Lebanon was a source of much anxiety for the Chalabi family. Everyone had an opinion, a side they supported, as they argued over the lunch table or tried to anticipate future developments. As the family discussed what the Americans would do next, say, if the Syrians responded to the Palestinians in this or that manner, the names of politicians and world leaders cropped up as often as references to food or the weather.

Bibi inevitably took the war very personally. She could not believe that this magical land had descended into violence and strife. The war

Bibi on a balcony in Beirut in the 1970s.

was a direct attack on her life and well-being, and she resented its disruption to her daily routine; she hated having to sit in the underground shelter, waiting for the fighting to abate. She was, however, very grateful for the white-haired former doorman Khalil's presence in the Building, imbuing him with almost supernatural powers which she believed would shield her from the fire of Palestinian Kalashnikovs from the nearby refugee camps. At the beginning of the war Khalil acquired a simple 9 mm gun for security, but he kept upgrading his weapons as the war turned more violent until he had acquired a .45 Magnum. Although he worked at my grandparents' house, Khalil drove me each day to my classes at the Collège Protestant, and was very much part of family life in the Building.

As rockets and *katyushas* fell from the skies, Bibi and Hadi became a source of increasing concern for their children. Both were in good health, but access to doctors was not easy in a war-torn city. They were less mobile than they had been, and could not reach shelter readily. Their presence in Beirut became more hazardous as the random nature of the fighting increased. Snipers became a new phenomenon, positioning themselves in strategic areas around the city, especially on either side of the 'Green Line' – the imaginary line that divided Beirut into east, with its Christian population, and west, where Muslims were the majority. The Building was not far from the Green Zone, and the neighbourhood changed dramatically during the war, attracting refugees, some of whom were armed. By the late seventies there was a gradual but steady decline in the quality of life, with a loss of public services and an influx of squatters and refugees who took over the homes of those who had fled the war.

The physical transformation of the city was dramatic. From a flourishing cultural cosmopolitan hub, Beirut became a carcass inhabited by displaced peoples. The elegant patio furniture on the balconies gave way to lines of washing. With the refugees came armed militia men, car bombs and explosions. Civil society still functioned, but schools struggled to stay open, with teachers running the gauntlet of shootouts in the streets in order to arrive on time.

My grandparents didn't leave Beirut overnight. For a long time Bibi only agreed to temporary absences in London, but as soon as there was a lull in the war she would persuade the family to let her return. But by the early 1980s she could no longer resist her children's worries for her safety, especially as most of them had already moved their own families out of Lebanon, and finally left for London. Hassan, Jawad and Hazem continued to use the Building as a home, leaving the city when the fighting became worse, and then returning as Bibi herself had done formerly.

Bibi didn't take well to the move, and continued to follow the war in Lebanon closely on the news. There was something absolute and traumatic about this departure. Beirut had long been a refuge for her and Hadi. To lose it at their advanced age was extremely distressing for them both.

After Bibi and Hadi had left with Rushdi and his family, Khalil and his family continued to live in the Building, protecting it from robbery and squatters and ferrying all its precious carpets, crystals and porcelain in a disused ambulance to a safe house. One day in September 1982, during the Israeli occupation of Beirut, Christian Phalange Lebanese forces seized Khalil, one of his sons and his two brothers, Aziz and Mahmud, while Israeli soldiers looted the Building. My father Ahmad's brothers were all out of Beirut at the time, some taking refuge in the mountains to the north, and it subsequently took Ahmad many months of investigation, of shifting through hearsay and lies, to discover that Khalil, his brothers and his sons had been killed at the Sabra and Shatila refugee camp that September. Thousands of Palestinians were killed, mostly women and children as the Palestinian combatants had already left the city under an international agreement.

Of the many episodes that occurred during the Lebanese War of 1975 to 1990, the Israeli occupation was the first that looked as if it might lead to some sort of culmination; it appeared that the battle that was raging all over the nation would finally lead to the emergence of a new Lebanon. The election of Bashir Gemayel as President in August 1982, even allowing for the blatant Israeli manipulation that achieved it, seemed to constitute for many of his Christian supporters the fulfilment of their wildest dreams. The letdown that followed his assassination less than a month later was channelled by these supporters, hungry for revenge, into the massacre at Sabra and Shatila.

Aside from the womenfolk in Khalil's family, only one of his sons survived the war, and that was because he was not in the Building on that fateful September day. News of Khalil's death was received with anguish and sorrow in London, especially by Bibi, who prayed for his soul, unable to imagine how his family could survive this tragedy.

38

Creased Maps

A Move to a Different Land
(1980s)

BIBI'S SECOND EXILE was as unwelcome as the first had been, over twenty years earlier. As they had done when she moved to Beirut, many of the family followed her back to London, Rushdi and Thamina's families among them. Her daughter Najla was there to help them to settle in, having spent the years since the 1958 revolution living in Surrey. She had weathered her difficult marriage to Abdul Latif, who had proved to be a ladies' man, and had worked hard to come to

Thamina and Bibi during a trip to Switzerland in 1974.

terms with her situation as an exile, and to reconcile her conservative Iraqi Muslim culture with life in Britain. She felt very vulnerable with respect to her four daughters, whose engagement with British life transgressed a multitude of sacrosanct social canons in her eyes. Whether it was the way they dressed and spoke, the friends they made or the way they addressed her, their behaviour challenged her understanding of the world. In this respect she was unwittingly her mother's daughter. Her attempts to instil in her daughters the values that she herself had grown up with made her a far more hands-on mother than her sisters Thamina and Raifa ever were. Her exile was in many respects the hardest.

In the late 1970s my uncles Rushdi, Hassan and Talal decided to found a bank in war-free Jordan. Their connection to the country was based solely on the family's inherited sentimentality towards the Iraqi royal family, which extended to the royal family's cousin, Jordan's Hashemite King Hussein. They asked their younger brother Ahmad, my father, to help them, and he took a one-year sabbatical from his teaching post in Beirut.

I don't recall much of our move to Jordan in 1980; I thought we were going there on holiday until I was taken to my first day at school in Amman. In my mind, Beirut remained home, our house was still there, and we returned fairly frequently, subject to the airport's being open – not always a given. The Iranian revolution had taken place a year earlier, taking the world by surprise. Some members of the family were shocked and saddened by the overthrow of the Shah, viewing it as further evidence of the fickleness of the West towards its friends, while others were intrigued by this dramatic revolution which brought with it a radical Islamic post-Marxist rejection of the West. The prosperity Iran had enjoyed during the 1960s and 1970s had not been shared by all; across Iranian society, social and economic resentment had been compounded by a feeling of political disenfranchisement that had its origins several decades earlier. The Shah had tried to channel some of the economic benefits to the people, but his coterie had benefited far more. The revolution represented the culmination of the successful mobilization of the masses; the presence of the cleric Ayatollah Khomeini at its helm raised tensions across the region.

My awareness of that other place called Iraq became far more acute when the Iran–Iraq war broke out on 22 September 1980. Supported by the West, Saddam Hussein declared war on Iran, which he labelled an anti-American theocratic state, thus pre-empting any potential alliance between Khomeini and a small core of Shi'a Islamists in Iraq. Until then the Lebanese Civil War had commanded most attention in our household, as many members of our family were still living in Beirut. I was familiar with the Civil War, with the names of the places I heard about on television, and with many of the images I saw on television and in my parents' newspapers. With the outbreak of the Iran–Iraq war, my father started to pore over detailed maps of the two countries in his efforts to follow the progress of the battles. However, neither my mother nor I was familiar with the places mentioned. The war would last eight years and would cost over a million lives, creating deep and insurmountable traumas in both countries.

My uncle Hassan and his new wife Jamila became regular visitors at our home in Amman. I knew they had been friends for a long time – which didn't surprise me, because they both seemed very old, being in their sixties – but I learned from my parents that they had only decided to get married earlier that year. I was a little taken aback, as I had always assumed they were already married. However, at the time I was more struck by Uncle Hassan's attachment to his radio: the small box rarely left his hands as he fiddled with the antenna to get better reception. Listening to the radio with him and to his and Jamila's conversations with my parents marked the first time I heard the names of Fao, Majnoon Island, Mehran and Abadan, the locations of the towns and islands where battles were being fought between the armies of Iraq and Iran. It was also the first time I heard the name Saddam Hussein – invariably accompanied by the grievances and frustrations aired by my father and his relations about the human rights abuses that were being committed in Iraq and overlooked by the rest of the world.

Saddam Hussein came from a rural tribal population in Tikrit, north-east of Baghdad. His was the first generation to be given the chance to advance educationally and socially in modern Iraq, and

Jamila Antoine and Hassan, Beirut, 1970s.

many among them sought radical change and were recruited by socialist and nationalist parties. Saddam himself came from a harsh background, growing up with a stepfather who was brutal by any
standards. The blend of his personal characteristics and his upbringing made him ripe for any number of roles within party politics, but
he was particularly drawn to the Ba'ath Party, which offered the opportunity for upward mobility while satisfying the populist desire for
'punishing the rich and the urban'. A daring activist and a violent
enforcer, by 1970 Saddam was part of the inner circle of the Ba'ath
Party leadership, and in a position to exterminate any potential opposition.

In 1979 the forty-two-year-old Saddam was able to force Iraq's
ailing and elderly President Ahmed Hassan al-Bakr to resign, and
assumed the leadership himself. His rule was characterized by extreme
ferocity and an unyielding demand for absolute loyalty. He created a
quasi-totalitarian state which was exceptional in its brutality, particu-

larly towards the Kurdish and Shi'a communities. Through sheer terror, Saddam commanded the loyalty of many in Iraq. What his regime lacked in efficiency, it made up in brutality.

One day at school a girl asked me why my father hadn't volunteered to fight the Persian enemies in Iraq. Through listening to the adults' conversations at home, I was vaguely aware that there were several Arab volunteer units fighting on Iraq's side, and that King Hussein of Jordan had been the first to fire a rocket at Iran as Saddam Hussein's guest of honour. But within my family there had always been a complete rejection of Saddam's rule. The girl's comment made me aware of my Iraqi heritage for the first time.

In 1980 Saddam enforced the deportation of hundreds of thousands of families from Iraq, including many Faili Kurds and Shi'a who were accused of being *taba'i* – fake Iraqis of Iranian origin. Their properties and assets were seized and their passports taken from them. Some of them came to Jordan, where many sought help from my father.

Even though he had originally only come to Jordan for a period of twelve months, my father continued to run and expand the bank, which had already become the second largest in the country. His animosity towards Saddam remained undiluted, although he was constrained by being in a country that was on friendly terms with Iraq. He did not belong to any of the outlawed Iraqi religious parties, and decided to bide his time, in the hope that he might eventually be able to expose Saddam through his financial dealings.

I remember one incident when I was a child that summarized the evil of Saddam for me. Muhammad Hadi Subaiti was an Iraqi electrical engineer who had fled his home because of his membership of the banned and targeted Shi'a Da'wa Party, and come to Jordan with his Lebanese wife and children. His wife Umm Hassan was from a well-known religious family in south Lebanon, and she wore a white *hijab*. I remember her as a proud and reserved, albeit frail, woman, whose young children always wore a look of fear on their faces. They were living in Zarqa, a service town outside Amman where many lower-income families lived, and had few resources besides the income from Subaiti's job.

One day, agents from the Jordanian intelligence services came to their house and took Subaiti with them, telling Umm Hassan that this was a routine procedure and he would be back in a few hours. He never returned. His wife was left alone with five children. She became a ghost of her former self, knocking on government door after government door, pleading for information about her husband.

Jordan was not the best place for dissidents such as Hadi Subaiti, particularly given the increasing closeness between King Hussein and Saddam; yet with Lebanon at war, the Subaitis had had little choice in their destination. After her husband disappeared, Umm Hassan received no assistance from the government. She would often arrive in tears at our house late at night to get whatever news about her husband she could from my father. Finally Umm Hassan gave up her search and returned with her children to her family in Tyre in southern Lebanon.

Some years later, it emerged that Hadi Subaiti had been drugged by the Jordanian intelligence agents and put on a plane to Baghdad, where Saddam's half-brother Barzan al-Tikriti had tortured him personally before executing him in the royal palace where the regicide had taken place, now renamed *Qasr al-Nihaya*, the Palace of the End.

My early relationship to Iraq was almost entirely defined and filtered through my family and our friends. It was portrayed to me politically by my father and more nostalgically by my grandmother Bibi. I had no other knowledge of the country, apart from Saddam's terror. Every time I listened to a family conversation, especially during our regular visits to Hadi and Bibi in London, it was always Saddam, Saddam, Saddam. They always pronounced his name with a dip at the end, *Sahd-dahm*. Thud-like. I didn't know then that his name was as violent as he was, coming from a root that meant 'to strike, to clash'.

I overheard gory details of the tortures that were taking place in Iraq. I learned how a scientist had been returned to his wife in pieces, but she had been forbidden from holding a funeral for him. The body of another victim had been returned to his family riddled with bullet holes and accompanied by an invoice for the ammunition. Saddam's brutality was harshest against the Kurds and Shi'a, but no community was spared. For me, Iraq meant only Saddam, and it was a deeply shameful connection. This exacerbated my sense of isolation, as the horrors in Iraq continued to be concealed from the international public.

I began to realize that there was a hierarchy of plights in the region: at the summit was the situation of the Palestinians and the Arab–Israeli conflict, followed by lesser confrontations such as the Lebanese Civil War. So far as the West was concerned, as long as Saddam carried on producing oil and fighting Iran, it didn't matter what he did in his backyard. The major oil-producing countries of the Gulf were the ones that mattered, foremost among them Saudi Arabia. While avoiding explicit expressions of sectarianism, the dominant Arab social culture had internalized the segregation that lay below its surface. Unnoticed by an unconcerned West, the Saudis promoted an exclusive and sectarian branch of Islam, Wahhabism, which considered the Shi'a to be infidels. Several times I was removed from my religious studies class in Jordan by a Saudi-educated teacher for challenging his claim that as perhaps one of a handful of Shi'a in Jordan, I was an infidel.

39

Lessons in Humility

The Loss of Everything Precious
(1980s)

ONE EVENING IN London in the early 1980s the family was holding a dinner party when their maid was suddenly heard screaming, 'Burglar, burglar!' Rushing upstairs as fast as her old legs could carry her, Bibi found the maid flailing her hands and pointing to the window. The girl, a fellow Iraqi who had come to work for them through a close recommendation, said she had discovered some men in Bibi's bedroom, but when they saw her they had escaped. An antique diamond ring, a large princess-cut emerald ring and three diamond-encrusted watches were found to be missing from Bibi's room, along with four crocodile handbags and £3,000 in cash. At Hadi's insistence the police were called. He and his daughters were convinced that the maid herself was behind the theft. Her story just didn't add up.

The police interviewed the girl, and eventually asked Bibi whether she wanted to press charges against her. She refused. Her daughters were exasperated, reproaching her for her part in what had happened. She only had herself to blame: by leaving items lying about her dressing room she had practically invited the maid to steal them. To Hadi, the theft represented another nail in the coffin of the old country, as he was convinced that it had been committed by the maid in cahoots with Iraqi accomplices.

When the police asked Bibi about the value of the jewels she gave them the arbitrary figure of £100, much to the outrage of Hadi and her daughters, who knew they were worth thousands of pounds. She didn't want to invite any more trouble, and was afraid that if she punished the Iraqi maid she might somehow expose her sons to trouble. The logic of this made sense only to Bibi, overcome as she was by superstition, paranoia and fear.

Hadi told the maid to leave the house. Bibi had not even thought of firing her.

Saddam committed many atrocious crimes during his career, but there was one which had a particular impact upon one of my cousins. The rise of Islamic opposition to the Ba'ath regime started in the 1960s, and by the 1970s the leading religious scholar Sayyid Muhammad Baqir al-Sadr had come to represent a strong voice of opposition. Connected as he was with the religious seminaries of Najaf, al-Sadr's appeal to the persecuted Shi'a meant that he posed a far more serious threat to Saddam than secular political opponents such as the Communists.

Saddam sentenced al-Sadr to life imprisonment in 1977, but released him two years later because of his undiminished popularity. Al-Sadr set about founding an Islamic political party, al-Da'wa, which appealed to vast numbers of the Shi'a middle class, which had been growing steadily since the foundation of the Iraqi state in the twenties. This group was against Saddam, having become disillusioned with secular parties such as the various nationalist groups and even the Communist Party, with which they had once felt a strong affiliation. Al-Sadr had a powerful intellect and made immense contributions to the field of political philosophy, writing many treatises on Marxism and economic theory, and propagating ideas of Islamic governance. His new party was immediately perceived by Saddam as a serious threat.

Al-Sadr's sister Bint al-Huda, who was based in Kazimiya, was an intellectual in her own right, and was deeply interested in the relationship between social issues and religious thought. She had a serene yet compelling personality, and women followed her weekly lectures in their thousands. Among them was Thamina's daughter Leila Bassam, the former child jewel-smuggler who had married and returned to Iraq.

Saddam's security forces arrested al-Sadr and his sister in the middle of the night. After a lengthy interrogation during which al-Sadr was asked to retract many of his statements but refused, he and Bint al-Huda were butchered by Saddam's half-brother Barzan al-Tikriti in front of each other, first the brother then the sister. Their burial site remains unknown; some say Saddam ordered their corpses to be burned. Even when their death became known, their family and followers were forbidden from holding any memorial for them, public or private. The outrage of their supporters was great, but the murders were quickly followed by the rounding up and killing of Da'wa Party members. As many as 10,000 people may have been slaughtered.

Deeply shocked by Bint al-Huda's murder, Leila took to wearing the *hijab*, which Bint al-Huda had also worn and advocated. It was a dramatic act that took everyone in the family by surprise, as Leila was a beautiful woman who had previously defined herself by her looks and her love of fashion. When Hassan heard of his niece's decision, he remembered how he had asked to be shown her when she was a newborn baby, and had gently run his hands over her soft infant face so he could imagine what she looked like.

During their second exile the family's headquarters in London had become a large flat behind the Albert Hall. My grandparents' lives continued to revolve around their large family, many members of

Talal's daughter Reem at her wedding in London, 1985, with Bibi
next to her and Thamina and her sister Peri standing behind her.

whom, together with other Iraqi friends, dropped in for lunch each day. Their contact with English people was limited to brief, functional encounters in shops, restaurants and taxis.

Whenever I saw her, Bibi showered me with affection and shared many stories of her past. By then she had abandoned her embroidery; her eyes were too weak. I would sit at her feet for hours, enthralled as she recited from her vast repertoire of poems, most of which I didn't understand, in classical Arabic and Persian. Omar Khayyam's *Rubaiyat* was her all-time favourite and she often spoke the verses aloud to no one in particular, simply taking pleasure in the sound of the words:

> And if the Wine you drink, the Lip you press
> End in what All begins and ends in – Yes;
> Think then you are Today what Yesterday
> You were – Tomorrow You shall not be less.

> Said one among them – 'Surely not in vain
> My substance of the common Earth was ta'en
> And to this Figure moulded, to be broke,
> Or trampled back to shapeless Earth again.'

She also knew an endless supply of Aesop's fables and the stories of the prophets from the Quran, telling me about Adam and Eve's fate, Noah's flood and Moses' crossing of the Red Sea. Most of all I loved her stories about her own life, when she was young and carefree in Baghdad, about parties at the Deer Palace, sumptuous banquets, games of cards and fragrant gardens:

Once upon a time, long, long ago, there was a small island in a river, which flowed in front of the Deer Palace garden. You remember me telling you about the Deer Palace, don't you? Anyway, in the springtime these small islands floated up above the river and the earth was so rich that people rushed to plant tasty things on them like lettuces and cucumbers. I used to have lovely lunch parties on one of the islands, and all my friends would come. Your busy grandfather always made sure that all the freshest produce was delivered from the farms in

*time for us to eat. I never had to worry about a thing. The
tables were set out underneath a canopy to shield us from the
sun. Even the ground was covered with carpets, so that the
island was transformed into a beautiful outside room. Like a
queen I just had to get dressed and receive people. My
favourite clothes always were polkadotted silks in blue ...*

Still intent on keeping busy, Hadi made a habit of travelling around
London by double-decker bus; he would sit upstairs and watch the
city go by before catching the same bus home. Hyde Park continued
to be a place of refuge for him, and he still enjoyed feeding the pigeons
there. His outings dismayed my grandmother; after almost seventy
years of marriage she was still a possessive wife, and often accused
him of going out to flirt with blonde young women. Hadi had been
known to have a roving eye in his youth, and Bibi still harboured some
jealousy towards him: as he took off his hat and hung up his walking
stick in the hallway after one of his outings, she would ask him how
many girls he had seen that day.

After their siesta, when she had decided to overlook what she
perceived to be the morning's grievances, they would play cards
together. Bibi was a bad loser, and if she lost a hand she would bring
up the subject of the blonde girls again. Hadi would just look at her
quizzically and say, '*Enti makhablah* – You're mad!' This was their
special way of teasing each other. They had been together for a lifetime, and for all their differences and disagreements they were very
attached to one another.

As I grew older, one of the most dreaded questions I had to face was
the seemingly innocuous 'Where do you come from?' asked by a
stranger. I was unsure of what to say, for fear of either being identified
with Saddam or of having to give an explanation that would last five
hours. I didn't feel satisfied with a simple answer that I was 'Iraqi',
'Lebanese', 'half-Lebanese' or 'half-Iraqi', as each seemed more like a
compromise than a true response; but a full explanation could not be
encapsulated within a single sentence.

By the late 1980s the Iraqi diaspora had vastly expanded, and a
reported 10 per cent of Iraqis were now living outside their homeland,

mainly in Iran but also in Europe, where they favoured Scandinavia, Holland and Britain. The majority of Iraqis who took refuge in Iran did so because of its proximity, but in some cases because of an ideological affinity with the country. The appeal of Britain – still seen as a Western role model – pre-dated the 1958 revolution but was still powerful. Migration to Europe was largely opportunistic: the more generous a country's treatment of refugees, the larger the community that settled there. Once an Iraqi community is settled the influx becomes a self-perpetuating process.

Wherever they made their homes, the Iraqi exiles brought with them the divisions of their own society, with its sectarianism, insecurities and biases. The terror of living in a police state also came with them. They were suspicious of one another and fearful of all agencies of the Iraqi state. They had good reason: all offices belonging to the Iraqi republic, such as the Iraqi Student Union, Iraqi Airways and the Iraqi Embassy, were fronts for Saddam's secret services. The regime's need to assert its power at home led to the monitoring of Iraqi nationals overseas. Exiles would often receive phone calls from relatives in Iraq who pleaded with them to cease their participation in activities deemed objectionable to the regime, on the clear understanding that if they did not comply with the request a cousin or brother who had been left behind might have to pay the price. Occasionally it was not only phone calls that exiles received, but videotapes containing disturbing images that revealed the fate of their relatives back home.

The Iraqi Embassies were a particular source of terror for Iraqis who had not succeeded in obtaining another nationality, and whose passports needed renewal or reissuing. This was not the standard bureaucratic process familiar to those in the West. Harrowing interrogations were sometimes followed by intimidation and even detention. The regime used the issue of passports as a means to reward loyal subjects and punish those who were seen as rebellious. With families at risk back at home, there was little that foreign-based Iraqis could do in terms of reporting their intimidation to the local authorities.

The prospect of renewing an Iraqi passport was almost as grave as going to Baghdad. For many this meant either a frenetic search for a means to acquire a new passport, or the risk of travelling with Iraqi papers that had been forged in Syria. It was said that Belize and Peru were offering Iraqis passports for those who invested a lump sum in the country. My uncles Rushdi and Talal and my cousin Ali decided to opt for Peru, and they were among the lucky ones who were able to find a solution, albeit a temporary one: Ali found himself arrested at Heathrow in 1988 for carrying what turned out to be forged Peruvian documents. In due course I learned that my aunt Jamila's expired Iraqi passport and related problems with her residency papers in Beirut were the key factors that had finally led to her marrying my uncle Hassan, more than forty years after she had first come for her job interview in Baghdad. By that time Hassan had acquired Lebanese citizenship, and she had the right to remain with him in Beirut as his wife.

40

The Mortality of Gods

Burials of the Banished

(1988)

MY GRANDFATHER HAD begun to age. He withdrew into an even more silent world, although he didn't suffer from any serious illnesses and insisted on smoking a cigarette a day, much to the consternation of his children. When I visited my grandparents in London during my holidays from Jordan I often saw him sitting in his dressing gown with his walking cane next to his armchair, silently staring at a silver box on the coffee table in front of him. The flat contained many of the valuable objects that had been rescued from Beirut and earlier from Baghdad, and they filled the rooms with colour and memories. Precious items nestled between fading black-and-white photographs of another world, of kings and horses, and magnificent feasts. Each item held a story.

The cover of the silver box was engraved with an intricate picture showing the Trooping of the Colour, while the four panels on each side were decorated with scenes of Buckingham Palace, St James's Palace, Westminster Abbey and the Houses of Parliament. Inside the lid in delicate engraved lettering were the words 'Abdul Hadi Chalabi, A Valued Friend, Lord Inverforth 1955'. This was the box from Aspreys that had been presented to Hadi by the Chief Director of Andrew Weir as a memento of his purchase of Latifiyyah. It was all he had left of the estate. Representing the cost of Dolphin Square, it had turned out to be one of the most expensive boxes imaginable.

* * *

One day in February 1988 Hadi woke up with a terrible pain in his abdomen. Hospital tests found a growth in his intestines which was blocking his digestive system and needed to be removed. Bibi couldn't bear to hear the word 'cancer'. She dreaded visiting him in hospital, and instead spent a lot of time biting her nails at home and praying for him. The surgery was successful, but Hadi's health continued to deteriorate, although he was still conscious and calm enough to bid his children farewell.

For a man whose life was so linked to Iraq, it was only natural that he wanted to be buried in the shrine city of Najaf near his father, mother, brother Abdul Rasul and Ni'mati, but he knew this would be highly complicated because of the political situation. Not wanting his children to risk their lives taking him there, he reminded them of an old saying: 'Man is buried with his deeds' – meaning that he would be remembered for his good acts, not for where he was buried.

Hadi died peacefully in hospital in London on 7 March 1988, surrounded by his children. Bibi had come to visit him for the last time on the day before he died, and had sat next to his bed, holding his

Shamsa and Hadi in London in the mid 1980s.

hand and weeping for him, for their old age. The impact of his death on the family was like a divine calamity, as if God had died. It unleashed a flood of sadness, shock, anger, regret and helplessness.

As it was impossible to return to Iraq, the family decided to lay Hadi to rest in Damascus, at the shrine of Sitt Zeinab, the Prophet's grand-daughter, Imam Ali's daughter and Imam Hussein's sister. The idea of burying him in England wasn't even considered. That would have been an eternal exile.

Grandchildren came from around the world to see the family patri-arch laid to rest, except for those who were prevented from travelling to Syria because of passport problems: once they left their countries of residence there were no guarantees that they would be allowed to return to them, as their birth nationality had made them pariahs. It was a poignant reminder of their greater orphaning through their loss of Iraq. As exiles, there was always an awareness of the homeland of which they had been deprived. All good things were projected onto the lost land, which became a focus of nostalgia and longing. Such attachment to one's roots, to the past, can become overwhelmingly painful.

Standing after midnight in the pitch darkness of Damascus airport with my mother, waiting for the plane from London to land with my relatives, I was filled with fear. How would I cope with seeing my grandfather's dead body? His was the first major death in my family that I had been aware of. My heart seized when a plane landed. Hardly able to breathe, I watched as its door opened and a group of young men in combat gear came out. These weren't my family. A man stand-ing beside us explained that these were Palestinian guerrillas back from their training in the Libyan desert. They walked silently past us, blink-ing sleepily as they headed towards the baggage hall. The plane from London finally landed, and my heart seized yet again. Here were my family: we hugged each other and cried together, and my fears subsided.

The funeral took place the next morning. We drove out early from our hotel to Sitt Zeinab shrine on the outskirts of Damascus. The city had become a centre for many fleeing Iraqis. The enmity between the Ba'ath regimes in Damascus and Baghdad was not ideologically based, but was more about the competition for power in the Arab world. Both regimes wanted to be seen as the legitimate embodiment of the 'Arab revolution'; however, the purpose of this revolution was to strengthen the regimes rather than to bring about any tangible change for their subjects.

This was the first time I had been to a Shi'a shrine, as the major ones were all in Iraq. Sitt Zeinab had a typical square interior court-yard, in the middle of which was a smaller building in which the sarcophagus was protected by perforated brace panels in the manner of the shrines at Najaf and Karbala. I followed my grandfather's coffin as it was carried around Zeinab's tomb by his male grandchildren before entering the cemetery outside. Guided by a cousin who gripped his elbow tightly, my uncle Hassan followed. Tears trickled underneath his dark glasses as he held out his hands in front of him to find his way in the crowd.

Although Muslim tradition means that the presence of women is not desired at burials, all Hadi's surviving female relatives were there, except for Bibi, who was too old to travel, and my aunt Najla who had stayed behind with her in London. The funeral procession included a host of uncles, aunts and cousins, second cousins, cousins once removed and great-uncles. When we finally arrived at the freshly dug plot, I saw that the procession had mushroomed to over five hundred people, including many unfamiliar recently exiled, suffering and home-less Iraqis. Everyone chanted repeatedly: 'Heaven, open your gate – Abdul Hadi is on his way.' It was a mighty sound. My father went down into the grave as it was his duty, as the youngest of his children, to lay the disc made from the clay of Karbala underneath Hadi's head. Throughout the service, Syrian government helicopters hovered over-head, watching us.

In accordance with tradition, several funeral services were held for my grandfather in different places around the world, reflecting the complex identity of his extended family. The most poignant service took place in London, where the family travelled to be with Bibi. Now in her late eighties, she looked very frail and confused; she had spent most of her life married to Abdul Hadi, and had never expected him to outdo her by dying first. She found great difficulty in adjusting to her new role as a widow. Her wit and life force started to fade. She missed her husband terribly, even after all the years of public bickering and unrelenting criticism of him. Death became the only thing on her mind.

She had previously declared to her children that she wanted to be buried in Najaf, as she was very taken by an old story about Imam Ali which said that he would take those buried near him by the hand and lead them to heaven. The story offered solace to Bibi, especially as she believed it had been corroborated by an incident many years earlier, when she had been a young girl. Her mother Rumia had made a pact with her friend Amira that whichever of them died first would visit the other in a dream to report on life beyond the grave. After Amira died she came to Rumia in a dream as promised, and told her that Imam Ali had indeed come to visit her and given her much comfort in her grave. When none of her children would commit to burying her in Najaf, Bibi turned to her granddaughter Leila, who promised that she would take her to the shrine city.

I was staying in Bibi's flat by the Albert Hall in 1988 when the news broke about Saddam's gassing of Iraqi Kurds in the town of Halabja. Kurdish groups in northern Iraq had been agitating for autonomy for decades, and had recently taken advantage of the Iraqi Army's engagement in Iran to seize control of large parts of their territory. When the Iran–Iraq War began to draw to a close, Saddam launched a genocidal campaign to recover his losses in the north and to punish the population that had supported the rebellion. Chemicals, including mustard gas and Sarin, which disrupts the nervous system, were shipped in from the West.

On 16 to 17 March 1988, Saddam Hussein experimented with chemical weapons on the civilians of Halabja, in an exercise whose long-term goal was the prevention of a potential Iranian invasion. The lack of response from the West indicated to him that he was at liberty to continue his retributive *al-Anfal* campaign against the Kurdish rebels. It is believed that, besides the devastating damage that was inflicted on the environment, as many as 182,000 people were killed.

The news reports were muddled, and seemed at times to assign responsibility for the attacks to Iran. As a result not all Western governments condemned the atrocities, although the visual evidence – footage of old men bent over dead babies in doorways – was highly incriminating. The gassing of the Kurds was preceded by the assassination of Sayyid Mahdi, the brother of Sayyid Muhsin al-Hakim, the Grand Ayatollah of Najaf.

It was absolutely clear that there was no possibility of my family's return to Iraq under Saddam's rule.

41

The Lost Talisman

When Everything is Taken
(1989–1992)

IT WAS 5 MARCH 1989, two days before the first anniversary of Hadi's death. As was the custom, a memorial service was to be held on the anniversary itself. This would include the recitation of the Quran, as well as readings of poems, panegyrics and prose passages. A large lunch would be served and alms given on behalf of the deceased to the poor. Bibi's daughters set about making the preparations, but she herself was upset and said she didn't want to attend the memorial, that it was a waste of time and they should give the money to the poor instead. She eventually settled down towards the end of the day, and went for a walk after dinner with Thamina and Raifa along the long communal hallway.

That evening Thamina put her mother to bed before going home to her own apartment across the street. She had barely taken her coat off when Bibi called to say that she wasn't feeling well. Quickly, my aunt put her coat back on and ran back. When Bibi told her she had severe pains in her stomach area, Thamina assumed it was of one of her usual imaginary ailments, but Bibi insisted that this time it was different. The doctor was called, but couldn't find anything wrong. However, Bibi began to suffer from spasms, and cried, 'I'm telling you, these are death pains.'

After the doctor had left all Bibi's children who were based in London came to be with her. Rushdi sat all night with her, while Thamina held her hand. Najla had a cold, so she sat outside the room so as not to infect her. Bibi continuously repeated to her children: 'Darlings, Najaf, Najaf, Najaf. Don't forget.' After midnight she grew quiet; her children sensed that she was fighting death. She quietly prayed and asked God for forgiveness, then repeated one last time: 'You have to take me to Najaf.'

Unlike Hadi, who died peacefully, and who didn't burden his children with the responsibility of where to bury him, Bibi had a difficult death. Yet she willed herself to die.

In fact, none of her children could fulfil her final wish. None of them could risk going to Iraq, and they couldn't bear to send her body off alone like a package. They decided to bury her temporarily next to Hadi in Damascus, until such a time as they could move her to Najaf.

Bibi's funeral was similar to Hadi's, but was rendered all the more poignant by her final request, which hung over the family. Leila was beside herself for failing to fulfil her promise to her grandmother, while Hassan spent days in solitude, weeping and forbidding anyone to talk to him. With Hadi and Bibi gone, the link to Iraq truly seemed to be slipping away.

Seven months before Bibi's death, in August 1988, the Iran–Iraq War ended in stalemate after eight years of fighting. Estimates put the death toll at between one and two million. Like the First World War, the conflict had involved torturous trench warfare and the use of mustard gas, which burned skin and blocked the respiratory tract, which if not immediately fatal caused severe long-term damage to the lungs, and cancer in many cases. The war achieved little besides strengthening the grip of both governments on their subjects.

Saddam Hussein presented himself to the world as a combination of two historical characters firmly rooted in the Iraqi psyche: the Babylonian king Nebuchadnezzar the Great, who captured Judah, and Saladin, the Sultan who won victories against the Crusaders – both empire builders who had fought enemies who threatened Iraq's integrity. They both came from the land that is present-day Iraq, although ethnically neither was an Arab, a somewhat ironic fact as Saddam placed great emphasis on his own Arab ethnicity.

Saddam exploited Arab nationalism and Ba'athism in very utilitarian ways. Conflating these ideologies with his ruthless and oppressive regime, he promoted a heightened sense of his own importance within the region, and of the region within the world. In accordance with his logic, any action undertaken by him assumed mythic, historic proportions. This is why he and his supporters called the Iran–Iraq War the *Qadissiya*, in reference to the historic Sassanian (Iranian) defeat at the hands of the Arab Muslims in the seventh century. Similarly, the first Gulf War was called *umm al ma'rek*, the 'mother of all battles'.

Saddam's ideology, which was really all about his own survival as an absolute leader, was grounded in his tribal background and an intense dose of machismo. While fellow tribesmen of a certain standing were respected by him, the peasantry were despised, while the urban classes – the bourgeoisie in particular – were resented. Saddam engaged in a comprehensive project of refashioning Iraq according to his own world view, tastes and fantasies. Under his rule, the urban charm of Baghdad was replaced by vulgar extravagance, manifested in the city's architecture and personality cults.

Although he had depleted the state's coffers in the Iran–Iraq War, more horror would soon follow with Saddam's invasion of Kuwait. During the first Gulf War in 1991, he brutally repressed a Shi'a uprising in the south which had been encouraged by George Bush Senior, who then looked away as Saddam's forces rounded the insurgents up. Tens of thousands are reported to have died. The impact this had on Iraqis abroad was tremendous, as their disappointment paralysed them. Again Saddam had survived against all odds, and with the assistance of the very country that had declared war on him.

When the family's fortunes dwindled after Bibi's death, her daughters observed that she had considered herself to be a lucky talisman, as her father-in-law Abdul Hussein had noted when she had married into the Chalabi family. My father recalled one of her superstitions: on her first sight of the new moon each month she would close her eyes, latch on to someone and ask them to lead her to a mirror for good luck. Before she died she gave her granddaughter Leila a photograph of herself and told her with absolute conviction, 'Keep this next to you; I'll bring you luck.' But soon after she died, any good luck that she had bestowed upon her tribe vanished as the banks established by my uncles ran into political and financial troubles.

My father's pioneering efforts in the spheres of banking and economic development had yielded impressive results. He brought world-class banking technology to Jordan, and through his entrepreneurial zeal helped to connect it to financial centres around the world. Under his leadership Petra Bank, which he had he founded, looked destined to become the top banking institution in the nation. His unusual style of conducting business – development banking, and supporting new projects and ventures beneficial to the country – won him a lot of friends and allies in the region. But it also earned him the ire of the established economic elite in Amman, who felt threatened by his success.

My father, who actively supported the opposition to Saddam Hussein, believed that Saddam's regime rested on three pillars. The first was terror and intimidation inside Iraq; he could do little about that. The second was his international support, which again it would be a mammoth task to tackle. The third was money. He decided to target the last, by undermining Saddam's creditworthiness among the international banking community. Specifically he focused on exposing the Iraqi regime's links to an Italian bank, Banco Nazionale del Lavoro, whose branch in Atlanta, Georgia, was lending illegally to the regime. My father's activities became known to Saddam in early 1989 when, following the Iran–Iraq War, Iraq found itself with serious financial issues to resolve.

Jordan had become increasingly dependent on Saddam's money and on Iraqi oil, and Saddam asked his ally King Hussein to put a stop to my father's activities. One way to achieve this would be to strip him of his chairmanship of Petra Bank, even though it was a private enterprise. On 2 August 1989 the Jordanian government passed an extra-judicial martial law under which Petra Bank was seized for the purpose of merging it with another local bank. The board of directors, including my father, were dismissed, and armed tanks surrounded the bank headquarters in Amman.

My father could not believe what had happened. He had many friends in various circles in Jordan, including politicians and members of the security services. A day after the raid on the bank some of these friends made him aware that Saddam was sending a team of his security forces to interrogate him and take him back to Baghdad. His arrest had already been approved by the Jordanian Prime Minister, General Zaid Shakir.

The rumours were soon confirmed. One of my father's friends, Abdul Hay Majali, the brother of the director of Jordan's security services, took my father to the house of another of his brothers, Abdul Wahab al-Majali, an eminent politician and the former Deputy Chairman of Petra Bank. When they arrived, my father was told that the Prime Minster would be joining them. He angrily said that he didn't want to meet him after what he had done in ordering the raid on the bank, but another of the brothers reproached him, telling him that bankers shouldn't talk like that. My father replied that he was no longer a banker after Shakir's actions. At that point the Prime Minister strode through the door. My father left the room and returned home.

The next day, several important people in the Jordanian government called to check whether he had been arrested, as they were aware of the orders that had been issued by the Prime Minister. This, combined with information about the Iraqi security team's presence in Amman, persuaded my father that it was time for him to get out of the country as soon as possible.

While these events were taking place, my mother, my siblings and I were on our summer holiday in London, waiting for my father to join us. We received his news by telephone, but none of us was able to grasp the magnitude of what had happened. We were convinced that a mistake had been made and that the King would surely fix it, given the good relations we had with the Jordanian royal family. Until my father arrived in London several weeks later, we heard little from him. Within a matter of days it became clear that we would not be returning to Jordan. Our entire lives were left behind in the country – our friends, our home, our personal belongings. We lost them all.

No charges were laid against my father until three years after the bank's seizure. When the military prosecutor reported to the Prime Minister that there was no basis for any charges, he was told to come up with something, then he was fired. Another prosecutor was brought in, and charges of misconduct and criminal negligence were levelled against my father. The same charges were made against his co-defendants, the employees of the bank. Every one of them was acquitted on all counts. With the exceptions of my father and my uncle Rushdi (who did not reside in Jordan), none of the other board members was ever charged with misconduct. Indeed, subsequently several of them and of the bank's key clients were granted government positions: Abdul Karim Kabariti, Samir Kawar and Laith Shubeilat became the Jordanian Prime Minister, a Cabinet Minister and a Member of Parliament respectively. My father was tried *in absentia*, and was prohibited from retaining counsel at the trial. Furthermore, he was not allowed to appeal against the charges in a higher court.

Before the sentences were passed in September 1992, my father met King Hussein for an hour in London, with the intercession of Jalal Talabani (the current President of Iraq). Jalal Talabani forcefully put his case to the King, who had little to say. Several more meetings followed with the King throughout the nineties until his death in early 1999, and he expressed a wish to solve the Petra Bank case. His deputies were at a loss as to how to do this, because over the years the bank's assets had been looted by influential people who had picked up shares and acquired real estate at cut-rate prices. The liquidation committee is still in operation today, over twenty years later, living off the assets of the bank.

The conspiracy against my father was rooted in a convergence between the Iraqi regime's political interests and the economic interests of the established Palestinian–Jordanian business elite. It had led to the subversion of Jordanian institutions and the exploitation of regulation, legislation and even the judiciary in order to undo the Petra Bank experiment and then lay the blame for its collapse on the very person who had lost the most from it, namely my father. The result was to tarnish his reputation, obliterate his wealth and cast him as an outlaw for years to come.

Far from confirming that the bank had been in deficit as they had alleged, the martial authorities authorized large transactions after the takeover on 2 August 1989. They later brought the international firm of auditors Arthur Andersen in to audit the bank. Arthur Andersen did not question the martial law committee's post-takeover transactions, nor the fact that the shareholders of the bank did not authorize their audit. Despite the fact that their report clearly stated that it was to be used for information purposes only, the authorities violated this request and published it, a flagrant abuse that has never been addressed.

My aunt Raifa told me that Bibi came to her in a dream a few months after her death and told her, 'Darling, I can't sleep. You're father's sleeping so comfortably next to me, but I keep tossing and turning, left and right.'

For nearly eight months after Bibi died, I kept dreaming of her. I was as upset as the rest of the family by the Najaf conundrum. Her fate, to be deprived of the final resting place she had longed for, sealed for me a great injustice and drove home my family's long exile. In my dreams I remember searching for signs of Bibi's disapproval as she talked to me about her mother's cooking, about a new dress she had bought me, about Hadi's gifts to her, about nothing in particular.

42

A Question of Identity

In Search of a Way to Be

(1990–2009)

EVEN BEFORE MY father decided to focus his efforts on building a unified opposition to Saddam Hussein's regime in 1991, the desire to make sense of what had happened to my family and to understand my own inheritance compelled me towards Iraq's history, which was unfamiliar to me, and to Saddam's terrible human rights abuses. At boarding school in Oxford I had aspirations to become an architect; I decided to write about the Gothic arch and its influence on Islamic architecture for my senior term paper. However, my architectural ambitions were soon abandoned, and I ended up writing on the Kurds of Iraq. The romance of the mountains may have had something to do with the fascination they held for me. Certainly their distinct foreignness appealed to my imagination, evoking a land full of different peoples and languages.

A few weeks after choosing my new assignment the news broke of Saddam's invasion of Kuwait on 2 August 1990. My first thought was that the invasion marked the anniversary of Petra Bank's takeover and my family's uprooting. From that day onwards, my relationship to Iraq became defined in terms of my opposition to the regime and a quest for an Iraq that was everything Saddam's tortured country was not. I became even more influenced by my father's stance when he committed himself full-time to opposition to Saddam's regime.

The invasion of Kuwait and the first Gulf War offered the world a fleeting glimpse of what Saddam's Iraq was really like, but it was the images of fleeing Kurdish refugees trapped in the mountains on the Turkish border that forced the international community to respond, enforcing a no-fly zone above the 32nd parallel. Iraq's sovereignty appeared to be eroding. Like many Iraqis I rejoiced when it seemed that the end might soon be approaching for the regime following news reports that the US military were near Baghdad, and like them I was devastated when Saddam survived, suppressing the uprising and killing tens of thousands.

I was on my summer holiday, about to start university in the United States, when the Iraqi National Congress (INC), the umbrella group of the Iraqi opposition, first met in Vienna in June 1992; to a large extent, it convened owing to my father's relentless efforts. As the Congress was in need of volunteers I asked to take part, and was ecstatic to be participating in such an event, to be involved in a collective effort to reclaim the Iraq I had only dreamed about. The assembly of Kurds and Arabs, Communists and Islamicists suggested once more to me a rich land full of diversity and possibility. A follow-up conference was convened in November 1992 on Iraqi soil, in the Kurdish town of Salahuddin under the no-fly zone, beyond Saddam's grasp. The university term meant that I couldn't go, despite my persistent lobbying to do so.

My father knew that he could not effectively oppose Saddam from an office in London, so after the INC's formation in 1992 he made his base in Iraqi Kurdistan, working closely with the Kurdish parties there. Communications were not as developed then as they are today, and I remember our anxiety whenever we were unable to reach him by telephone, or if a few days passed without any news. Despite the no-fly zone, he was not far from Saddam's forces, and the fear of something happening to him haunted us constantly. When the two main Kurdish factions, the Kurdistan Democratic Party and the Patriotic Union of Kurdistan, both of which were founding members of the INC, started fighting each other in 1994, my father took upon himself the task of mediating, often going to the battleground to implement ceasefires.

I finally got my wish to visit Iraq in the winter of 1994. Flying from Istanbul to Diyarbakir in eastern Turkey, I nervously considered the fact that I would soon be arriving in Iraq. For days I had been building up this moment in my head, wondering how I would feel when I finally stepped onto Iraqi soil.

Following a three-hour car journey, the party I was travelling with finally arrived at the Iraqi border. A Turkish guard took our passports, slowly leafing through each one. After what seemed an eternity, a man came out of the guards' hut and explained in broken Arabic that I wasn't allowed to cross into Iraq because of new restrictions on journalists and foreigners. 'But I'm not a foreigner,' I explained. 'I'm going to see my father.' The man apologised, but said he couldn't see how I had any connection to Iraq: my passport was Lebanese, and I had been born in Beirut. He advised me to go to Ankara and obtain special permission from the Foreign Ministry.

I couldn't believe this was happening. The border supervisor whose name my father had given me in case of any problems was not there, and there seemed to be no way to resolve the impasse. I knew that Abu Muhammad, my father's contact, who I had never met before, was waiting for me on the other side of the border, but we were separated by stretches of no-man's land on either side of a river, and the border was guarded by Turkish soldiers. All I had was a verbal description of Abu Muhammad to help me identify him. I felt that some sinister force was preventing me from fulfilling my destiny.

I am not sure what seized me at that moment. Perhaps it was a lack of sleep, or the dramatic landscape – an expansive open sky with infinite space in every direction – that made me react in the way I did. My passport was still with the officer inside the hut, who must have assumed that I was trying to make arrangements to return to Ankara. I spoke to a friend in the group, and agreed that we would stage a goodbye; then the rest of the party would drive across the border, and would meet me on the other side. I simply had to get across on my own.

Ahead, all I could see were rifles and military helmets. I was hardly inconspicuous in my green jeans and colourful jumper, but I took a deep breath and started walking. I must have walked for half an hour, crossing the metal bridge under the curious gaze of a few young Turkish soldiers who thankfully did nothing, before I reached the Iraqi–Kurdish border, where there was a lot of commotion. I told myself that I just had to find Abu Muhammad among the crowd, and then everything would be fine. There were hardly any women in the throng, I realized; I must certainly have stood out.

As I searched the people around me, I spotted a wiry man with his head cocked to one side. He came towards me and said: 'Tamara? I'm Abu Muhammad.'

I beamed at him; I was exhausted, but I had made it to Iraq. I had crossed the border illegally. My passport was still at the border control. The adrenaline that had pushed me into crossing was gone, and I suddenly felt depleted. The mountains overwhelmed me.

On the long drive to Salahuddin, Abu Muhammad and I quickly established a rapport that remains to this day. He struck me as the very essence of what I had imagined Iraq to be, a collage of disparate elements. Although he was an Iraqi Turkmen whose first language was Turkish, he also spoke Persian, Kurdish and Arabic, badly but with great poetic flair. Like Ni'mati, he mixed genders and switched verbs around when he talked. Anyone who had heard him address me would have been sure he was speaking to a man. He asked me why I spoke Iraqi like an Armenian, which I thought was rich coming from him. As the car drove into the Zagros mountains I was very struck by their majesty and their colour, a dramatic combination of purples and greys running up towards the sky. Only after some time did I realize that the irregularly-shaped stones that covered the mountainsides were in fact rough headstones. Was this a type of grave particular to this region, I asked myself. Or were these makeshift graves for fallen victims? They were all identical, and I wondered how anyone could ever find their loved ones.

I was in northern Iraq for two weeks. I stayed with my father, who I had not seen for nearly a year, in his rented house in the former summer resort of Salahuddin, near Arbil. He was in charge of the INC, working with the Kurds, recruiting people – many of whom had fled to the no-fly zone – collecting information and running an anti-regime television and radio station broadcasting to the rest of Iraq. I was thrilled to see him. He was in good spirits, and his efforts and those of the people around him inspired me. I could see the risks they were taking, and security was a prime concern at all times, although they attempted to lead as normal lives as possible. Nevertheless, the limitations were many. The threat of Saddam was ever-present, looming from the south near Mosul, and among the people I met were Iraqis who had escaped from Baghdad, many of whom had undergone severe physical and psychological trauma.

I was finishing my undergraduate thesis, and needed to research Iraqi schoolbooks and propaganda pamphlets. But at the same time I was surprised by everything around me. The people and the landscape were all so foreign to me that I made a point of writing down everything I saw. I filled my diary with ideas for the future, mostly concerned with healing the pain I saw.

When it was time to leave, I had to hide in the car as we crossed the border, because I had never officially left Turkey. My passport had been picked up after I had crossed into Iraq by someone who told the border guards that I had forgotten it and had gone back to Ankara.

As the struggle against Saddam's regime continued, my father spent most of his time in northern Iraq with the INC. In 1996 there was a huge setback when Saddam's forces entered the northern city of Arbil. Over a hundred people working for the INC were killed, while many others were left stranded, either in hiding or fleeing towards the Turkish border. Many of these, including Abu Muhammad, were granted asylum in the United States with my father's help.

The pressure on Saddam had to come from a different direction now. For the next two years my father devoted all his energies to drafting a law committing the USA to support a democratic Iraq, and lobbying Congress in Washington to adopt this law. Both houses passed the Iraq Liberation Act (ILA) with large majorities, and it was enacted into law by President Clinton on 31 October 1998. The Liberation Act stipulated that the United States would support the efforts of Iraqi opposition groups to remove Saddam's regime, and promote the emergence of a democratic system. This support would take a variety of forms: logistical, broadcasting, humanitarian and military. It was a great achievement for the Iraqi opposition to have the USA pass such an unprecedented law. Although it was not explicitly stated, it was assumed that Saddam would be toppled by a popular uprising. The possibility of the US engaging in a war with Iraq was not even contemplated in Washington.

In the event, only a fraction of the clauses in the Iraq Liberation Act were ever implemented as inter-agency squabbling between the State Department, the Pentagon and the CIA undermined the Iraqi opposition – an unfortunate outcome that was to endure to Iraq's detriment even after Saddam's removal in 2003.

* * *

I was in America, pursuing a PhD in History at Harvard University, on 9/11. Like everyone else, I was shaken by what I saw on live TV. I was also taken aback by the transformation in the language of the US government and media concerning the Middle East, which triggered confusion and fear of the American authorities among my fellow students and friends who had links with the region. America's reaction to 9/11 gathered an unstoppable momentum, culminating in the 'war on terror'. To me it seemed that Iraq's story became an ideological battleground between liberals and conservatives struggling to define the new America, and by extension the new global landscape. The decades-long struggle against Saddam's regime by Iraqis took on a different aspect when seen through the lens of President George W. Bush's simplistic language. While many Iraqis rejoiced at the overt recognition of Saddam's Iraq as part of the 'axis of evil', it also seemed as if the debate ignored them, and was speaking to another audience: it was an embodiment of American angst.

A large conference was held in London in December 2002 for the Iraqi opposition to discuss Iraq's future with American and British officials, in an attempt to organize and unify the anti-Saddam forces. Despite many disagreements, a successful communiqué was issued, planning a larger conference on Iraqi soil in February 2003, at which the establishment of a provisional government would be discussed in greater detail.

I was visiting my family in London after submitting my PhD thesis in early January 2003 when my father announced that he was shortly to be a member of a delegation that was going to Iran to have discussions with members of the Iraqi opposition and government officials. I immediately decided to accompany him.

After ten days in Tehran, it became clear that my father and the rest of the delegation were not planning to return to London, but to go to northern Iraq and prepare for the conference set for late February with the Kurdish forces on the ground, ironing out the details of the conference's aims, which included the nature of a future democratic government in Baghdad and discussions on federalism.

We flew to Urumia, in north-western Iran. This was followed by a long car journey through deep snow and high mountains populated only by smugglers on horseback to the Iraqi border at Hajji Umran, which we crossed on foot. This time I entered the country without difficulty. In fact there was a welcoming committee. Various opposition groups were gathering in Iraqi Kurdistan to discuss the creation of a provisional Iraqi government. Our first port of call was Salahuddin, the INC's former headquarters, near Arbil, where the KDP was in control. After Saddam's incursion to this area in 1996, and his subsequent withdrawal, the Kurds settled their territorial disputes by dividing the Kurdish area into two spheres: the one in the west run by Mustafa Barzani, the head of the KDP (Kurdistan Democratic Party); and the one in the east by Jalal Talabani, head of the PUK (Patriotic Union of Kurdistan). There was a rotating presidency, and a parliament in Arbil.

What I most remember from those days is the uncertainty and frustration, but also the fragile hope that the dictator's end might be nearing. Following the huge anti-war protests in many Western cities in mid-February, I listened to bewildered Kurds who wondered why the international community loved Saddam so much. Walking around the bazaar in Arbil, I could see a calm resignation as men stacked up sandbags and sold gas masks, anticipating another chemical attack from Saddam. All expectation was that the war would start from the north.

At the planned conference in Salahuddin, Iraqi opposition and US representatives debated the creation of a transitional government. But the anticipated American blessing never came, as the various US agencies each supported their own chosen Iraqi faction on the ground. I experienced the build-up to war as surreal, and am not clear at what point the idea of an Iraqi-led internal uprising gave way to an all-out American attack on Saddam. After all the years the opposition had been trying to remove Saddam, it took some time to come to terms with the fact that what had been hoped for for so long was actually happening.

Over the next few weeks we moved from Salahuddin to Lake Dokkan, a beautiful spot near Suleymania surrounded by mountains and clouds, under PUK control. Many young Iraqi volunteers had gathered there, enough to make a battalion, and they were training and marching every day in preparation for engaging with Saddam's forces. I accompanied my father on several trips to Iran and Turkey during this time, as he took part in talks with the governments in both countries in anticipation of the imminent war. As we travelled on roads that traced the footsteps of the rulers of the ancient empires, Assyrian,

Persian and Ottoman, it struck me once again how intermingled Iraq was with these other, older narratives and identities.

On 20 March 2003 I was at Ankara airport with my father and an Iraqi delegation waiting for a flight to Mardin, near where my distant ancestors started their Mesopotamian venture many centuries ago, when it was announced on a television in the airport lounge that coalition forces had entered southern Iraq. I was gripped with anticipation and uncertainty; I didn't know what to think, other than that Saddam's days in power were surely numbered.

Everyone was waiting to see what would happen. Having devoted years of their lives to arguing for action against Saddam, many leaders of the Iraqi opposition now seemed confused about their roles. The Americans seemed deaf to all their concerns, and were clearly working to a prearranged plan, whether it corresponded to the reality on the ground or not. The military machine had been switched on, and was now moving forward with unstoppable momentum. Opposition forces such as the Kurdish Peshmerga and the INC had expected to engage with Saddam's forces, but this was not a priority for the Americans.

Arriving back in Dokkan, I sensed a very different atmosphere from that I had left behind only a few days earlier. Frustration was increasing among the opposition groups, and even when a US military liaison officer, Colonel Seale, arrived to coordinate action with the INC, it was unclear what there was to coordinate, despite his level-headedness and his willingness to engage in joint activity. The sense of powerlessness that ran through opposition circles did not, however, discourage my father from seeking a more visible and active role for Iraqis.

Resistance from Saddam's militias in the south was greater than expected, and my father felt there was a need for Iraqis to be involved in the fight there, rather than simply to wait until the US forces arrived in Baghdad. I am certain that many in the US military considered this request at best a nuisance, and at worst a potential danger. But the opposition's position could not be preserved unless it actively took part in fighting Saddam's forces, rather than merely looking on from afar. The pressure exerted on Washington by my father and his group finally bore fruit, and a volunteer battalion named the Free Iraqi Forces (FIF) was authorized to fly south from Dokkan.

Over three days several batches of men were transported in US C10 planes from Harir airport near Arbil to Nassiriya. When the first group arrived in the south they called to complain that they had been left in an abandoned hangar in the middle of the desert without any facilities, food or water. My father was to fly south on the second day, and I insisted on going too, despite much resistance from him. I would be the only female to accompany the FIF apart from Marie Colvin of the *Sunday Times*, who was 'embedded' with us.

In the car driving to Harir, my father received a call from Lieutenant General Abizaid of Central Command (CENTCOM) asking him to delay his trip, to which there was considerable opposition in Washington. My father insisted that he could not put his men in the line of fire and not join them. The exchange showed that even at this late hour disagreement and indecision still pervaded the highest levels of the coalition, adding yet greater uncertainty to our journey. Abizaid's call did not make my father change his mind, and he held my hand as we boarded the dark plane to fly into the unknown.

Like the previous FIF arrivals, when we arrived in Nassiriya we were left in the middle of nowhere, in an abandoned hangar miles outside the city and surrounded by parched desert on all sides. We had little in terms of provisions, infrastructure or transport. After a few hours the ever-practical Abu Muhammad decided to go into Nassiriya to get some food and water. He found a *yashmak* that he wrapped around his head in an attempt to look more local, and asked me to pray that he was not captured by the Fidayin Saddam, who were still active in the area. After two hours during which we anxiously waited for his return, he returned with several kilos of overripe tomatoes, the only things he could find, such was the dearth in the city.

With little choice but to adapt to the situation in which we found ourselves, we started unpacking our equipment, which included a satellite dish for communication and a carpet on which to eat and sleep. Protesting telephone calls were made to Washington, and efforts were made to contact the local Marine forces in order to establish some mode of operation. A few raids were undertaken on nearby villages, which secured ammunition and captured several Fidayin, but the reluctance of the Americans to rely on local support was strong. I remember that when a group of them came to talk to us they were stunned to discover that I was an Iraqi, as all they had seen previously was a smattering of heavily covered local women. I was much concerned by their ignorance of the region and their cultural misconceptions.

Within two days of our arrival we received a visit from a group of Nassiriya notables, tribal leaders who came to pay their respects and enquire about the general situation of the war, about which they had little information. Although they had a lot of information about Saddam supporters who were still fighting, there was no mechanism for them to communicate it to the US military, who were suspicious of everyone: the locals were afraid to approach them for fear of being arrested or misidentified as Saddam supporters. Through the help of Colonel Seale, some of their information was relayed, but often little was done about it.

As the days passed, many of Nassiriya's tribal notables invited us for lunches in their homes. They were anxious to discuss the future, and were looking for leadership. I was heartened by the clarity with which these men saw the steps by which Iraq could make the transition out of a dictatorship. Two recurrent themes were their inability to communicate with the Americans, and their disconnection from the progress of the war. The fact that I was the only woman present at these lunches led several of our hosts to assume that I must be an American adviser who was well versed in Arabic; they could not imagine that I might be Iraqi.

In Nassiriya, I was struck by the evidence of years of wasted resources. So much had been funnelled into statues and images of Saddam, which were still to be seen everywhere, while the people lacked the most basic services. Standing in the desert not far from the town was the ancient Sumerian site of Ur, which according to the Book of Genesis was the birthplace of Abraham. Next to it, Saddam had built a military airbase.

Living conditions barely improved for us. The sand blew everywhere, including into our noses and mouths. I was adapting to the lack of showers and a toilet, using a makeshift metal shack tucked behind the big hangar. This was comparative luxury: the men had to make do with the open air. More people started gathering at the camp, including officials from Baghdad who were looking for help to secure government assets such as the central bank from looting, and could not find a way to communicate this to the Americans.

When Baghdad fell on 9 April, everyone wanted to go there, but we had to wait for cars to reach us from Kuwait by road. When they finally arrived, it was made clear to us by the US military that they would not assist us in our journey: we were going at our own risk. We taped signs to the roofs of the cars and carried white flags to identify us as non-hostile, and headed out at dawn on 15 April, in the early stages of a sandstorm. Our convoy of twenty-five cars drove north to Baghdad through primitive villages, with no water or electricity. As I sat silently watching the landscape pass by, my father received a telephone call from someone in CENTCOM asking what his intentions were in travelling to Baghdad. I remember his bemused response: 'What do you mean? I am going home.' I doubt the officer was satisfied with it.

North of Samawa, the cars had to slow down because of the large number of people walking in a long line down the highway. All of them, men, women and children, were wearing black. Some were carrying flags, others water. I had never seen anything like this. There was total silence, except for the sound of footsteps and the murmuring of the lush palm groves by the side of the road, beyond which the Tigris was hidden. We were near Hilla, two hours south of Baghdad and only a few kilometres from Babylon. The thought of being so near to this mythical place set my imagination loose. I imagined the city Alexander the Great would have built, a meeting of civilizations, and projected onto it the possibilities suggested by that story for Iraq's future. Then I see the derelict mud houses and parched land in the distance, and realize that change may not be possible at the pace I imagine.

I am told the people on the road are walking to Karbala, the resting place of the martyred Imam Hussein, a pivotal figure in Shi'a Islam. They are commemorating the fortieth day of his death, and this march is called the Fortieth, *Arba'in*. By walking there, from all over the country, they atone for their sins and become closer to heaven. The reason there are so many of them is that this is the first time in years they have been able to do this, since the *Arba'in* was forbidden under Saddam.

It was dark when we arrived in Baghdad many hours later. Our destination was the Hunt Club, a British-colonial-style private club in the middle-class Mansour district, with a large garden. Until recently it had been a playground for Saddam's son 'Uday and his cronies – a den of sleaze, from the look of its dark, filthy red velvet rooms. A large stash of passports was found in a storeroom, of women of various nationalities. If their passports had been discarded like this, what had happened to the women themselves? Hovering over the club grounds was the sinister shell of a vast concrete mosque, surrounded by inactive cranes, that Saddam had been building. It was to be the mother of all mosques, while a few streets away drinking water was a luxury.

In Baghdad I did not feel the comfort of returning home, but the unease of arriving in a foreign, deeply wounded and dilapidated city. I missed Beirut. In the weeks that followed, as dreams were realized and dashed in equal measure by Saddam's removal and the declaration of US occupation, I heard that a mass grave had been discovered near Hilla. I decided to go and investigate. Before I had even reached the site, the combination of the heat and the stench of the dead formed a toxic potion that made me throw up. Three old mechanical diggers were planted in the middle of a reeking, muddy field. A man walked by with a plastic bag containing a bone, half a human skull and a shred of red cloth. He handed it to a grief-stricken man and woman. This, they were told, was the son they had lost in 1991.

The truth was that nobody knew, or would ever know, if it was indeed their son, since almost no effort or technology would be applied to identifying the bodies. When I raised this matter with the new US-led Coalition Provisional Authority in Baghdad, a mid-Western junior lawyer handed me a piece of paper and said that a form had to be filed before the contents of a mass grave could be dealt with. Meanwhile, local volunteers were filling plastic bags with pieces of different human beings that were then taken to Najaf for burial. Soon there would be no evidence to investigate. The lawyer could not understand my anger. The victims of the mass graves have still not been given their due. Their numbers remain unknown, not all the graves have been uncovered, and little has been done to attempt to identify them.

The years since Saddam's removal have been a time of emotional turmoil for me. I was relieved when he was caught, but troubled by the circumstances of his capture – how could such a monster be found hiding in a hole? The myth and the reality clashed.

This sensation was confirmed when I spent a day at the criminal court on 15 December 2005, on the first occasion that Saddam was to meet his victims from Dujail. Sitting in the gallery with a group of people including a woman who had lost her father to Saddam, I couldn't help feeling deeply depressed. In addition to having serious reservations about the conduct of the trial, I couldn't banish from my mind the banality of what I saw in front of me: Saddam and his cronies, pathetic old men playing a game of survival to the very end, mired in paternalistic macho language and jests, yet still able to taunt their audience, their victims. I felt, what a waste, what an infinite waste of life.

The main reactions I had when I saw the images of Saddam's messy, unresolved execution were anger and shame. I felt it to be a betrayal of all those people who had struggled for so many years against him. It reduced his crimes and Iraqis' suffering. I could not believe that it was conducted with such a lack of professionalism and solemnity, and before some of his major crimes had been exposed, most particularly those relating to the Kurds. The argument that his execution afforded his victims an opportunity for revenge didn't touch me; I had hoped for more dignity in the proceedings. A particular Shi'a group monopolized the event, which I felt robbed the Iraqi people of their collective redemption. In his cruelty, Saddam was pathological. In its clumsiness, his execution was pathetic, leaving the collective trauma of Iraq unresolved.

My disappointments are many, but chief among them are the wasted possibilities of what could have been. With hindsight, the cycle of violence that gripped Iraq in the aftermath of Saddam's fall may appear to have been inevitable. But it did not have to be. Perhaps my hopes for a rapid transition towards a representative government after Saddam's fall were unrealistic, but they should not be abandoned. Those age-old fears – sectarian and ethnic, an innate distrust of the state in addition to unfulfilled expectations after years of suffering – remain unresolved, causing division and discord.

Iraq still challenges me, even when I attempt to pull apart the fabric of who I am and examine it one layer at a time: as a woman, as someone with a cosmopolitan multicultural upbringing, as an Arab, as a person with a Muslim Shi'a heritage, as a secular individual, as an exile. What persists is my identity as a human being, beyond borders and frontiers, but still with dreams for the land of my fathers.

30 JANUARY 2005, ELECTION DAY IN BAGHDAD

I wake up with a jolt, wondering if I am going to die. A bomb has exploded nearby, making the two-storey house shake from top to bottom. It is just after 6 a.m. on the day of the first election in Iraq after the fall of Saddam.

A few hours later we gather in the hallway with our ID cards. We are going to walk to the polling station a few streets away. The silence all around us is eerie. What if a bomb has been planted at the voting station? There have been so many threats in these last few days. Although I indulge these dark thoughts, I wouldn't want to be anywhere else at this moment. I simply have to vote. I want to belong, and this is the test. This is going to be my first vote ever, and I am casting it in Iraq after a brutal dictatorship has come to an end after so many years.

The elation I feel after I have voted remains with me me all day as I watch unforgettable scenes on television of people across the country risking their lives to have their say in the future of Iraq. Old women, widows and mothers with dead children, orphans and young men, cast their votes sombrely and with defiant dignity. I am so proud. The euphoria is overwhelming.

Epilogue

After fifteen years, Bibi finally got her last wish. In early 2004 her children decided to move her to Iraq as soon as they could, despite the continuing lack of security in the country. Her eldest granddaughter Leila, who still felt guilty for failing to fulfil her final request, travelled to Damascus to supervise Bibi's exhumation. As the tomb also held the remains of Hadi and, by this time, Aunt Najla, Leila was concerned that the right body be removed. But when the workman removed the tiles, Bibi's face was still recognizable, and her hair remained. She was still waiting. Leila smiled and said to her, 'Bibi, I'm here to take you home.'

A Lebanese sheikh travelled with Bibi through the Syrian desert to Baghdad. She spent a night in her birthplace of Kazimiya, and visited the shrine as was the custom with the dead. Then she was taken to Karbala for a similar visit before the funeral party went on to Najaf.

Spanish military forces had control of Najaf and, confused by the convoy with the coffin, stopped the party from continuing. Then someone remembered that one of Bibi's grandchildren lived in Madrid and spoke Spanish. Once they had called him and explained the situation, he talked to the Spanish soldier who swiftly let them through.

My father was waiting for Bibi at the shrine, where he had the thankless task of reburying her, a small bag of bones. It took some time for the party to find the plot that belonged to her family in the Valley of Peace, as it is called, as so many had been destroyed and paved over in the last decade, but at long last Bibi was laid to rest near Imam Ali.

Jamila, my uncle Hassan's wife, died a year later. She had asked to be buried in her old family church in Baghdad, so she made a similar journey to Bibi's, leaving Hassan alone and inconsolable. Now that she was gone, he expressed his love for her without holding back. His loss was compounded by his growing disillusionment with what was happening in Iraq in the wake of Saddam's defeat. He had long nurtured hopes that he would be involved in the creation of Iraq's new constitution, but it was not to be.

On the day the constitution was published we were in a car on our way to a classical concert in London. Hassan particularly liked the violin, and we were going to hear a performance of Bach's violin and oboe concerto; he seemed equivocal about the oboe. The journey would be long enough for me to read out to him the text of the new constitution in Arabic, which I had printed out that morning. He had called me three times to remind me to bring it along.

And so I began to read, my tongue tripping every so often on the multisyllabic words, exposing my faltering command of literary Arabic. My uncle sat next to me on the back seat, listening intently, his hands flat on his knees. He picked up every mistake I made, correcting every mispronunciation and conjugating the word properly back to me. He was still a teacher as well as a lawyer at heart.

When I reached the end he looked thoughtful. 'This is not a good document,' he said. 'It's very weak and quite dangerous.' He said that the second clause – which stated that Iraq was an Islamic county, with Islam being the foundation of its legislation, which also abided by the principles of democracy, human rights and freedoms – represented a major legal contradiction, that the whole constitution process was illegal and that the document had no legitimacy because it was based on another text written under occupation. He energetically slapped his knee to express his frustration as he made each point. Then he went silent and brooded until we arrived at our destination.

In some respects, my family's long journey has come full circle. Yet of all of Bibi's children only my father lives in Iraq. But it is no longer forbidden territory: now we can come and go as we please. When

Hassan visiting the shrine of Imam Hussein in Karbala, 2010.

Hassan went to Baghdad for a visit he wept when he arrived at the Sif with its palm grove. His tears were of joy and pain. He wept again when he was reunited with old friends such as the singer Afifa Eskandar.

In a long conversation my uncle guided me through my family's past, explaining the significance of Iraq to it and his own ambivalent attitude towards the country. What he couldn't observe with his eyes, he has seen through the lens of history and memory. Like the image of a mirror in a painting, the reflections in his family's story capture what reality may not. Theirs is not a history of ideas and social patterns, but of individuals, of people who lived their lives in the best way they could.

I searched long and hard in Iraq, amidst broken lives and continuing pain, to find that country of dreams mythologized in my family by the power of longing and exile. I discovered faint traces of it, in the old houses and the tales of old men, hinting at another place, other possibilities. But these clues were few and far between. The Deer Palace too was perhaps a dream.

I still find myself harbouring mixed feelings of belonging and not belonging to Iraq. Aren't there other places that are mine too? Beirut was home until I was told it wasn't. London, the city in which my family sought exile from exile, has now become my home. And then there are all the places I've lived in between.

Some weeks after Iraq's first election, in the course of a casual conversation in London my cousin Nadia asked me, 'So does this mean we are still exiles or not any more?' I couldn't answer her.

Does exile ever really end? Rather than being a physical separation from a place, I believe that it is essentially a state of mind. It grows and evolves, taking on a life of its own. To have an inheritance of exile is a never-ending journey between myth and reality. Part of my coming to terms with Iraq entails accepting a reality that was built on an old dream; the dream of another home.

Glossary of Iraqi Terms

Clothing:

Abaya: long loose over-garment worn in public

Boyana: cloth that covers part of the head and is held by an *usbah*, for women

Charawiya: cloth tied around the head, similar to a turban, for men.

Dishdasha: informal long dress made of light cloth, worn at home by both men and women

Faisaliya: almond-shaped headdress for men similar to the Nehru cap, but stiffer and more upright, named after King Faisal I; also called *sidara*

Hijab: head cover for women; also used to mean custom-made talismans

I'gal: black cord that sits on top of the head holding the *yashmak*, for men

Keshida: conical hat with a cloth wrapped around its base, for men

Usbah: band covering the front of the head, holding a scarf underneath, for women

Zuboun: waistcoat for men, worn over a *dishdasha*

Food and Drink:

'Amba: type of rice

Arak: alcoholic drink made from aniseed

Bagila: broad beans, usually cooked with rice and dill

Baklava: filo pastry layered with almonds, walnuts, sugar and cinnamon

Burma: filo pastry stuffed with a mix of walnuts, cardamom and sugar

Chai: tea

Claytcha: soft round biscuits stuffed with mashed dates

Dolma: vine leaves stuffed with meat and rice

Fesanjoon: stew made of chicken, walnuts and pomegranate

Gahwa: coffee

Gargari: boiled and salted lupin beans

Gaymar: cream of buffalo milk

Halawa: flour-based pudding flavoured with saffron

Kubbah: stuffed meatballs mixed with bulgur wheat or rice

Kubbat hamud: stew of meatballs mixed with rice in a turnip and dried lime soup

Mann al-sima: sweet made of the boiled bark of trees, mixed with pistachios and covered in icing sugar

Mihalabi: rice pudding made with milk and flavoured with orange blossom essence

Namlet: locally produced flavoured soda water

Nargilleh: water pipe

Numihilu: lemon-flavoured sweets

Patcha: stew made of boiled sheep's head and entrails

Quzi: grilled whole lamb stuffed with rice and almonds

Sabzi: spinach and fenugreek stew

Shakar borek: baked sugared biscuits, rich and thick

Shalgham: boiled turnips with sweet molasses, a popular street food in winter

Simatch masguf: fish with a spicy tomato filling, baked on an open wood fire, unique to Iraq

Timan: rice

Turshi: pickle

Za'faran: saffron

Zarda: rice pudding flavoured with saffron

General:

'Adadah: professional female mourner

Amiriye: Ottoman primary school

Andaroun: harem, women's quarters

Al-Anfal Campaign: Iraqi forces' operation against the Kurds in 1998

Ardahaltchi: writing clerk found outside government offices, usually sought by poorly educated or illiterate people

Assalamu alaikum: general term for saying hello, literal meaning 'may peace be upon you'

'Azimah: party or meeting, usually centring around food, either lunch or dinner

Baghwan: gardener

Bibi: grandmother

Bulbul: nightingale.

Çelebi (Chalabi): Ottoman honorific title for a man with several meanings: gentleman, sage and, originally, prince

Dawakhana: men's reception area in a house

Diram: walnut-based lipstick

Effendi: Ottoman Turkish title of respect and courtesy for a man

Eid: Muslim festival; there are two main ones: Eid al-Fidr and Eid al-Adha

Farhud: the 'Great Loot', in reference to the 1941 Baghdad Farhud

Hafafa: beautician, sugaring lady

Hajji: title given to men who have performed the *hajj*, pilgrimage to Mecca

Hakawati: professional storyteller

Hamam: Bath house.

Hashemite: a member of the Hashem clan that originated from Mecca and whose members led the Arab Revolt during World War I

'Idiyah: gift, usually money, given during the Eid

Ingliz: the English, but also used to describe the British

Inqilab: *coup d'état*

Istikan: small hourglass-shaped teacup

Jiddo: grandfather

Jizrah: small islet that appears in the Tigris River in the summer time, on which vegetables were grown and parties were held

Khatun/Khatuna: darling, a term of endearment for a woman

Kursidar: sitting room

Kutab: informal primary school in a shrine or a mosque, where boys are taught by a mullah

Madarban: corridor

Mahir: official wedding ceremony

Mesopotamia: ancient Greek name for Iraq, 'the land between two rivers'

Miri/amiri: state-owned lands

Mu'akhar: usually a sum of money agreed before marriage to be paid to the woman in case of divorce

Mulukiye: Ottoman secondary school

Nawab: an honorific Mughal title held by Indians, some of whom came to live in Iraq

Nidir: a wish or plea

Ottoman: Relating to the Turkish dynasty of the house of Osman, which ruled for 800 years

Qabul: weekly ladies' get-together

Razil: rascal

Safarbarlik: Ottoman Turkish term for military conscription

Sarifa: shack

Sawt al-Arab: 'The Voice of the Arabs' radio station, which emerged from Egypt under Abdul Nasser; Arab nationalist and anti-monarchist in orientation

Shanashil: ornate wooden lattices that cover windows

Sibdaj: blusher paste

Siraj: sesame oil, usually used for kosher cooking

Soug: market

Tawli: backgammon

Thawra: revolution

Tinkhawa: mineral hair conditioner

Ukhut: sand fly whose bite can transmit *Leishmaniasis*, also called the 'Baghdad Boil', which can leave a distinctive mark on the body

Za'im: leader

Zaffa: wedding procession, involving music and singing

Zanabil: large containers made of reeds

History and Religion:

Ali: cousin and son-in-law of the Prophet Muhammad; the fourth caliph (successor to Muhammad); also considered the first of the Twelve Imams by the Shi'a, his followers

'Alim (pl. ulama): man learned in Islamic legal and religious studies

Ashura: first ten days of Muharram, when the Shi'a commemorate the death of Imam Hussein at Karbala in AD 680

Abbasids: relating to the second of two dynasties that ruled the Islamic Empire from AD 750 to AD 1258 with Baghdad as its capital; the Abbasid period was known as a golden age in which learning and culture flourished

Ahl al-Bayt: descendants of the Prophet Muhammad through his daughter Fatima and her husband, who is also the Prophet's cousin Ali

Arba'in: religious observation that marks the fortieth day after Imam Hussein's death; it commemorates the end of the month of mourning

Caliph: successor to Muhammad in leading the Muslims; the term later came to denote religious and civil rulers of the Muslim community

Fatimid: dynasty that took its name from Fatima, the Prophet's daughter, from whom they claimed descent; based in North Africa, the dynasty expanded to the Middle East AD 909–1171.

Fatwa: formal judgement or decision of a mufti, a scholar who interprets and expounds Islamic law

Ghadir: festival commemorating the farewell speech of the Prophet Muhammad, during which the Shi'a believe that he designated his cousin and son-in-law Ali as leader of the Islamic community in his stead

Imam: in Shi'a Islam, spiritual successors to Muhammad and his descendants, similar to saints; there are twelve Imams

Jihad: war in the name of religion against unbelievers

Karbala: location where the battle for the Caliphate was fought in 680 and Imam Hussein was killed; today a medium-sized city in Iraq

Maghsal: special area in mosque for washing the dead

Muharram: the second month of the Islamic lunar calendar, during which Imam Hussein and his family were killed by Umayyad forces

Mullah: learned Muslim man with religious authority

Sayyid (pl. Sadah): descendant of the Prophet

Shi'a: group of Muslims who believe in the right of Ali and his descendants (*Ahl al-Bayt*) to lead the Muslim community

Sunni: often referred to as orthodoxy, position that accepted the authority of the first generation of leaders to follow the Prophet, in contrast to Shi'a beliefs; majority group among Muslims, who have also long held political power

Umayyad: first dynasty to rule the Muslim Empire AD 661–750; headed by a Meccan tribal family, who established Damascus as their base; in their assertion of power they fought many prominent Muslim figures, most significantly Imam Hussein

Wakf: religious endowment, usually property

Music and Entertainment:

Chopi: light popular songs, in comparison to the traditional *maqam*; often accompanied by dancing

Daqbuli: male servants in the brothels in the *kallachiya* of Baghdad

Dunbuk: Iraqi drum made of stretched skin over a clay base

Kallachiya: old red light district of Baghdad

Maqam: form of classical music with a strict structure that cannot be changed by the singer, influenced by Indian ragas; traditionally, the majority of *maqam* instrumentalists in Iraq have been Jews

Murshid: master of ceremonies during a *zorkhana* wrestling match

Oud: type of lute or mandolin from the Middle East

Zorkhana: literally 'house of force'; traditional wrestling clubs popular in Iraq and Iran

Titles and Positions:

Mejlis/Majlis: council, legislative body, parliament

Mukhtar: head of a village or town

Pasha: high-ranking Ottoman political appointee

Qa'immaqam: Ottoman term for the deputy governor of a provincial region

Wali: governor

Wilayet: province (during Ottoman era)

Transport:

'Arabana: general term for all horse-drawn carriages, later used for cars

Gharri: horse-drawn carriage

Guffa: circular boats used to transport people and goods on the Tigris, propelled with a paddle; they are mentioned by Herodotus in the 5th century BC.

Rabbil: small horse-drawn carriage

Reyll: Iraqi colloquial term for rail

Takhtarawan: wooden palanquin

Tenta: convertible car

Trammai: Iraqi colloquial term for tramway

List of Illustrations

ALL PICTURES ARE the author's own, with the following exceptions. While it is believed that these images are in the public domain, the publishers would like to apologise for any omissions and will be pleased to incorporate missing acknowledgments in any future editions:

p. 13   The shrine of the Imam Musa al-Kazim and Muhammad Jawad in Kazimiya was renowned for its two domes and four minarets.

p. 20   A *guffa* on the Tigris in Baghdad, circa 1914.

p. 41   A woman in an *abaya* walks by the river.

p. 64   British troops entering Baghdad, March 1917.

p. 80   Men in a popular café, sitting on traditional high wooden benches.

p. 97   Rashid Street, one of Baghdad's main thoroughfares, in the late 1920s.

p. 143 Samau'al Street, Baghdad.

p. 168 The Haidarkhana mosque in Baghdad, late 1930s.

p. 170 Salima Murad, aka Salima Pasha, one of Iraq's most famous singers, in the early 1930s.

p. 174 Afifa Eskandar, another renowned Iraqi singer, in the late 1930s.

p. 207 Rashid Street in the 1940s.

Acknowledgements

THIS BOOK WOULD not have come to life without the help of many people, books, articles, photographs, letters and songs. It is simply not possible to thank you all, but I would like to give particular thanks to the following. Where I have quoted directly from a source I have included the page numbers on which the citation appears.

Verse 71 in Ali, A. Y. (tr.), *The Holy Qu'ran* (New York, 1987) appears on p. 196; Allaf, A. K., *Baghdād al-Qadimah* (Baghdad, 1960); Allaf, A. K., *Qian Baghdad* (Baghdad, 1969); Allawi, A.A., *Tajjarub wa Dhikrayat* (London, 1999); Jawahiri, 'My Brother Jaafar' in Badawi, M. M., *A Critical Introduction to Arabic Poetry* (Cambridge, 1975) appears on p. 231; Baghdadi, A., *Baghdad fi al-'Ishriniyat* (Beirut, 1999); Barker, A. J., *The Neglected War: Mesopotamia 1914–1918* (London, 1967); '1958 Revolution Proclamation No. 1' in Battatu, H., *The old social classes and the revolutionary movements of Iraq: a study of Iraq's old landed and commercial classes and of its Communists, Ba'thists, and Free Officers* (Princeton, 1978) appears on p. 265; Bell, G., *The Letters of Gertrude Bell*, 2 v. (London, 1927), p. 802; Bell, G., letters dated 8/7/1921 and 4/12/1922 reproduced on *http://www.gerty.ncl.ac.uk* appear on p. 115; Caractacus, *Revolution in Iraq* (London, 1959); *Chicago Daily Tribune* newspaper (15 July 1958); Chiha, H. K., *La province de Bagdad: son passé, son présent, son avenir* (Cairo, 1908); Cooper, A.,

Cairo during the War 1939–1945 (London, 1989); *Christian Science Monitor* (14 July 1958); Dakkash, Laure, 'Amantu Billah' (song) on YouTube apperas on p. 167; *Dijlah* newspaper (1921–1925); Falle, S., *My Lucky Life* (Sussex, 1996); Verses 42 & 84 in Fitzgerald, E., *The Rubaiyat of Omar Khayyam* (Oxford, 1981); Gailani al Werr, L., 'A Museum is Born' in Polk, M., and Schuster, A. (eds), *The Looting of the Iraq Museum* (New York, 2005); Geniesse, J. F., *Passionate Nomad: The Life of Freya Stark* (London, 2000); British Colonial Office, *'Report on Iraq administration'* (London, 1922); British Naval Intelligence Division, *A Handbook of Mesopotamia 1916–1917*, 4 v. (London, 1917); Haidar, R., *Mudhakkirāt Rustum Haidar* (Beirut, 1988) appears on p. 9; Hassani, A. R., *Tarikh al Wuzarat al Iraqiyya*, 10 v. (Baghdad, 1988); Ireland, P., *Iraq: A Politcal Study* (London, 1937); Khalili, J., *Mawsuat al-Atabat al-Muqadasah*, vols 9–10 (Beirut, 1987); Kojaman, Y., *The Contemporary Art Music of Iraq* (London, 1978); *LA Times* newspaper (14 July 1958); *Layla* women's monthly magazine (Baghdad, 1925); MacMillan, M., *Peacemakers: The Paris Conference of 1919 and Its Attempt to End War* (London, 2001); *al-Mada* newspaper (2003–2007); Madfa'i, Ilham, 'Chalchal Alaya al-Rumana' (song) on *Ilham Madfa'i* (2004) appears on p. 62; 'Mali Shughul Bil Soug' (Iraqi folk song), reproduced on *http://www.ilhamalmadfai.com/lyrics.htm*, appears on p. 69; Litvak, M., 'Money, Religion, And Politics: The Oudh Bequest In Najaf And Karbala, 1850–1903' in *International Journal of Middle East Studies*, 33:1:1–21 (Cambridge, 2001); *Murrays' Handbook for Travellers in Turkey and Asia* (London, 1878); PRO BW 39/1; PRO E 2515/1090/93; PRO E 6899/78/79; PRO FO 317/E6899/78/93; PRO FO 371/40014; PRO FO 371/34/98; PRO FO 371/5071, 5072, 5073, 5076, 5078, 5079, 6350, 6351, 6352, appears on p.p. 88–9; Faisal's Coronation Speech recorded in Iraq Intelligence Report No. 19, 1 September 1921, in PRO 371/6353-W10532/100/93 p. 71, 6355, 133067, 133069, 133090, 134256; PRO FO 930/278; PRO FO 481/1; PRO FO 624/1; PRO 624/30 appears on p. 107; 'Proclamation of Lieutenant-General Sir Stanley Maude at Baghdad', reproduced in *http://www.harpers.org/archive/2003/05/0079593*, appears on p. 65; MR97-269.5, US State Department; Qattan, A. R., *Mudhakarat min Janub al-Iraq* (London, 2005); *National Geographic Magazine*

(December 1914, February 1916, April 1922); Qazwini, A. J., *Tarikh al-Qazwini fi a'lam al musinin wal ma'rufin min a'lam al-Iraq wa ghayrahum*, 1900–2000, 4 v. (Beirut); Rizk Khoury, D., *The Ambiguities of the Modern: The Great War in the Memoirs and Poetry of the Iraqis* (forthcoming); Rusafi, M., *Al-Armalah al-Murdhi'a* reproduced on *http://www.aliraqi.org/forums/archive/index.php/t-55932.html* appears on p. 34; *al-Sabah* newspaper (2004–2008); 'Allahu Akbar' (Sawt al-Arab song) reproduced on *http://www.ebnmasr.net/forum/t90788.html* appears on pp. 263–4; Qazzaz, Said, speech reproduced in *al-Sharq al-Awsat* newspaper (17 March 2001) appears on pp. 312–3; Shahbandar, M., *Dhikrayat Baghdadiya* (London, 1993); Sheikh Ali, F., *Mudhakarat Warithat al 'Arsh* (London, 2002); Lionel Smith Papers, GB165-0266 ALF Smith (St Antony's College, Oxford); Stark, F., *Dust in the Lion's Paw* (London, 1961); Stillman, N., *The Jews of Arab Lands in Modern Times* (Philadelphia, 1991); *New York Times* (14 & 17 July, 19 November 1958); Storrs, R., *Orientations* (London, 1945); Ta'i, J., *Al-Zorkhana al-Baghdadiya* (Baghdad, 1986); *Time Magazine* (17 June 1957); *The Times* newspaper (14, 15, 16 & 30 July 1958); Tripp, C., *A History of Iraq* (Cambridge, 2005); 'Imnahu al Jad la al'abaya' in Al-Uzri, A. H., *Diwan al-Hajj Abdul Hussein al-Uzri* (Beirut), p. 52 appears on p. 104; Wallach, J., *Desert Queen: The Extraordinary Life of Gertrude Bell: Adventurer, Adviser to Kings, Ally of Lawrence of Arabia* (London, 1997); al-Wardi, A., *Lamhat Ijtima'yah min Tarikh al-Iraq*, 6 v. (Beirut, 2005); *Wall Street Journal* newspaper (17 & 18 July 1958); *Washington Post* and *Times Herald* newspapers (16 June, 16, 17 & 30 July 1958); Wilson, J., *Lawrence of Arabia* (London, 1988); al-Yasin, M. H., *Tarikh al-Kazimiya* (Baghdad, 1970); *Azzaman* newspaper (2003–2007); Zubaida, S., 'Entertainers in Baghdad 1900–1950' in Rogan, E. (ed.), *Outside In, On the Margins of the Modern Middle East* (London, 2002); Zubaidi, F. (ed.), *Baghdad min 1900–1934* (Baghdad, 1990).

I am grateful to many individuals in different parts of the world who gave me their time, knowledge and friendship. In no particular order, thank you to Lamia Gailani al-Werr, Abbas Kelidar, Sami Zubaida,

Hala Fattah, Haitham Hadid, Abdul Razzaq al Safi, Tamara Daghistani, Nasser Saadoun, Louay Suwaidi, Salwa Suwaidi, Nibraz Kazimi, Dina Risk Khoury, Betjullah Destani, Safa Killidar, the staff of the London Library, Lamys Araktingi, Muhammad Biyara, Silvia Kedourie, Fatima Mudamgha, Houchang Chehabi, Kevin Conroy-Scott, Muhammad Hassan al-Musawi, the archivist at Downing College, Cambridge, Ali Bahruluum, the late Hussein Ali Mahfouz, Fadhil Chalabi, Yiorgos Borovas, Jonathan Foreman, Assaad Eskandar and the staff of the Iraq National Library and Archives in Baghdad, Rima Osseiran, Megan Ring, Issam Ibrahim, Lauren Rizzo, Alex Selim. Special thanks to Hassan Mneimneh, Fabio D'Andrea, Amr Shalakany, Elif Uras, Michael Soussan, Terence Coleman, Joumane Chahine. A big thank you to Justine Hardy for her infinite wisdom and kindness and to Fouad Ajami. I would like to thank my sister Mariam for her unswerving encouragement, my father Ahmad, and my cousin Nadia Chalabi in particular, who indulged my idea for a book long before it became one and helped me in its early stages and shared with me her many insights and her memories.

At HarperCollins, I would like to thank Arabella Pike and Tim Duggan for commissioning the book, Annabel Wright, Robert Lacey and Sophie Goulden. I would like to thank Sue Lascelles for her wonderful support, Andrew Wille, Richard Kelly and Sarah O'Reilly. I would also like to thank my agent Elizabeth Sheinkman, and Jonny Geller and Felicity Blunt at Curtis Brown.

To several characters in this book who kindly recalled their histories in Iraq and elsewhere, I thank them for trusting me: my cousins Ghazi, Ali and Zina, Leila, Mahdi, Issam, Mohammad, Ali and Sarah, my great uncle Saleh, my uncles Jawad, Talal and Hazem. A special thank you to my aunts Thamina and Raifa, who walked me through their lives with enthusiasm, drawing on all details of their childhoods in the Deer Palace and their time in Iraq, from the furniture to the flowers, and answering every query I had, however farfetched. Finally, I would like to thank my uncle Hassan, who took this book very seriously and personally. He guided me through the labyrinth of Iraq's history and the family's with passion, patience and dedication. He held my hand every step of the way. It is to him that I dedicate this book, with love.

Index

9/11 375
Abbas, Sheikh 133
Abdiya, Princess 268
Abdul Ilah, Prince Regent
  becomes Regent 191
  flees to Transjordan 202–3
  returns to Baghdad 205, 206–7
  orders executions for treason 208, 234
  relationship with Nuri Pasha 220–1,
    258–9
  dislike and unpopularity of 226–7, 247
  and Anglo-Iraqi Treaty 229, 230
  asks for money to be transferred offshore
    259, 313
  murder of 262, 264, 266, 268–9, 322
Abdul Wahab 166
Abizaid, Lieutenant General 378
Abu, Ahmad 186
Abu Ghraib prison 273, 309
Abu Hanifa shrine 197
Abu Muhammad 372–3, 374, 378
Abul Timman family 91
'Abussi, Abdul Sattar 268
Afnan, Sayyid Hussein 106
Agarguf 11
Agha Jaafar, Abdul Latif 237, 313
Agha Jaafar, Jaafar 295, 302
Agha Jaafar, Najla Chalabi 130, 150, 159
  and the Eid festival 163, 165, 166
  dislike of modernising ways 194
  on holiday in Broumana 194
  and death of Rumia 219–20
  unwitting visit to the kallachiya 223,
    225
  and the military coup 277

children sent to London 295
arranges for food to be sent to Ahmad at
  school 304
attempted escape from Baghdad 309–11
difficult marriage of 343–4
exile in Britain 343–4
and death of Bibi 363–4
Agha Jaafar, Zeinab 295, 302
Akbar, Ali 130, 151–2, 205
Akhuwat-i-Iran 82–3
al Tai, Yunis 321–2
al-Abidin, Zein 115, 116, 151
al-Anfal campaign 361–2
al-Arab, Nabih al-Din 328–9
al-Arab, Shehab al-Din 328
al-Askari, Jaafar Pasha 139, 60
al-Bakr, Ahmed Hassan 346
al-Bassam, Aminah 55
al-Bassam, Fahima 42–3, 55, 61
al-Bassam, Kuku 235, 276, 295, 296, 326
al-Bassam, Leila 220, 252, 274, 295, 351,
  352, 364, 385
al-Bassam, Mahdi 220, 252, 262, 264,
  276, 302, 303
al-Bassam, Murtada 266, 274
al-Bassam, Rumia Kazimi (Postforoush)
  family background 36
  and choosing a husband for her daughter
    37
  description of 37, 62
  meets her daughter's future in-laws 40–4
  arranges her daughter's wedding day
    44–5
  and the mahir 46–50
  and her daughter's trousseau 50–2

and her daughter's wedding 54, 55
worries about the Chalabis' lifestyle
  59–60
and British occupation of Baghdad 82,
  83, 91
preparations for King Faisal's visit 100–1,
  103, 105
reassures Bibi on her fertility problems
  108
and blindness of her grandson Hassan
  123
visited by her grandchildren 161
known by the telephone operator 162
death of 219–20
makes pact with her friend Amira 361
al-Bassam, Saleh 195, 197–8, 201, 218,
  247, 261–3, 273–4, 276
al-Bassam, Sayyid Hassan 36, 38–40
al-Duwayh, Rahim 126–7
al-Duwayh, Zahra 126–7, 131, 134, 244
al-Falaki, Adawiyah 'Girl of the Bridge'
  230–1
al-Farun–Imad 185–6
al-Hakim, Sayyid Muhsin 362
al-Hashemi, Yassin 152–3
al-Hassani, Sayyid Muhammad Jawa al-
  Tag 179
al-Hayat 331
al-Husri, Sati' 113–14, 116, 157, 183–4
al-Jawahiri, Jaafar 231
al-Jawahiri, Muhammad Mahdi 231
al-Karkhi, Mulla Abboud 141, 167,
  224–5
al-Karmali, Pere Anastate 29
al-Kazim, Imam Musa 14, 155
al-Khalisi, Abdul Rasul (see also Khalisi,
  Sheik Mahdi) 290, 307
al-Lampachi, Dawud 224
al-Madfa'i, Ilham 69–70
al-Madfa'i, Jamil 152–3, 208
al-Majali, Abdul Wahab 367
al-Malaika, Nazik 251
al-Nayif, Abdel Razzaq 335
Al-Qotob, 'the Pivot' 40, 52, 269
al-Rashid, Haroun 14, 314
al-Sadr, Bint al-Hudda 351–2
al-Sadr, Sayyid Muhammad 84, 87, 91, 94,
  231, 232, 333–4
al-Sadr, Sayyid Muhammad Baqir 351–2
al-Sadr, Sayyid Musa 335–7
al-Said, Nuri Pasha
  relationship with Hadi 156
  as Prime Minister 199, 200, 220–1
  untenable position of 200–1
  replaced by Rashid 'Ali 202–3
  relationship with Prince Regent 220–1,
    258

Communist vendetta 221, 253
rapport with Rushdi 226, 249
replaced by Saleh Jabr 229
and Arab–Israeli war 233
policies and plans 253–4
as official face of Iraqi politics 255
and the Suez crisis 257
takes refuge in Thamina's house 264–6
and murder of his son 269
murder of 269–70
al-Said, Sabbah 269
al-Thawra 322
al-Tikriti, Barzan 348, 351
al-Uzri, Abdul Hussein 24–5, 29, 83, 104
al-Uzri, Amira 25, 30
al-Uzri, Issam 270, 271, 272, 286, 287
Ali (clerk) 133
Alia, Queen Mother 210
Allawi, Abdul Amir 219, 257, 266–7
Allawi, Ali 252, 276, 294–5
Allawi, Ghazi 220, 252, 276, 292, 293–4,
  302, 303
Allawi, Jaafar 241
Allawi, Raifa Chalabi
  memories of her mother 74–5
  and the Baghdad markets 75–6
  birth of 123, 124
  fascinated by chandelier in the dining
    room 131
  childhood of 149–50
  and the Eid festival 163, 165
  witnesses parental argument 173
  shopping in Beirut 194
  visits Cairo 218
  marriage of 219
  birth of children 220
  and her mother's charity-giving 244
  and murder of Prince Regent 266
  exile in London 313
  dreams of Bibi 369
Allawi, Zina 73, 326–7
Allerton, John 304
Alto, Alvaar 250
American University (Beirut) 122, 178,
  337
Amir Abdullah 87
Amir, Umm Abdul (see also Istrabadi, Bibi)
  300
Amira (friend of Rumia) 361
Amman 344, 366, 367
Anglo-Iraqi Treaty (1930) 147–8, 182,
  202, 203, 229, 275
Anglo-Persian Oil Company 235
Arab Legion 232–3
Arab Revolt (1916) 87, 89
Arab–Israeli war (1967) 329
Arab–Israeli war (1948) 232–3

Arafat, Yassir 338
Arbil 373, 374, 376, 378
Arif, Abdul Salam 263, 265, 267, 278,
    305, 322, 323
Arif Agha family 91
Arthur Andersen 368
Ashura festival xxviii, 78–9, 243, 279–80
Aswân Dam 257
Ayubi, Ali Jawdat 323
A'zamiya 197, 235–6, 241, 263, 277, 299,
    320
Azerbaijan 36
Aziz (brother of Khalil the doorman)
    341–2

Bab al-Mua'dham (Baghdad) 154
Bab al-Murad 4
Baban, Ahmad Mukhtar 323
Baban, Jamil 156
Babylon 380
Badiya, Princess 211, 261, 268
Baghdad
  history of xxx–xxxi
  attempts at modernization of 10–11
  morale in 28
  arrival of German military in 31–2
  treatment of army deserters in 32
  military casualties in 33
  British occupation of 63–7, 77–84
  beautiful memories of 74
  markets in 75–6, 120, 161, 206
  café culture in 79–80, 81–2
  opposition to British rule 88, 91–2
  King Faisal's arrival at 102, 106–7
  education in 113, 114
  bustling streets of 142–4
  unrest in 151–2, 154, 160, 191, 202–10
  life and culture in 161, 166–7, 168,
    169–72
  nightlife 173–4
  severe flooding of region 196
  seized by Iraqi Army 202–3
  Jews in 205–8
  great uprising in 230–2
  modernizing of 236–7, 245, 250–3
  military coups in 260–9, 276–8, 300,
    309–10, 312–13
  and murder of the royal family 268–9
  British Embassy attacked 274
  Mahdawi courts 307, 308, 312–13
  overthrow of Saddam 377, 380–1
Baghdad College 212–13, 277
Baghdad Modern Art Group 252
Baghdad Opera House 189–90
Baghdad Pact (1955) 255–6, 260, 266, 290
Baghdad Radio 167, 252
Baghdad Stock Exchange 142

Baghdad Times 120
Baghdad Vilayet 7
Bahri, Yunis 183
Balfour, Frank 86
Balkans 7
Banco Nazionale del Lavoro 366
Baquba 144
Barzani, Mulla Mustafa 335, 376
Bashir, Jamil 169–70, 213–14
Basra xxx, 17, 20, 26, 63, 78, 89, 95, 96,
    102, 113, 118, 144, 204, 277
Basra railway 96–8
Bassam, Thamina Chalabi 218
  memories of her mother Bibi 73, 74–5
  and the markets of Baghdad 75–6
  as favourite of her grandfather 131
  childhood of 132, 133, 149–50
  meets Crown Prince Ghazi 137
  description of 150–1, 157
  witnesses parental argument 159
  and the Eid festival 163, 165
  education of 193
  shopping in Beirut 193–4
  marriage of 195, 197–8, 201
  friendship with Princess Badiya 211
  birth of children 220, 235
  unwitting visit to the kallachiya 223,
    225
  and her mother's charity-giving 244
  and the Saudi Royal family 247
  rescued from mob violence 262–3
  gives refuge to Nuri Pasha 265–6
  in fear of her servants 276
  arrest and return of Fahima 277
  daughters sent to London 295
  moves to Beirut 319
  exile in London 343
  and death of Bibi 363–4
Bassim, Zaki 234
Batul (friend of Saeeda) 234
Bazargan family 91
Bazzaz, Abdul Rahman 289
BBC 166, 252, 329
Bedouin 7–8, 63, 87, 106, 119
Beirut xxvii, 122, 138, 176, 178, 191–5,
    226, 252, 254, 300, 316–20, 322–3,
    325–6, 334, 337, 338–41, 356
Beirut College for Women 218
Bell, Gertrude 95–6, 105, 107, 115–16,
    183, 185–6, 389
Berlin-Baghdad railway 10, 23
Betahon (maid) 326
Bevin, Ernest 229
Beydoun family 319
Blomberg, Axel von 204
Bombay 17
Broumana 191–5

'the Building' (Beirut) 325–6, 329–32, 339, 341–2
Bush, George Snr 365
Bush, George W. 375

Cairo 213, 215–17, 218–19, 252, 253–4
Cairo Conference (1921) 95
Cairo Zoo 217
Cambridge University 122–3, 127–8, 250
Central Treaty Organization (CENTO) 256
Chadirchi family 91
Chaigahwa, Habib 118, 132–3, 215
Chalabi, Abdul Ghani 8, 25, 32, 111, 112, 203
Chalabi, (Abdul) Hadi
  and the stone deer xxvii
  befriends Ni'mati 15, 16, 18, 19, 148
  father's fears for 24, 26
  works for the Turkish 6th Army 26–30, 33, 61
  description of 27–8
  sees his first aeroplane 30–1
  observes East–West interactions 32
  wife chosen for 35–45
  wedding of 53–6
  and British occupation of Baghdad 66–7, 77
  and the Ashura procession 78–9
  enjoys café life 81–2
  and post-War politics 87
  and preparations for King Faisal's visit 99–101, 103
  moves into the Deer Palace 112
  business enterprises 118–20, 153, 159–60, 247–9, 254
  and death of his brother Abdul Rasul 138–41
  political career 142–6, 153
  and killing of girl and her husband 145
  arrest of 151–4
  imprisonment and release 154–6, 157
  relationship with Bibi 158–9, 354
  incensed at Rushdi's nightly escapades 173–5
  agrees to send Rushdi to London 176
  agrees to send Rushdi to Beirut 178
  moves to Sif Palace 179–80
  and marriage of Thamina 195, 197–8
  and assassination of Haidar 198–200
  political and family tensions 203, 205–6
  provides refuge for Jewish families 205–6
  loyalty to Kazimiya 227
  as owner of dogs and horses 227
  supports Jabr 231–2
  and oil revenues 235–6
  buys jewels for Bibi 238–9
  takes his family to Paris 238–41
  moves into his new house 241
  declines to buy Dolphin Square, London 248, 315
  visit to Turkey 257
  and military discontent 258–9
  asked to deposit money offshore for Prince Regent 259, 313
  and the military coup 275
  exile in London 285–8, 306–7, 313–15, 352–4
  welcomes Ahmad and Ghazi to London 293–4
  organizes his grandchildren's lives 296–7
  moves to Beirut 316–17, 319, 322–3, 324, 326
  returns to Baghdad 320–2
  returns to London 323
  love of carpets 330–1
  and death of Ni'mati 332
  illness and death 357–9
  funeral of 359–61
Chalabi, Abdul Hussein
  appearance of 5–6
  rigorous daily schedule 5–6
  listens to grievances and requests 6–12, 133
  administration of his lands 7–8
  and the locust story 8–10
  summoned to the shrine 12–16
  agrees to help with gift of carpets 16
  visits his sister and brother-in-law at the Deer Palace 16–19
  and the pot of earth 19–22
  and the outbreak of War 23–6
  and marriage of his son Hadi 35, 50
  meets Bibi's grandfather 44
  relationship with Bibi 59, 61–3, 129–30
  and defeat of the Ottomans 67
  and British occupation of Baghdad 77–8, 83–5
  involvement in politics 86–8, 91–2, 94, 95
  meets and entertains King Faisal 96–101, 103–5
  at King Faisal's installation 106
  as Minister of Education 109–10, 113–14, 127, 153
  banned from the Kazimiya shrine 110–11, 287–8
  moves into the Deer Palace 112–13, 114–16
  fashionable appearance of 115
  visited by Gertrude Bell 115–16, 127
  re-admitted to the shrine 116
  encourages Abdul Rasoul to go abroad 121–2

attends Gertrude Bell's funeral 129
loved by his grandchildren 131
and death of Abdul Rasul 140, 141
progressive attitude of 146–7
saddened at death of King Faisal 148
and arrest of Hadi 149–51, 154, 156
relationship with his wife 158
plays down his political role 159–60
and the Eid festial 164
listens to the radio 167
domestic authority of 175
moves to Sif Palace 179–80
death of 181–2, 191
Chalabi, Abdul Jabbar 250
Chalabi, Abdul Rasul 285
  hears about the first aeroplane over
    Baghdad 31
  meets Bibi for the first time 53–4, 55
  celebrates Ashura 78
  and political self-determination 87, 94
  moves to Deer Palace 112
  education of 114, 118, 120–3
  attends Cambridge University 122–3,
    127–8
  returns to Baghdad 136–8
  illness and death 138–41
  elegy written for 225
Chalabi, Ahmad
  birth of 220
  devoted to Saeeda 232, 241, 267, 292–3
  fascinated by the Marsh women 237
  visits Hassan in Paris 238, 239–40
  precociousness of 239–40
  love of popular culture 252
  escapes to Sheikh Jamil farm 270–2
  bravery of 271–2, 277
  called a traitor at school 276, 292
  in exile in London 282–4
  attends Seaford College, Sussex 302,
    303–6, 329
  and death of Saeeda 306
  studies mathematics in America 334–5
  forges lifelong friendship with Mulla
    Mustafa Barzani 335
  marriage of 336–7
  and death of Khalil and his family 342
  moves to Amman 344
  and death of Hadi 360
  and seizing of his bank 366–9
  charged with misconduct 367–8
  organises opposition to Saddam Hussein
    370, 371, 373, 374, 375–81
  and re-burial of Bibi at Najaf 385–6
  lives in Iraq 387
Chalabi, Ali Jnr 325, 356
Chalabi, Ali Snr 8–10, 117
Chalabi, Amira 35, 36–7, 41–3, 62, 244

Chalabi, Bashar 325
Chalabi, Bibi al-Bassam
  memories of Baghdad xxxvii
  as prospective wife of Hadi 36–7
  refuses to marry a distant cousin 37–8
  childhood 38–9
  and death of her father 39–40
  meets her future in-laws 40–4
  agrees to marry Hadi 44–5
  and the mahir 46–50
  preparations for the wedding 51–3
  wedding of 53–6
  and inspection of her trousseau 56–7
  married life 58–60
  worries about conceiving 60–1, 108–9
  relationship with her in-laws 61–3
  and fall of Baghdad 64
  family memories of 74–5
  pregnancies 81, 85, 109, 123, 130, 138,
    220
  worries about Hadi 81–2
  reaction to British occupation of Baghdad
    82–5, 91
  and King Faisal's visit 100–1, 103–5
  moves into the Deer Palace 112, 113
  meets Gertrude Bell 116
  attends qabuls in Deer Palace 118, 123
  dislikes Hadi's scrapes with the Bedouin
    119
  and blindness of her son Hassan 123–5
  social skills, ambition and curiosity
    129–30
  hosts parties at the Deer Palace 137
  attends concerts 138, 146–7
  and killing of girl and her husband 145
  and monthly family gatherings 149–51
  arrest and release of Hadi 151, 154–6
  relationship with Hadi 158–9, 354
  and the Eid festival 163–6
  love of music 167–8, 169
  worries about Rushdi's nightly escapades
    173–5
  encourages Rushdi to go to London
    175–8
  moves to Sif Palace 179–80
  love of fashion and shopping 192–5
  considered a snob 195–6
  and marriage of Thamina 195, 197–8,
    201
  and flooding of Sif Palace 196–7
  vehemently opposed to Hitler 205
  not impressed with Freya Stark 210–11
  monthly visits to the royal household 211
  has her fortune told in Cairo 218–19
  unwitting visit to the kallachiya 223, 225
  visits Paris 233, 238–41
  cards and gambling 242–3, 354

household duties 242–3
charity-giving 243–4
religious inclinations and superstitions
    243–4, 291, 302, 365
and the Communist party 246–7
and Egyptian coup 246
anti-Arabist views 253
as supporter of Nuri Pasha 254
escapes to Sheikh Jamil farm 267, 270–3
and military coup 269
and death of Nuri Pasha 270
sends Ahmad to London 293
secretes jewels in Leila's coat 295
pleads for Rushdi's release 298–9
exile in London 300–2, 306, 307,
    313–15, 343–4, 352–4
moves to Beirut 316–19
serious illness of 324
as matriarch 325–8
and the Lebanese civil war 339–41
and death of Khalil 342
and theft of jewellery 350–1
entertains her grandaughter Tamara with
    stories and verse 353–4
and death of Hadi 358–61
death and funeral of 363–4
exhumation and re-burial at Najaf 385–6
Chalabi, Burhan see Nawab, Burhan
    Chalabi
Chalabi, Hadi 337
Chalabi, Hashim 337
Chalabi, Hassan 109, 112
and sense of loss xxvii–xxix
blindness of 123–5, 133–5, 239
childhood and education 132–5
mimics his great aunt Munira 150
arrest and release of his father 151–2,
    157
politicised education of 157
and the Eid festival 163–6
impact of radio on 167, 345
love of music 167–70, 326, 386
attends a wrestling match 171–2
and death of his grandfather 181
studies law 212–17
meets Jamila 214–16
moves to Cairo 216–17
persuades Jamila to join him in Paris 228
illness and recovery 237–8
visited by his family in Paris 239–40
and Egyptian coup 245–6
dislike of Nasser 253–4
and democratic progress 254
escape from military coup 267
escapes to Sheikh Jamil farm 270–1
keeps up family morale 276
at the Ashura ceremony 280

returns to Law Faculty 289–90
says goodbye to Ahmad and Ghazi 293
refuses to escape from Baghdad 308,
    311
makes a case for Rushdi's innocence 311
visits Bibi in Beirut 316–17
returns to Baghdad 319–20
possible arrest of 322, 323
leaves Iraq for Beirut 333–5, 341
forges lifelong friendship with Mulla
    Mustafa Barzani 335
decides to found a bank in Jordan 344
visits Ahmad and his family in Amman
    345
learns of his niece Leila's decision to wear
    the hijab 352
marriage of 356
and death of Hadi 360
and death of Bibi 364
comment on new Iraqi constitution 386
returns to Iraq 387
Chalabi, Hazem
and the Eid festival 163, 165
enchanted by tales told by Rumia 220
trip to Paris 238, 239
at Cambridge University 250
and development of Baghdad 250
exile in London 286
sends children to Lysses School in
    Hampshire 303
marriage of 318–19
children of 325
in Beirut 341
Chalabi, Hussein 235, 276, 299, 325
Chalabi, Ibrahim 112, 137, 191
Chalabi, Ilham Agha Jaafar 226, 235, 267,
    270, 299–300, 309–11, 313
Chalabi, Jamila Antoine 311
meets and works for Hassan 213–16
agrees to join Hassan in Paris 228
alerts Bibi of Hassan's illness 238
visited by the Chalabi family in Paris 239
has a hysterectomy 247
protects Hassan from student attack 290
visits Beirut with Hassan 316–17
returns to Baghdad with Hassan 319
leaves Baghdad for Beirut 334
visits Ahmad and his family in Amman
    345
marriage of 356
death of 386
Chalabi, Jamila al-Uzri 30
and choosing a wife for her son Hadi 35
meets and likes Hadi's prospective wife
    40–3
describes Bibi to her son 53
hosts post-wedding tea party 56–7

relationship with Bibi 60, 61–2
alarmed at increased political
  developments 91
gives birth to final child 108
moves into the Deer Palace 112
holds *qabuls* at Deer Palace 117–18,
  146–7
and female emancipation 121
concerned at her son's Western costume
  127
and death of Abdul Rasul 138, 140–1
family gatherings at the Deer Palace 149
relationship with her husband 158
death of 181
Chalabi, Jawad
  birth of 138
  slapped for singing a song from a *Tarzan*
    film 155–6
  overeats when upset 156–7
  and the Eid festival 163–4
  enjoys his visit to Baghdad playground
    165
  witnesses Hadi's anger at Rushdi 173
  and engagement of his sister Najla 237
  family takes refuge with 263
  escapes to Sheikh Jamil farm 266–7, 271
  accompanies Rushdi back to Baghdad
    273
  keeps family business afloat 291–2, 324
  attempted escape from Baghdad 308–11
  returns to Baghdad 320
  possible arrest of 322
  leaves Baghdad 333
  in Beirut 341
Chalabi, Khadja Malaika 30, 328
  chooses a wife for her grandson Hadi
    35–7, 40–3, 44
  and wedding of grandson 55
  relationship with Bibi 59
  stays with her son Abdul Ghani 112
Chalabi, Leila Osseiran 336–7, 367
Chalabi, Mariam 337
Chalabi, Mohamed 271, 313, 325, 329
Chalabi, Muhammad Ali 31, 54, 112, 177,
  197, 230, 317
Chalabi, Munira *see* Nawab, Munira
  Chalabi
Chalabi, Nadia
  birth of 73, 235
  nostalgia for the past 75–6
  education of 276–7, 299
  journey into exile 295
  attempts to escape from school 302–3
  obsessed with cats 325
  fashion conscious in Beirut 326–7
  and lewdness of Nabih al-Arab 329
  and question concerning exile 388

Chalabi, Najla *see* Agha Jaafar, Najla
  Chalabi
Chalabi, Peri 325
Chalabi, Raifa *see* Allawi, Raifa Chalabi
Chalabi, Reem 325
Chalabi, Rushdi 73
  birth of 85, 108
  moves to the Deer Palace 112
  as favourite of his grandfather 131
  attends the *dawakhana* 132
  arrest and release of his father 151–2
  politicised education of 157
  and Eid festival 163–5
  impact of radio on 166
  attends a wrestling match 171–2
  enjoys Baghdad nightlife 173–5, 225–6
  as pampered and spoilt 175
  unhappy visit to London 175–8
  attends American University in Beirut
    178
  joins the *Ikhwan al-Hurriya* 209–10
  describes Jamila to Hassan 215
  political career 217–18, 227
  marriage of 226
  temperament of 227
  and oil revenues 235–6
  reaction to Hadi's refusal to buy Dolphin
    Square 248–9, 315
  appointed Deputy Minister of Agriculture
    249–50
  and the Communists 253–4
  and army discontent 258–9
  arrest and imprisonment 271–2, 275–6,
    278, 290–1, 298
  daugher Nadia sent to London 295
  released from prison 299–300
  attempted escape from Baghdad 308–11
  travel ban lifted 312, 313
  exile in London 313–14, 343
  moves to Beirut 319
  manages family's overseas assets 324
  listens to BBC World Service 329
  decides to found a bank in Jordan 344
  and acquisition of a passport 356
  and death of Bibi 363–4
Chalabi, Saleh 108, 112, 132, 134, 135,
  238
  attends a wrestling match 171–2
Chalabi, Salem 326
Chalabi, Salim 203, 222, 246–7
Chalabi, Sarah 325
Chalabi, Shamsa 64, 112, 138, 146, 149,
  150, 182, 270
Chalabi, Shaouna 35, 36–7, 41–3, 62, 244
Chalabi, Talal 149
  and Eid festival 163, 165
  plays cards with Raifa 194

enthralled by Rumia's tales 220
enjoys trip to Paris 238
taken to house of family friend 267
escapes to Sheikh Jamil farm 270–2
attempted escape from Baghdad 309–11
marriage of 318–19
children of 325
decides to found a bank in Jordan 344
and acquisition of a passport 356
Chalabi, Tamara
  first visit to Baghdad xxvi–xxvii
  search for family connections
    xxvii–xxxix, xxxii–xxxiii, 3–4, 69–70,
    73–6, 189–90, 386–8
  finds Gertrude Bell's grave 185–6
  watches re-enactment of Battle of
    Karbala 279–80
  and rebuilding of Iraq 283–4
  entertained by stories by Bibi 353–4
  dreams of Bibi 369
  opposition to Saddam Hussein 370–82
  votes in first Iraq elections 383
  and burial of Bibi in Najaf 385–6
  and Iraq constitution 386
Chalabi, Thamina see al-Bassam, Thamina
  Chalabi
Churchill, Winston 95, 96
Clemenceau, Georges 90
Colvin, Marie 378
Cornwallis, Kinahan 98
Costain (Richard) Ltd 248
Cox, Sir Percy 102, 106, 107

Dabbakhana 83
Daccache, Laure 167
Daftari family 91
Daghistani, Ghazi 278
Damascus 94, 176, 192, 218, 359
Dardanelles 67
Dayyah Saadah 112, 197
Deer Palace xxvii, 203, 314, 387
  naming of 17, 18, 19, 20
  requisitioned by Turkish military 26–7,
    33
  Abdul Hussein's move to 112–13
  gardens of 114–15, 131
  visitors to 115–16, 131, 132, 136, 146
  qabuls at 117–18, 137
  daily life in 130–3
  monthly family gatherings at 149–51
  childhood lives in 161–7
  radio introduced into 166–7
  move from 179–80
  parties at 353–4
Deutsche Bank 235
Dhour el-Choueir 220
Diala province 119

Diwaniya 142
Diyarbakir 371
Dokkan 377
Downing College, Cambridge 122–3,
  127–8
Doxiadis, Constantinos 250, 279

Eden, Anthony 257
Eid festival 163–6
Eisenhower, Dwight D. 257
Eskandar, Afifa 174, 387

Fahima (nanny to Thamina's children)
  261, 266, 277
Fahima (wife of Ni'mati) 112, 148
Faili, Hadji Abbas 329–31
Falle, Sam 274
Falluja 196, 308, 309
Farhud (Great Loot) 207–8
Farouk, King 216–17, 253
Fath Ali Shah 36
Fatima 14
Fatima (bread-maker) 131
Fatimid dynasty 105
Fattah family 91
Faisal I, King 87, 142, 198
  attends Versailles Peace Conference
    89–90
  involvement in the Arab revolt 89
  in Damascus 94, 95
  as putative King of Iraq 95–9
  meets Abdul Hussein 98–9
  arrival in Baghdad and Kazimiya 102–3
  lunches with the Chalabi family 103–5
  coronation speech 106–7
  administration of 109–10, 113–14
  relationship with Gertrude Bell 116
  misses Gertrude Bell's funeral 129
  death of 147
Faisal II, King 191, 211, 247, 253, 267,
  268, 269, 276, 277
Fermor, Patrick Leigh 217
Fidayin Saddam 378
First World War
  recruitment and conscription 23–4, 33
  debates concerning 24–6
  and call for jihad 26
  effects of 28–33
  casualities of 33–4
  and the fall of Baghdad 63–7
Free Iraqi Forces (FIF) 377–8

Gailani, Rashid 'Ali 152, 156, 159, 202–5,
  206, 220
Gemayel, Bashir 342
Geneva 238
George, Zakiya 167

INDEX

Germans, Germany 10
Ghadir, festival of the 106
Ghaffuri (driver) 205
Ghazi, Crown Prince 137, 147–8, 191
Gilgamesh epic xxix, 251
Goebbels, Joseph 183, 184
Golden Square (nationalist group) 202, 204, 208
Goltz, Colmar Freiherr von der 31–2
Grobba, Herr 182
Gropius, Walter 250
Groppi coffee house (Cairo) 217
Gulbenkian, Calouste 'Mr 5 Per Cent' 235–6
Gulf War (1990–1) 365, 370–1

Habbaniya 204, 206
Habib, Master Yusuf 168–9
Hadi City (Madinat al-Hadi) 248
Hadi, *Hadji* (cook) 132–3, 169, 170–2
Hadid, Muhammad 159, 274
Haidar, Rustum 106, 121, 136, 142, 198–200
Haifa 218
Hajji Umran 375–6
Halabja 361–2
Hammudi the Arabanchi ('tram-man') 170–2
Harir 378
Hasouni family 321
Hejaz 115
Hellenic Stores (Greek Street, London) 294
Hidat (maid) 326
Hilla 144, 380
Hitler, Adolf 182, 183, 203, 204, 205, 213
Hiyam, Princess 268
HMS *Northbrook* 98
Hunt Club (Baghdad) 380–1
Huntington House school (Surrey) 302–3
Hussein, King 267, 313, 322–3, 348, 366, 367, 368
Hussein, Saddam xxxii, 279, 283, 319–20, 333, 334, 335, 345–8, 351–2, 355, 361–2, 364–5, 366–7, 370–1, 373, 374, 375–81, 382
Hussein, Taha 217
Husseini, Mufti Amin 203Hyderabad 17

Ibrahim (estate manager) 7–8
  and locust story 8–10
Ibrahim, Naji 294
Ikhwan al-Hurriya (Brotherhood of Freedom) 209–10
Imam Ali (1st Imam) 93, 99, 106, 141, 182, 361, 386

Imam Hussein (3rd Imam) xxviii, 78, 99, 105, 278
Imam Musa al-Kazim (7th Imam) 14, 60–1, 81, 155
Independence Guard (*Haras al Istiqlal*) 91–4
Inverforth, Lord 248, 249, 357
Iran, Shah of 322, 335, 344
Iran-Iraq War (1980–8) 345–8, 364
Iranian revolution (1979) 344
Iraq
  history of xxxiii–xxxi
  post-Saddam view of xxxii–xxxiii
  formation of 89–94
  ambivalence during Second World War 182–4
  cultural history 190
  mounting tensions in 202–10
  British *de facto* occupation of 203–4, 208–9, 210
  Communist ideology in 221–2
  anti-British sentiment in 229–32
  anti-Jewish feelings in 232–3
  political instability in 232–3
  executions in 234
  oil revenue 235–6
  development of 250–4
  culture in 251–3
  and the Suez crisis 257–8
  military coups in 260–9, 276–8, 291–2, 300, 309–10, 312–13, 322–3
  martial law declared 278
  military courts 278, 307, 308, 312–13
  and the 'expired epoch' 283–4, 307
  Ba'ath coup in 333–4, 335
  under Saddam's regime 345–8, 349, 361–2, 364–5
  and Iraqi exiles 354–6
  financial state of 366
Iraq Liberation Act (ILA) 374
Iraqi Airways 355
Iraqi Communist Party (ICP) 203, 221–2, 234, 246–7, 253–4, 289, 290, 307, 309–10, 351
Iraqi Development Board 236–7, 250, 253–4
Iraqi Embassies 355
Iraqi Military coup 260–9
Iraqi National Congress (INC) 371, 373, 374, 376, 377
Iraqi National Museum 127, 183–4, 389
Iraqi Petroleum Company 235
Iraqi Red Crescent 227
Iraqi stock exchange 160, 163
Iraqi Student Union 355
Israel 338
Israel Radio 329

Istanbul 10, 181, 293, 371
Istrabadi, Bibi (*see also* Amir, Umm Abdul) 52, 124, 269–70
Istrabadi family 321
Istrabadi, Sadiq 154, 267, 269

Jaafar, Dhia 271
Jaafar, Omar 270
Jabr, Muhammad 231
Jabr, Saleh 229–30, 231–2, 255
Jamali, Fadhil 255, 266–7, 271, 272, 273, 290
Jawad, *Hadji* 83–4
Jawad, Muhammad 14
Jawahiri family 321
Jeddah 98
Jews xxxi, 205–8, 222, 232–3
Jurf al-Sakhir 119

Kabariti, Karim 368
kallachiya (Baghdad) 223–5
Kalthoum, Umm 146, 166, 167
Kanaan family 321
Karaj 279
Karbala 99, 102, 199, 231, 278, 321, 360, 380, 385
Karbala, Battle of xxvii, 279–80
Karim (driver) 133, 140, 151–2, 158, 205
Kasbani, Amin 106
Kawar, Samir 368
Kayseri (ancient Caesaria) 29, 104
Kazim (friend of Hassan's) 228, 238
Kazimi, Abdul Raouf 82–3
Kazimi family 321
Kazimiya xxxi, 3–4, 8–22, 26, 28, 58, 63–4, 66, 79, 91, 145–6, 148, 161, 197, 222, 227, 250, 266, 267, 287, 306, 321, 351, 385
Kazimiya, Mayor of 88, 269
Kazimiya shrine 12–16, 60, 92, 103, 110–11, 116, 154–5, 182, 280, 287, 385
Kazimiya Tramway 61
Kelidar (shrine overseer) 14–16, 61, 92
Khalil family 319
Khalil Hamadani (Baghdad shoe shop) 163
Khalil (Palestinian doorman) 318, 326, 341–2
Khalil Pasha 63, 64
Khalisi, Sheikh Mahdi 110–11, 116, 287
Khomeini, Ayatollah 344, 345
King Fouad University (Cairo) 215, 216, 253
Kirkuk 235–6
Kramer, Herr 171
Kubba family 91

Kufa 93, 333
Kurdistan Democratic Party (KDP) 371, 376
Kurds xxxi, 78, 92, 105, 121, 156, 253, 255, 319, 335, 347, 348, 361–2, 370, 373, 376, 382
Kut 32, 63, 67, 119, 144, 234
Kuwait 98, 370
Kuwaiti, Daoud 169
Kuwaiti, Saleh 169

Lake Dokkan 376
Laleh (maid) 101
Latifiyyah Estate 248, 248–9, 250, 287, 320
Lawi, Gurji 205–6
Lawrence, T.E. 89, 95, 96
Layla 120–1
Le Corbusier 250
League of Nations 107, 229
Lebanese Civil War (1975–1990) 338–42, 349
Lebanon 121
Lemnos 67
Lloyd George, David 90
Lodge, Henry Cabot 272
London 139, 141, 176–8, 248–9, 285–8, 293–4, 296–7, 300–2, 306–7, 341, 352–4, 360–1, 367, 375, 388
Lysses School (Hampshire) 303

Madam Adel's school (Baghdad) 276–7
Mahdawi, Colonel 278, 307, 308, 312
Mahmud (brother of Khalil the doorman) 341–2
Majali, Abdul Hay 367
Mallowan, Max 217
'Manifesto of the association Against Imperialism' 222
March, Walter 184
Mardin 377
Marseilles 138, 141, 176
Marsh people 237
Massachusetts Institute of Technology (MIT) 334
Maude, Sir Frederick 64–7, 129, 186, 274
Maysaloun, Battle of (1920) 106
Mecca 7, 96, 222, 243
Menderes, Adnan 255, 256–7
Mesopotamia xxix–xxxi, 10, 13–14, 16, 67, 78, 89, 96, 190, 377
Midhat Pasha 11
Milan 238
Millfield College (Somerset) 303
Moore, Henry 252
Mosul 78, 113, 121, 144, 160, 183, 204, 213, 235–6, 268, 309, 373

Mudros 67
Mughaysil cemetery (Kazimiya) 151–2
Muhammad, Abu 279–80
Muhammad (refugee) 283–4
Murad IV, Sultan xxx
Murad, Salima 169
Murjan, Abdul Wahab 257, 290, 307
Muruwa, Kamil 331–2

Nabil 205–6
Nafisa, Queen 211, 268
Najaf 92, 93, 99, 102, 109, 141, 157, 182,
    199, 231, 257, 293, 306, 321, 332,
    351, 360, 361, 364, 381, 385
Najafi, Mahin 131
Nasser, Jamal Abdul 245–6, 252–3, 254,
    257, 289, 315, 316, 323, 328, 329,
    331–2
Nassiriya 63, 234, 378–9
NATO (North Atlantic Treaty
    Organization) 255, 256
Nawab, Agha Muhammad 7, 16–19, 26
Nawab, Burhan Chalabi 18–19, 60, 151
Nawab, Munira Chalabi
    visited by her brother 7, 18–19
    culinary skills 18, 99, 100–1, 103
    complains about the statue of the deer
        19
    moves out of the Deer Palace 26
    and choosing a wife for her nephew 35,
        36–7, 41–3
    relationship with Bibi 62
    and visit of King Faisal 99–101, 103–4
    sells Deer Palace to Abdul Hussein 112
    monthly visit to the Deer Palace 149–51
    visits Bibi in her new house 244
Ni'mati
    found and given a home by Hadi 15, 16,
        18, 19
    accompanies Hadi on his political errands
        26, 27
    prepares a horse for Bibi 53
    and café gossip 79–80
    accompanies Bibi to her mother's house
        101
    moves to the Deer Palace 112
    accompanies Hadi on his business travels
        119
    and blindness of Hassan 123
    looks after the Chalabi children 131
    death of his wife Fahima 148
    and the Eid festival 163, 165
    disapproves of Bibi abandoning the
        abayas 192
    and use of Arabic language 205
    criticizes Jabr 231–2
    worried over son's safety 234

visits Rushdi in prison 276, 291
greets Hadi on his return to Baghdad
    320–1
death of 332
Ni'mati, Muhammad 134–5, 234
Numah, Lilu 81
Numah, Rosa 81
Nur al-Din Beg 28, 29

Orosdi Back (department store) 201
Osseiran, Adel 336, 337, 339
Ottomans xxx–xxxi, 3, 7, 8, 10–11, 24–6,
    32, 33, 63–7, 89, 96, 106
Oudh Bequest 16

Palestine 67, 89, 178, 206, 229, 230,
    232–3
Palestine Liberation Organization (PLO)
    338
Paris 141, 176, 228, 238–41
Pasha, Nuri Fattah 159
Patriotic Union of Kurdistan (PUK) 371,
    376
Persia, Persians 3
Persian Gulf Treaty (1975) 335
Petra Bank 366–9
Philby, St John 98
Picasso, Pablo 252
Pitcher (maid) 326
Ponti, Gio 250
Postforoush family 36

Qassim, Abdul Karim 267, 273, 274–5,
    278, 287, 298, 305, 309–10, 320, 321,
    322
Qassim, Hamed 249–50, 291–2, 294, 298,
    299, 311, 312, 320, 322
Qassim, Umm Adnan 298–300
Qazzaz, Said 312–13
Qian 82
Qishla 30
Qotob family 36–7

Rangoon 12
Rashid Camp 204
Rashid, Haroun 251
Rashidiye Law School 120
Republican Guards 323
Revolutionary Council 278
Rishan (maid) 326
Royal Berkshire Regiment 106
Royal Dutch Shell 235
Rusafi, Ma'ruf 33–4

Saadoun (friend of Hadi's) 308
Sabra and Shatila refugee camp 342
Sadr City 279–80

Saeeda (nanny)
soothes the young Bibi 40, 60–1
finds scarce produce in the markets 41
loyalty and temperament of 51, 81, 124
and marriage of Bibi 54
superstitious beliefs 61, 155
informs Bibi of Hadi's café visits 81
leaves Rumia's service and joins Bibi 85
looks after Rushdi 101
moves to Deer Palace 112
failed marriage 124–5, 234
uncomfortable with modern ways 194–5
and death of Rumia 220
devoted to Ahmad 232, 241, 267, 292–3
important role of 232, 233–4
dislike of violence and chaos 234
escapes from military coup 267
death of 306
Sahla Mosque (Kufa) 93
St Leonard's Forest School (Sussex) 303
Salahuddin 371, 373, 376
Salim, Jawad 252
Salman, Abdul Razzaq Ali 269
Samachi, Fattuma 224
Samarra 64
Samawa 63
Samerai, Saleh 335
San Remo 91
Sattar (steward) 12, 14, 16, 19, 21, 27
Sawt-al-Arab (radio station) 252–3, 266, 305, 329
Sayyid Nassir (Bibi's grandfather) 44
Seaford College (Sussex) 302, 303–6
Seale, Colonel 377, 379
Second World War 173, 180, 182–4, 195, 202–10, 220
Shakir, Zaid 367
Sharif Hussein, Emir of Mecca 87, 89, 91, 96
Sha'shou, Munashi 206
Sheikh Abdul Qadir Mosque 108
Sheikh Jamil farm 266–7, 270–3
Sheikh Sandal mosque (Baghdad) 91–2
Shibibi, Hussein Muhammad 234
Shirazi, Grand Ayatollah 93, 109
Shosa 134–5
Shubbar, Kadhim 334
Shubeilat, Laith 368
Siddiq, Yehuda 234
Sidqi, Bakr 160
Sif Palace 179–80, 196–7, 241
Sitt Zeinab shrine (Damascus) 359–60
Somerville, Mr 140
Souq al-Saray (Baghdad market) 75–6
Souq Hanoun market (Baghdad) 206

Souq Istrabadi 79
Spanish Civil War 166
Stark, Freya 209–11
Subaiti, Muhammad Hadi 347–8
Subaiti, Umm Hassan 347–8
Suez Crisis (1956) 257, 306
Suleymania 376
Suwaidi family 91
Suwaidi, Tawfiq 290, 323
Sykes, Sir Mark 65
Sykes–Picot agreement (1916) 91, 95
Syria 89

Talabani, Jalal 319, 368, 376
Talib (member of Revolutionary Council) 278
Tarzi, Salim 248, 285
Ta'sisiyah school 276–7
Tehran 275
Tikrit 333, 345
Tokatelian, Madame 201
Tudeh Communist Party (Iran) 221
Turkey 255–7
Turkish Sixth Army (Ottoman 6th Army) 24, 28

Ugaili family 321
Ummayyad dynasty 105
United Arab Union 274
United Nations 229, 257, 272, 274
University of Chicago 335
Ur 379
Urumia 375

Versailles Treaty (1919) 89–90
Vienna 138, 371
Voice of America 329

Wahhabism 349
Wathbah (great uprising) 230–2
Waugh, Evelyn 217
Weir (Andrew) & Co. 142, 144, 198, 218, 247–9, 285, 357
Wilhelm, Kaiser 23
Wilson, A.T. 88
Wilson, President 87, 88
Wright, Frank Lloyd 189–90, 250

Young Turks 10, 65
Yunis, Thabet 267
Yussif, Sultana 167
Yusuf, Yusuf Salman, 'Fahd' 234

Zagros mountains xxx, 373
Zilkha, Salman 206
Zionism 232–3